Machine Learning

Theory and Practice

Machine Learning

Theory and Practice

M N Murty
Honorary Professor, Department of Computer Science and Automation, Indian Institute of Science, Bengaluru, India

Ananthanarayana V S
Professor, Department of Information Technology, National Institute of Technology Karnataka, Surathkal, Mangaluru, India

Universities Press

MACHINE LEARNING: THEORY AND PRACTICE

UNIVERSITIES PRESS (INDIA) PRIVATE LIMITED

Registered Office
3-6-747/1/A & 3-6-754/1, Himayatnagar, Hyderabad 500 029, Telangana, India
info@universitiespress.com; www.universitiespress.com

Distributed by
Orient Blackswan Private Limited

Registered Office
3-6-752 Himayatnagar, Hyderabad 500 029 Telangana, India

Other Offices
Bengaluru / Chennai / Guwahati / Hyderabad / Kolkata
Mumbai / New Delhi / Noida / Patna

© Universities Press (India) Private Limited 2024

Cover and book design:
© Universities Press (India) Private Limited 2024

ISBN 978-93-93330-69-7

Typeset in Adobe Garamond Pro 11 points *by*
Cameo Corporate Services Limited, Coimbatore

Printed in India by
Shree Maitrey Printech Pvt. Ltd., Noida

Published by
Universities Press (India) Private Limited
3-6-747/1/A & 3-6-754/1, Himayatnagar, Hyderabad 500 029, Telangana, India

502026

Preface

This book deals with machine learning (ML), an important topic of current interest. Some key areas of artificial intelligence (AI) and ML are:

- **Search**: In practical applications of current interest, the data is typically high-dimensional. These applications include image classification, information retrieval, problem-solving in AI, biological and chemical structure analysis and social network analysis. Searching for an ML model with appropriate parameters is a challenge in such high-dimensional spaces. The main problem is that most of the popular tools like the k-nearest neighbor classifier, decision tree classifier and clustering algorithms that depend on inter-pattern distance computations fail to work well. So, representing the data in a lower-dimensional space is inevitable.

- **Representation**: While learning ML models from the data using a machine, it is important to represent the data in a form suitable for effective and efficient machine learning. In this book, we propose to cover a wide variety of representation techniques that are important in both theory and practice. Specifically, we cover:

  ○ *Feature selection schemes*: Here, an appropriate subset of the given feature set is identified and used in learning.
  ○ *Feature extraction schemes*: Here, linear or non-linear combinations of the given features are used in learning. Some of the popular linear feature extractors are based on principal components (PCs), random projections (RPs) and singular value decomposition (SVD). We cover all these techniques in the book.

- **Prediction**: We discuss models for classification and regression based on neighborhood, decision trees, posterior probabilities, frequent itemsets, compression based on clustering, linear discriminants and neural networks. Various deep learning (DL) architectures for prediction are also discussed. We present experimental results on some real-world benchmark data sets to illustrate various ideas.

This book is meant for both undergraduate students and teachers and also helps practitioners in implementing ML algorithms. It is intended for senior undergraduate and graduate students and researchers working in machine learning, data mining and pattern recognition. We present the material in such a manner that it is accessible to a wide variety of readers with some basic exposure to undergraduate-level mathematics. The presentation has intentionally been made as simple as possible to make the reader feel comfortable.

This book is organized as follows: Chapter 1 deals with a generic introduction to machine learning and various concepts including feature engineering, model selection, model estimation, model validation and model explanation. Two important tasks in ML are classification and clustering. Chapter 2 deals with the representation of KNNC and its variants. Several state-of-the-art ML models are based on decision trees. These include random forest and gradient boosting. We discuss these models in Chapter 3. It is possible to characterize an optimal predictor based on Bayes' rule. We discuss the related classifiers in Chapter 4. It has been shown that frequent itemsets are discriminative. So, Chapter 5 deals with ML based on frequent itemsets. The important problem of representation in ML is considered in Chapter 6 while clustering algorithms for grouping data items are considered in Chapter 7. Perceptron is the most basic building block for DL architectures. Further, optimal predictors based on separation of the pattern space such as SVM and logistic regression are discussed in Chapter 8, along with MLP. Chapter 9 provides an introduction to DL architecture, including autoencoders, CNNs, RNNs and GANs.

M N Murty
Ananthanarayana V S

Acknowledgements

We would like to thank Madhavi Sethupathi, Managing Editor at Universities Press, for excellent editorial support. She has provided innumerable editorial tips and corrections to improve the quality of the material in the book. We would also like to thank Kallol Das, Senior Acquisitions Editor at Universities Press, for his critical inputs on earlier versions of the material.

We would like to thank our collaborators over the past four decades for providing us an excellent platform to appreciate several aspects of machine learning.

M N Murty
Ananthanarayana V S

Contents

Preface   *v*

Acknowledgements   *vii*

List of Acronyms   *xiv*

1   Introduction to Machine Learning   **1**

Evolution of Machine Learning | Paradigms for ML | Learning by Rote | Learning by Deduction | Learning by Abduction | Learning by Induction | Reinforcement Learning | Types of Data | Matching | Stages in Machine Learning | Data Acquisition | Feature Engineering | Data Representation | Model Selection | Model Learning | Model Evaluation | Model Prediction | Model Explanation | Search and Learning | Explanation Offered by the Model | Data Sets Used

2   Nearest Neighbor-Based Models   **21**

Introduction to Proximity Measures | Distance Measures | Minkowski Distance | Weighted Distance Measure | Non-Metric Similarity Functions | Levenshtein Distance | Mutual Neighborhood Distance (MND) | Proximity Between Binary Patterns | Different Classification Algorithms Based on the Distance Measures | Nearest Neighbor Classifier (NNC) | K-Nearest Neighbor Classifier | Weighted K-Nearest Neighbor (WKNN) Algorithm | Radius Distance Nearest Neighbor Algorithm | Tree-Based Nearest Neighbor Algorithm | Branch and Bound Method | Leader Clustering | KNN Regression | Concentration Effect and Fractional Norms | Performance Measures | Performance of Classifiers | Performance of Regression Algorithms | Area Under the ROC Curve for the Breast Cancer Data Set

3   Models Based on Decision Trees   **49**

Introduction to Decision Trees | Decision Trees for Classification | Impurity Measures for Decision Tree Construction | Properties of the Decision Tree Classifier (DTC) | Applications in Breast Cancer Data | Embedded Schemes for Feature Selection | Regression Based on Decision Trees | Bias–Variance Trade-off | Random Forests for Classification and Regression | Comparison of DT and RF Models on Olivetti Face Data | AdaBoost Classifier | Regression Using DT-Based Models | Gradient Boosting (GB) | Practical Application

4   The Bayes Classifier   **81**

Introduction to the Bayes Classifier | Probability, Conditional Probability and Bayes'
Rule | Conditional Probability | Total Probability | Bayes' Rule and Inference | Bayes' Rule
and Classification | Random Variables, Probability Mass Function, Probability Density
Function and Cumulative Distribution Function, Expectation and Variance | Random
Variables | Probability Mass Function (PMF) | Binomial Random Variable | Cumulative
Distribution Function (CDF) | Continuous Random Variables | Expectation of a Random
Variable | Variance of a Random Variable | Normal Distribution | The Bayes Classifier
and its Optimality | Multi-Class Classification | Parametric and Non-Parametric Schemes
for Density Estimation | Parametric Schemes | Class Conditional
Independence and Naïve Bayes Classifier | Estimation of the Probability Structure |
Naïve Bayes Classifier (NBC)

5   Machine Learning Based on Frequent Itemsets   **109**

Introduction to the Frequent Itemset Approach | Frequent Itemsets | Frequent Itemset
Generation | Frequent Itemset Generation Strategies | Apriori Algorithm | Frequent
Pattern Tree and Variants | FP Tree-Based Frequent Itemset Generation | Pattern Count
(PC) Tree-Based Frequent Itemset Generation | Frequent Itemset Generation Using
the PC Tree | Dynamic Mining of Frequent Itemsets | Classification Rule Mining |
Frequent Itemsets for Classification Using PC Tree | Frequent Itemsets for Clustering
Using the PC Tree

6   Representation   **131**

Introduction to Representation | Feature Selection | Linear Feature Extraction |
Vector Spaces | Basis of a Vector Space | Row Vectors and Column Vectors |
Linear Transformations | Eigenvalues and Eigenvectors | Symmetric Matrices |
Rank of a Matrix | Principal Component Analysis | Experimental Results on Olivetti
Face Data | Singular Value Decomposition | PCA and SVD | Random Projections

7   Clustering   **157**

Introduction to Clustering | Partitioning of Data | Data Re-organization | Data
Compression | Summarization | Matrix Factorization | Clustering of Patterns |
Data Abstraction | Clustering Algorithms | Divisive Clustering | Agglomerative
Clustering | Partitional Clustering | K-Means Clustering | K-Means++ Clustering |
Soft Partitioning | Soft Clustering | Fuzzy C-Means Clustering | Rough Clustering |
Rough K-Means Clustering Algorithm | Expectation Maximization-Based Clustering |
Spectral Clustering | Clustering Large Data Sets | Divide-and-Conquer Method

8   Linear Discriminants for Machine Learning   **201**

Introduction to Linear Discriminants | Linear Discriminants for Classification |
Parameters Involved in the Linear Discriminant Function | Learning w and b | Perceptron
Classifier | Perceptron Learning Algorithm | Convergence of the Learning Algorithm |
Linearly Non-Separable Classes | Multi-Class Problems | Support Vector Machines |
Linearly Non-Separable Case | Non-linear SVM | Kernel Trick | Logistic Regression |
Linear Regression | Sigmoid Function | Learning w and b in Logistic Regression |
Multi-Layer Perceptrons (MLPs) | Backpropagation for Training an MLP | Results on
the Digits Data Set

9   Deep Learning   **233**
Introduction to Deep Learning | Non-Linear Feature Extraction Using Autoencoders |
Comparison on the Digits Data Set | Deep Neural Networks | Activation Functions |
Initializing Weights | Improved Optimization Methods | Adaptive Optimization | Loss
Functions | Regularization | Adding Noise to the Output or Label Smoothing |
Experimental Results on the MNIST Data Set | Convolutional Neural Networks |
Convolution | Padding Zero Rows and Columns | Pooling to Reduce Dimensionality |
Recurrent Neural Networks | Training an RNN | Encoder–Decoder Models | Generative
Adversarial Networks

Conclusions   *277*

Appendix – Hints to Practical Exercises   *A1*

Index   *325*

List of Acronyms

AI	artificial intelligence
AUC	area under the curve
CNN	convolutional neural network
DL	deep learning
DTC	decision tree classifier
FP tree	frequent pattern tree
GAN	generative adversarial network
GB	gradient boosting
GRU	gated recurrent unit
HD	Hamming distance
JC	Jaccard coefficient
KLD	Kullback–Leibler divergence
KMA	k-means algorithm
KNNC	k-nearest neighbor classifier
KNNs	k-nearest neighbors
LR	logistic regression
LSTM	long short-term memory
MAE	mean absolute error
MDC	minimal distance classifier
MI	mutual information
ML	machine learning
MLE	maximum likelihood estimate
MLP	multi-layer perceptron
MND	mutual neighbor distance
MSE	mean square error
NBC	naïve Bayes classifier
NNC	nearest neighbor classifier
PC tree	pattern count tree
PCA	principal component analysis
RF	random forest
RNN	recurrent neural network
ROC	receiver operating characteristic
SMC	simple matching coefficient
SVD	singular value decomposition
SVM	support vector machine

CHAPTER 1
Introduction to Machine Learning

Learning Objectives

At the end of this chapter, you will be able to:

- Give a brief overview of machine learning (ML)
- Describe the learning paradigms used in ML
- Explain the important steps in ML, including data acquisition, feature engineering, model selection, model learning, model validation, model explanation, representation, and search and explanation

1.1 EVOLUTION OF MACHINE LEARNING

Machine learning (ML) is the process of learning a model that can be used in prediction based on data. Prediction involves assigning a data item to one of the classes or associating the data item with a number. The former activity is classification while the latter is regression. ML is an important and state-of-the-art topic. It gained prominence because of the improved processing speed and storage space of computers and the availability of large data sets for experimentation. Deep learning (DL) is an offshoot of ML. In fact, perceptron was the earliest popular ML tool and it forms the basic building block of various DL architectures, including multi-layer perceptron networks, convolutional neural networks (CNNs) and recurrent neural networks (RNNs).

In the early days of artificial intelligence (AI), it was opined that mathematical logic was the ideal vehicle for building AI systems. Some of the initial contributions in this area like the General Problem Solver (GPS), Automatic Theorem Proving (ATP), rule-based systems and programming languages like Prolog and Lisp (lambda calculus–based) were all outcomes of this view. Various problem solving and game playing solutions also had this flavour. During the twentieth century, a majority of prominent AI researchers were of the view that *logic is AI and AI is logic*. Most of the reasoning systems were developed based on this view. Further, the role of artificial neural networks in solving complex real-world AI problems was under-appreciated.

However, this view was challenged in the early twenty-first century and the current view is that *AI is deep learning and deep learning is AI*. The advent of efficient graphics processing units (GPUs), platforms like TensorFlow and PyTorch along with the demonstrated success stories of convolutional neural networks, gated recurrent units and generative models based on neural networks have impacted every aspect of science and engineering activities across the globe.

So, ML along with DL has become a state-of-the-art subject. Artificial neural networks form the backbone of DL.

A high-level view of AI is shown in Fig. 1.1. The tasks related to conventional AI are listed separately. Here, ML may be viewed as dealing with more than just pattern recognition (PR) tasks. Classification and clustering are the typical tasks of a PR system. However, ML deals with regression problems also. Data mining is the efficient organization of data in the form of a database.

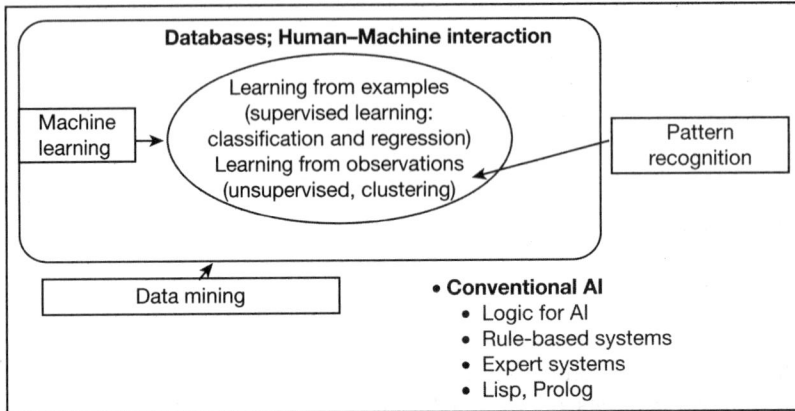

FIG. 1.1 A high-level view of AI

The typical background topics of AI are shown in Fig. 1.2.

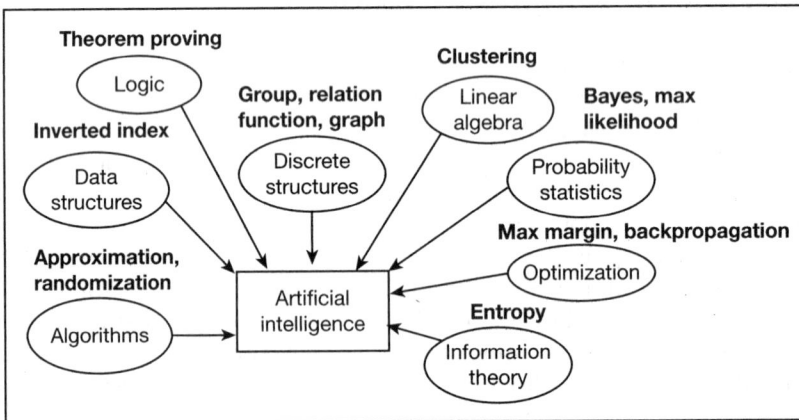

FIG. 1.2 Background topics of AI

Note that data structures and algorithms are basic to both conventional and current AI systems. Logic and discrete structures played an important role in the analysis and synthesis of conventional AI systems. The importance of other background topics may be summarized as follows:

- In ML, we deal with vectors and vector spaces and these topics are better appreciated through **linear algebra**. The data input to an ML system may be viewed as a matrix, popularly called the data matrix. If there are n data items, each represented as an l-dimensional vector, then the corresponding data matrix A is of size $n \times l$. Linear algebra is useful in analysing the weights associated with the edges in a neural network. Matrix multiplication and eigen analysis are important in initializing the weights of the neural network and in weight updates. It can also help in weight normalization. The whole activity of clustering may be viewed as data matrix factorization.

- The role of **probability and statistics** need not be explained as ML is, in fact, statistical ML. These topics help in estimating the distributions underlying the data. Further, they play a crucial role in analysis and inference in ML.
- **Optimization** (along with calculus) is essential in training neural networks where gradients and their computations are important. Gradient descent–based optimization is an essential ingredient of any DL system.
- **Information theoretic concepts** like entropy, mutual information and Kullback–Leibler divergence are essential to understand topics such as decision tree classifiers, feature selection and deep neural networks.

We will provide details of all these background topics in their respective chapters.

1.2   PARADIGMS FOR ML

There are different ways or paradigms for ML, such as learning by rote, learning by deduction, learning by abduction, learning by induction and reinforcement learning. We shall look at each of these in detail.

1.2.1   Learning by Rote

This involves memorization in an effective manner. It is a form of learning that is popular in elementary schools where the alphabet and numbers are memorized. Memorizing simple addition and multiplication tables are also examples of rote learning. In the case of data caching, we store computed values so that we do not have to recompute them later. Caching is implemented by search engines and it may be viewed as another popular scheme of rote learning. When computation is more expensive than recall, this strategy can save a significant amount of time. Chess masters spend a lot of time memorizing the great games of the past. It is this rote learning that teaches them how to 'think' in chess. Various board positions and their potential to reach the winning configuration are exploited in games like chess and checkers.

1.2.2   Learning by Deduction

Deductive learning deals with the exploitation of deductions made earlier. This type of learning is based on reasoning that is truth preserving. Given A, and *if A then B* $(A \to B)$, we can deduce B. We can use B along with *if B then C* $(B \to C)$ to deduce C. Note that whenever A and $A \to B$ are *True*, then B is *True*, ratifying the truth preserving nature of learning by deduction. Consider the following statements:

1. It is raining.
2. If it rains, the roads get wet.
3. If a road is wet, it is slippery.

From (1) and (2), we can infer using deduction that (4) the roads are wet. This deduction can then be used with (3) to deduce or learn that (5) the roads are slippery. Here, if statements (1), (2) and (3) are *True*, then statements (4) and (5) are automatically *True*.

A digital computer is primarily a deductive engine and is ideally suited for this form of learning. Deductive learning is applied in well-defined domains like game playing, including in chess.

1.2.3   Learning by Abduction

Here, we infer A from B and $(A \to B)$. Notice that this is not truth preserving like in deduction as both B and $(A \to B)$ can be *True* and A can be *False*. Consider the following inference:

1. An aeroplane is a flying object ($aeroplane \to flying\ object$).
2. A is a flying object.

From (1) and (2), we infer using abduction that A is an aeroplane. This kind of reasoning may lead to incorrect conclusions. For example, A could be a bird or a kite.

1.2.4   Learning by Induction

This is the most popular and effective form of ML. Here, learning is achieved with the help of examples or observations. It may be categorized as follows:

- **Learning from Examples:** Here, it is assumed that a collection of labelled examples are provided and the ML system uses these examples to make a prediction on a new data pattern. In supervised classification or learning from examples, we deal with two ML problems: classification and regression.

    1. **Classification:** Consider the handwritten digits shown in Fig. 1.3. Here, each row has 15 examples of each of the digits. The problem is to learn an ML model using such data to classify a new data pattern. This is also called **supervised learning** as the model is learnt with the help of such exemplar data. It may be provided by an expert in several practical situations. For example, a medical doctor may provide examples of normal patients and patients infected by COVID-19 based on some test results. In the case of handwritten digits, we have 10 class labels, one class label corresponding to each of the digits from 0 to 9. In classification, we would like to assign an appropriate class label from these labels to a new pattern.
    2. **Regression:** Contrary to classification, there are several prediction applications where the labels come from a possibly infinite set. For example, the share value of a stock could be a positive real number. The stock may have different values at a particular time and each of these values is a real number. This is a typical regression or curve fitting problem. The practical need here is to predict the share value of a stock at a future time instance based on past data in the form of examples.

- **Learning from Observations:** Observations are also instances like examples but they are different because observations need not be labelled. In this case, we cluster or group the observations into a smaller number of groups. Such grouping is performed with the help of a clustering algorithm that assigns similar patterns to the same group/cluster. Each cluster

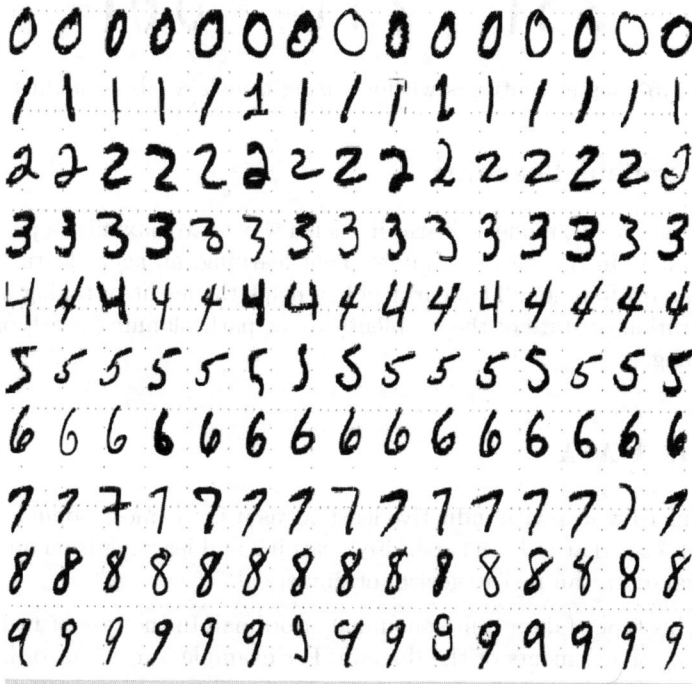

FIG. 1.3 Examples of handwritten digits labelled 0 to 9

could be represented by its centroid or mean. Let $x_1, x_2, \ldots, x_p$ be p elements of a cluster. Then the centroid of the cluster is defined by

$$\frac{1}{p} \sum_{i=1}^{p} x_i$$

Let us consider the handwritten digit data set of 3 classes: 0, 1 and 3. By using the class labels and clustering patterns of each class separately, we obtain 3 clusters that give us the 9 centroids shown in Fig. 1.4.

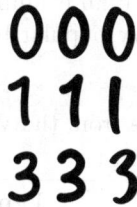

FIG. 1.4 Cluster centroids using the class labels in clustering

However, when we cluster the entire data of digits 0, 1 and 3 into 9 clusters, we obtain the centroids shown in Fig. 1.5. So, the clusters and their representatives could differ based on how we exploit the class labels.

$$3\,3\,|\qquad 3\,7\,1\qquad 0\,0\,|$$

FIG. 1.5 Cluster centroids without using the class labels in clustering

1.2.5   Reinforcement Learning

In supervised learning, the ML model is learnt in such a way as to maximize a performance measure like prediction accuracy. In the case of reinforcement learning, an agent learns an optimal policy to optimize some reward function. The learnt policy helps the agent in taking an action based on the current configuration or state of the problem. Robot path planning is a typical application of reinforcement learning.

1.3   TYPES OF DATA

In this book, we primarily deal with inductive learning as it is the most popular paradigm for ML. It is important to observe that in both supervised learning and learning from observations, we deal with data. In general, data can be categorical or numerical.

- **Categorical:** This type of data can be nominal or ordinal. In the case of **nominal data**, there is no order among the elements of the domain. For example, for colour of hair, the domain is {brown, black, red}. This data is of categorical type and the elements of the domain are not ordered. On the contrary, in **ordinal data**, there is an order among the values of the domain. For example, the domain of variable *employee number* could be $\{1, 2, \ldots, 1011\}$ if there are 1011 employees in an organization. Here, ordering among the elements of the domain is observed, indicating that senior employees have smaller employee numbers compared to junior employees; the most senior employee will have employee number 1.
- **Numerical:** In the case of numerical data, the domain of values of the data type could be a set/subset of integers or a set/subset of real numbers. For example, in Table 1.1, a subset of the features used by the Wisconsin Breast Cancer data is shown. The domain of *Diagnosis*, the class label, is a binary set with values *Malignant* and *Benign*. The domain of *ID Number* is a subset of integers in the range $[8670, 917897]$ and the domain of *Area_Mean* is a collection of floating point numbers (interval) in the range $[143.5, 2501]$. It is possible to have binary values in the domain for categorical or numerical data. For example, the domain of *Status* could be {Pass, Fail} and this variable is nominal; an example of a binary ordinal type is {short, tall} for humans based on their height. A very popular binary numerical type is $\{0, 1\}$; also in the

TABLE 1.1 Different types of data from the Wisconsin Breast Cancer database

Feature Number	Attribute	Type of Data	Domain
1	Diagnosis	Nominal	$\{Malignant, Benign\}$
2	ID Number	Ordinal	$[8670, 917897]$
3	Perimeter_Mean	Numerical	$[43.79, 188.5]$
4	Area_Mean	Numerical	$[143.5, 2501]$
5	Smoothness_Mean	Numerical	$[0.05263, 0.1634]$

classification context, the *class label* data can have the domain $\{-1, +1\}$ where -1 stands for the label of the negative class and $+1$ stands for the label of the positive class.

Typically, each pattern or data item is represented as a vector of feature values. For example, a data item corresponding to a patient with ID 92751 is represented by a five-dimensional vector $(Benign, 92751, 47.92, 181, 0.05263)$, where each component of the vector represents the corresponding feature shown in Table 1.1. *Benign* is the value of feature 1, *Diagnosis*; similarly, the third entry 47.92 corresponds to feature 3, that is *Perimeter_Mean* and so on. Note that *Diagnosis* is a nominal feature and *ID Number* is an ordinal attribute. The remaining three features are numerical.

Here, *Diagnosis* or the class label is a dependent feature and the remaining four features are independent features. Given a collection of such data items or patterns in the form of five-dimensional vectors, the ML system learns an association or mapping between the independent features and the dependent feature.

1.4 MATCHING

Matching is an important activity in ML. It is used in both supervised learning and in learning from observations. Matching is carried out by using a **proximity measure** which can be a distance/dissimilarity measure or a similarity measure. Two data items, u and v, represented as l-dimensional vectors, match better when the distance between them is smaller or when the similarity between them is larger.

A popular distance measure is the **Euclidean distance** and a popular similarity measure is the **cosine of the angle between vectors**. The Euclidean distance is given by

$$d(u, v) = \sqrt{\sum_{i=1}^{l} (u(i) - v(i))^2}$$

The cosine similarity is given by

$$cos(u, v) = \frac{u^t v}{||u||||v||},$$

where $u^t v$ is the dot product between vectors u and v and $||u||$ is the Euclidean distance between u and the origin; it is also called the Euclidean norm.

Some of the important applications of matching in ML are in:

- **Finding the Nearest Neighbor of a Pattern:** Let x be an l-dimensional pattern vector. Let $\mathcal{X} = \{x_1, x_2, \ldots, x_n\}$ be a collection of n data vectors. The nearest neighbor of x from $\mathcal{X}$, denoted by $NN_x(\mathcal{X})$, is x_j if

$$d(x, x_j) \leq d(x, x_i), \ \forall x_i \in \mathcal{X}$$

  This is an approximate search where a pattern that best matches x is obtained. If there is a tie, that is, when both $x_p \in \mathcal{X}$ and $x_q \in \mathcal{X}$ are the nearest neighbors of x, we can break the tie arbitrarily or choose either of the two to be the nearest neighbor of x. This step is useful in classification and will be discussed in the next chapter.

- **Assigning to a Set with the Nearest Representative:** Let $C_1, C_2, \ldots, C_K$ be K sets with $x^1, x^2, \ldots, x^K$ as their respective representatives. A pattern x is assigned to C_i if

$$d(x, x^i) \leq d(x, x^j), \ for \ j \in \{1, 2, \ldots, K\}$$

This idea is useful in clustering or learning from observations, where C_i is the ith group or cluster of patterns or observations and x^i is the representative of C_i. The centroid of the data vectors in C_i is a popularly used representative, x^i, of C_i.

1.5   STAGES IN MACHINE LEARNING

Building a machine learning system involves a number of steps, as illustrated in Fig. 1.6. Note the emphasis on data in the form of training, validation and test data.

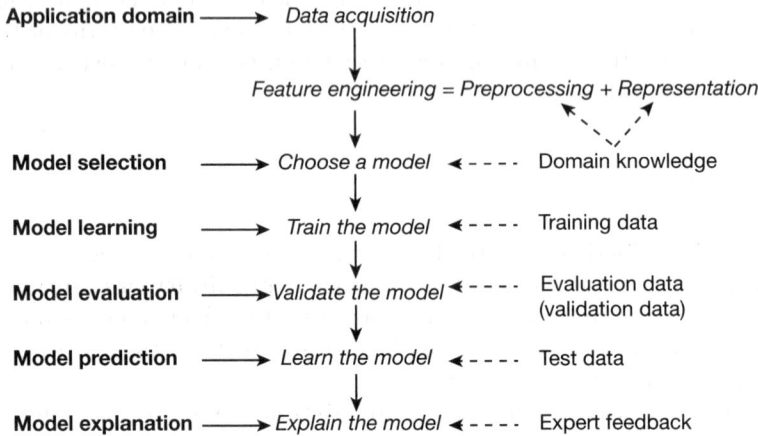

Application domain ⟶ *Data acquisition*

Feature engineering = Preprocessing + Representation

Model selection   ⟶ *Choose a model* ◄ - - - Domain knowledge

Model learning   ⟶ *Train the model* ◄ - - - Training data

Model evaluation   ⟶ *Validate the model* ◄ - - - Evaluation data (validation data)

Model prediction   ⟶ *Learn the model* ◄ - - - Test data

Model explanation   ⟶ *Explain the model* ◄ - - - Expert feedback

FIG. 1.6 Important steps in a practical machine learning system

Typically, the available data is split into training, validation and test data. Training data is used in model learning or training and validation data is used to tune the ML model. Test data is used to examine how the learnt model is performing. We now describe the components of an ML system.

1.5.1   Data Acquisition

This depends on the domain of the application. For example, to distinguish between *adults* and *children*, measurements of their *height* or *weight* are adequate; however, to distinguish between *normal* and *COVID-19-infected* humans, their *body temperature* and *chest congestion* may be more important than their *height* or *weight*. Typically, data collection is carried out before feature engineering.

1.5.2   Feature Engineering

This step involves a combination of data preprocessing and data representation.

Data Preprocessing

In several practical applications, the raw data available needs to be updated before it can be used by an ML model. The common problems encountered with raw data are missing values, different ranges for different variables and the presence of outliers. We will now explain how to deal these problems.

1. **Missing Data:** It is likely that in some domains, there could be missing data. This occurs as a consequence of the inability to measure a feature value or due to unavailability or erroneous data entry. Some ML algorithms can work even when there are a reasonable number of missing data values and, in such cases, there is no need for preprocessing. However, there are a large number of other cases where the ML models cannot handle missing values. So, there is a need to examine techniques for dealing with missing data. Different schemes are used for dealing with the prediction of missing values:

   - *Use the nearest neighbor:* Let x be an l-dimensional data vector that has its ith component $x(i)$ missing. Let $\mathcal{X} = \{x^1, x^2, \ldots, x^n\}$ be the set of n training pattern vectors. Let $x^j \in \mathcal{X}$ be the nearest neighbor of x based on the remaining $l - 1$ (excluding the ith) components. Predict the value of $x(i)$ to be $x^j(i)$, that is, if the ith component $x(i)$ of x is missing, use the ith component of $x^j = NN_x(\mathcal{X})$ instead.
   - *Use a larger neighborhood:* Use the k-nearest neighbors (KNNs) of x to predict the missing value x_i. Let the KNNs of x, using the remaining $l - 1$ components, from $\mathcal{X}$ be $x1, x2, \ldots, xK$. Now the predicted value of $x(i)$ is the average of the ith components of these KNNs. That is, the predicted value of $x(i)$ is

$$\frac{1}{K} \sum_{j=1}^{K} xj(i)$$

   EXAMPLE 1: Consider the set data vectors

$$(1,1,1), (1,1,2), (1,1,3), (1,-,2), (1,1,-), (6,6,1)$$

   There are 6 three-dimensional pattern vectors in the set. Missing values are indicated by $-$. Let us see how to predict the missing value in $(1,-,2)$. Let us use $K = 3$ and find the 3 nearest neighbors (NNs) based on the remaining two feature values. The three NNs are $(1,1,1), (1,1,2)$ and $(1,1,3)$. Note that the second feature value of all these three neighbors is 1, which is the predicted value for the missing value in $(1,-,2)$. So, the vector becomes $(1,1,2)$.

   - *Cluster the data and locate the nearest cluster:* This approach is based on clustering the training data and locating the cluster to which x belongs based on the remaining $l - 1$ components. Let x with its ith value missing belong to cluster C^q. Let μ^q be the centroid of C^q. Then the predicted value of $x(i)$ is μ_i^q, the ith component of μ^q. We will explain clustering in detail in a later chapter; it is sufficient for now to note that a clustering algorithm can be used to group patterns in the training data into K clusters where patterns in each cluster are similar to each other and patterns belonging to different clusters are dissimilar.

   EXAMPLE 2: Consider the following data matrix. It has 5 data vectors in a four-dimensional space.

$$\begin{bmatrix} 5.1 & 3.5 & 1.4 & 0.2 \\ 4.9 & 3.0 & 1.4 & 0.2 \\ 4.7 & 3.2 & 1.3 & 0.2 \\ 4.6 & 3.1 & 1.5 & 0.2 \\ 5.0 & 3.6 & 1.4 & 0.2 \end{bmatrix}$$

Let us imagine that two values are missing to obtain

$$\begin{bmatrix} NA & 3.5 & 1.4 & 0.2 \\ 4.9 & 3.0 & 1.4 & 0.2 \\ 4.7 & 3.2 & 1.3 & 0.2 \\ 4.6 & NA & 1.5 & 0.2 \\ 5.0 & 3.6 & 1.4 & 0.2 \end{bmatrix}$$

If we assume that this data is from the same cluster, we can compute the mean in each case and use it to predict the missing value. The mean of the available 4 values in the first column is 4.8; similarly the mean of the available 4 values in the second column is 3.325. So, using these mean values to replace the missing values in the matrix gives us the updated matrix

$$\begin{bmatrix} \mathbf{4.8} & 3.5 & 1.4 & 0.2 \\ 4.9 & 3.0 & 1.4 & 0.2 \\ 4.7 & 3.2 & 1.3 & 0.2 \\ 4.6 & \mathbf{3.325} & 1.5 & 0.2 \\ 5.0 & 3.6 & 1.4 & 0.2 \end{bmatrix}$$

We can compute the **mean squared error (MSE)** with respect to the predicted values based on the deviations from the original values. The computation of MSE may be explained as follows: Given the n true (target) values to be $y_1, y_2, \ldots, y_n$ and the predicted values to be $\hat{y}_1, \hat{y}_2, \ldots, \hat{y}_n$, MSE is defined as

$$\frac{\sum_{i=1}^{n}(y_i - \hat{y}_i)^2}{n}$$

In the above example, we have predicted two missing values based on the mean of the remaining values of the respective feature. In the first case, instead of 5.1, our estimated value is 4.8. Similarly, in the second case, for the value 3.1, our estimate is 3.325. So, the MSE here is

$$\frac{(5.1 - 4.8)^2 + (3.1 - 3.325)^2}{2} = \frac{0.09 + 0.00050625}{2} = 0.04525$$

EXAMPLE 3: Consider three clusters of points and their centroids:

a. Cluster1: $\{(1,1,1), (1,2,3), (1,3,2)\}$ Centroid 1: $(1,2,2)$
b. Cluster2: $\{(3,4,3), (3,5,3), (3,3,3)\}$ Centroid 2: $(3,4,3)$
c. Cluster3: $\{(6,6,6), (6,8,6), (6,7,6)\}$ Centroid 3: $(6,7,6)$

Consider a pattern vector with a missing value given by $(1, -, 2)$. Its nearest centroid among the three centroids based on the remaining two features is Centroid 1, $(1,2,2)$. So, the missing value in the second location is predicted to be the second component in Centroid 1, that is, 2. So, the pattern with the missing value becomes $(1,2,2)$.

We will now illustrate how the KNN-based and mean-based schemes work on a bigger data set. We consider 20,640 patterns of the California Housing data set. It has 8 features and the target variable is the median house value for California districts, expressed in hundreds of thousands of dollars ($\$100,000$). This is a regression problem. Some values in the data set are removed to create missing values. The missing values are imputed using the KNN scheme and the mean-based scheme. Now the regressor (a function to predict the target) is used on the whole data set without missing values, on the KNN-based imputed data set and the mean-based imputed data set. The resulting mean squared error of the predictions of the regressor is shown in Fig. 1.7.

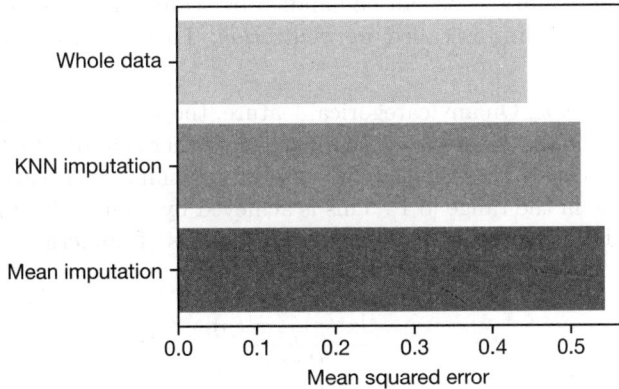

FIG. 1.7 MSE of the regressor on data imputed using KNN and mean

It is easy to observe that the regressor performs best when the whole data is available. However, when prediction is made by removing some values and guessing them, the performance of the regressor suffers; this is natural. Note that between the KNN-based and the mean-based imputations, the former made better predictions leading to smaller MSE. This is because the KNN-based scheme is more local to the respective point x and so is more focussed.

2. **Data from Different Domains:** The scales of values of different features could be very different. This would bias the matching process to depend more upon features that assume larger values, toning down the contributions of features with smaller values. So, in applications where different components of the vectors have different domain ranges, it is possible for some components to dominate in contributing to the distance between any pair of patterns. Consider for example, classification of objects into one of two classes: *adult* or *child*. Let the objects be represented by *height* in metres and *weight* in grams. Consider an adult represented by the vector $(1.6, 75000)$ and a child represented by the vector $(0.6, 5000)$, where the heights of the adult and the child in metres are 1.6 and 0.6, respectively, and the weights of the adult and the child in grams are 75000 and 5000, respectively. Assume that the domain of *height* is $[0.5, 2.5]$ and the domain of *weight* is $[2000, 200000]$ in this example. So, there is a large difference in the ranges of values of these two features.

Now the Euclidean distance between the adult and child vectors given above is

$$\sqrt{1.6 - 0.6)^2 + (75000 - 5000)^2} = \sqrt{1 + 4.9 \times 10^9} \approx 70000$$

Similarly, the cosine of the angle between the adult and child vectors is

$$\frac{0.96 + 375 \times 10^6}{\sqrt{25000000.36} \times \sqrt{5625000002.56}} \approx 1.0$$

Note that the proximity values computed between the two vectors, whether it is the Euclidean distance or the cosine of the angle between the two vectors, are dependent largely upon only one of the two features, that is, *weight*, while the contributions of *height* are negligible. This is because of the difference in the magnitudes of the ranges of values of the two features. This example illustrates how the magnitudes/ranges of values of different features contribute differently to the overall proximity. This can be handled by scaling different components differently and such a process of scaling is called *normalization*. There are two popular normalization schemes:

- *Scaling using the range:* On any categorical feature, the values of two patterns either match or mismatch and the contribution to the distance is either zero (0) (match) or 1 (mismatch). If we want to be consistent, then in the case of any numerical feature also we want the contribution to be in the range $[0,1]$. This is achieved by scaling the difference by the range of the values of the feature. So, if the pth component is of numerical type, its contribution to the distance between patterns x^i and x^j is

$$\frac{|x^i(p) - x^j(p)|}{Range_p},$$

  where $Range_p$ is the range of the pth feature values. Note that the value of this term is in the range $[0,1]$; the value of 1 is achieved when $|x_p^i - x_p^j| = Range_p$ and it is 0 (zero) if patterns x^i and x^j have the same value for the pth feature. Such a scaling will ensure that the contribution, to the distance, of either a categorical feature or a numerical feature will be in the range $[0,1]$.

- *Standardization:* Here, the data is normalized so that it will have 0 (zero) mean and unit variance. This is motivated by the property of standard normal distribution, which is characterized by zero mean and unit variance.

  EXAMPLE 4:   Let there be 5 l-dimensional data vectors and let the the qth components of the 5 vectors be 60, 80, 20, 100 and 40. The mean of this collection is

$$\frac{60 + 80 + 20 + 100 + 40}{5} = 60$$

  We get zero-mean data by subtracting this mean from each of the 5 data items to obtain $0, 20, -40, 40, -20$ for their q^{th} components. Note that this is zero-mean data as these values add up to 0. To make the standard deviation of this data 1, we divide each of the zero-mean data values by the standard deviation of the data. Note that the variance of the zero-mean data is

$$\frac{0 + 20^2 + (-40)^2 + 40^2 + (-20)^2}{5} = 800$$

  and the standard deviation is 28.284. So, the scaled data is $0, 0.707, -1.414, 1.414, -0.707$. Note that this data corresponding to the qth feature value of the 5 vectors has zero mean and unit variance.

3. **Outliers in the Data:** An outlier is a data item that is either noisy or erroneous. Noisy measuring instruments or erroneous data recordings are responsible for the presence of such outliers. A common problem across various applications is the presence of outliers. A data item is usually called an outlier if it

   - Assumes values that are far away from those of the average data items
   - Deviates from the normally behaving data item
   - Is not connected/similar to any other object in terms of its characteristics

Outliers can occur because of different reasons:

- *Noisy measurements:* The measuring instruments may malfunction and may lead to recording of noisy data. It is possible that the recorded value lies outside the domain of the data type.
- *Erroneous data entry:* Outlying data can occur at the data entry level itself. For example, it is very common for spelling mistakes to occur when names are entered. Also, it is possible to enter numbers such as salary erroneously as 2000000 instead of 200000 by typing an extra zero (0).
- *Evolving systems:* It is possible to encounter data items in sparse regions during the evolution of a system. For example, it is common to encounter isolated entities in the early days of a social network. Such isolated entities may or may not be outliers.
- *Very naturally:* Instead of viewing an outlier as a noisy or unwanted data item, it may be better to view it as something useful. For example, a novel idea or breakthrough in a scientific discipline, a highly paid sportsperson or an expensive car can all be natural and influential outliers.

An outlying data item can be either out-of-range or within-range. For example, consider an organization in which the salary could be from $\{10000, 150000, 225000, 300000\}$. In this case, an entry like 2250000 is an out-of-range outlier that occurs possibly because of an erroneous zero (0). Also, if there are only 500 people drawing 10000, 400 drawing 150000, 300 at 225000 and 175 drawing 300000, then an entry like 270000 could be a within-range outlier.

There are different schemes for detecting outliers. They are based on the density around points in the data. If a data point is located in a sparse region, it could be a possible outlier. It is possible to use clustering to locate such outliers. It does not matter whether it is within-range or out-of-range. If the clustering output has a singleton cluster, that is, a one-element cluster, then that element could be a possible outlier.

1.5.3   Data Representation

Representation is an important step in building ML models. This subsection introduces how data items are represented. It also discusses the importance of representation in ML. In the process, it deals with both feature selection and feature extraction and introduces different categories of dimensionality reduction.

It is often stated in DL literature that feature engineering is important in ML, but not in DL because DL systems have automatic representation learning capability. This is a highly debatable issue. However, it is possible that, in some application domains, DL systems can avoid the representation step explicitly. However, preprocessing including handling missing data and eliminating outliers is still an important part of any DL system. Even though representation is not explicit, it is implicitly handled in DL by choosing the appropriate number of layers and number of neurons in each layer of the neural network.

Representation of Data Items

The most active and state-of-the-art paradigm for ML is **statistical machine learning**. Here, each data item is represented as a vector. Typically, we consider addition of vectors in computing the mean or centroid of a collection of vectors, multiplication of a vector by a scalar in dealing with operations on matrices, and the dot product between a pair of vectors for computing similarity as important operations on the set of vectors. In most of the practical applications, the dimensionality of the data or correspondingly size of the vectors representing data items, L, can be very large. For example, there are around 468 billion Google Ngrams. In this case, the dimensionality of the vectors is the vocabulary size or the number of Ngrams; so, the dimensionality could be very large. Such high-dimensional data is common in bioinformatics, information retrieval, satellite imagery, and so on. So, representation is an important component of any ML system. An arbitrary representation may also be adequate to build an ML model. However, the predictions made using such a model may not be meaningful.

Current-day applications deal with high-dimensional data. Some of the difficulties associated with ML using such high-dimensional data vectors are:

- Computation time increases with the dimensionality.
- Storage space requirement also increases with the dimensionality.
- *Performance of the model:* It is well-known that as the dimensionality increases, we require a larger training data set to build an ML model. There is a result, popularly called the **peaking phenomenon**, that shows that as the dimensionality keeps increasing, the accuracy of a classification model increases until some value, and beyond that value, the accuracy starts decreasing.

This may be attributed to the well-known concept of **overfitting**. The model will tend to remember the training data and fails to perform well on validation data. With a larger training data set, we can afford to use a higher dimensional data set and still avoid overfitting. Even though the dimensionality of the data set in an application is large, it is possible that the number of available training vectors is small. In such cases, a popular technique used in ML is to reduce the dimensionality so that the learnt model does not overfit the available data.

Well-known dimensionality reduction approaches are:

- **Feature selection:** Let $\mathcal{F} = \{f_1, f_2, \ldots, f_L\}$ be the set of L features. In the feature selection approach, we would like to select a subset F_l of $\mathcal{F}$ having $l (< L)$ features such that F_l maximizes the performance of the ML model.
- **Feature extraction:** Here, from the set $\mathcal{F}$ of L features, a set $\mathcal{H} = \{h_1, h_2, \ldots, h_l\}$ of l $(< L)$ features is extracted. It is possible to categorize these schemes as follows:

1. *Linear schemes:* In this case,

$$h_j = \sum_{i=1}^{L} \alpha_{ij} f_i$$

That is, each element of $\mathcal{H}$ is a linear combination of the original features. Note that feature selection is a specialization of feature extraction. Some prominent schemes under this category are:

a. *Principal components (PCs):* Consider the data set of n vectors in an L-dimensional space; this may be represented as a matrix A of size $n \times L$. The covariance matrix Σ of size $L \times L$ associated with the data is computed and the eigenvectors of Σ form the principal components. The eigenvector corresponding to the largest eigenvalue is the first principal component (PC). Similarly, the second largest eigenvalue provides its corresponding eigenvector as the second PC. Finally, the eigenvector corresponding to the lth largest eigenvalue is the lth PC. Both the original feature vectors and PCs are sufficiently powerful to represent any data vector. So, PCs may be viewed as linear combinations of the given features.

b. *Non-negative matrix factorization (NMF):* Even when the data is non-negative, it is possible that PCs have negative entries. However, it is useful to have representations using non-negative entries; NMF is such a factorization of $A_{n \times L}$ into a product of $B_{n \times l}$ and $C_{l \times L}$. Its use is motivated by the notion that NMF can be used to characterize objects in an image represented by A. In NMF, the columns of B can be viewed as linear combinations of the columns of A because of linear independence.

We will examine, in detail, the concepts of eigenvalue, eigenvector and linear independence in later chapters.

2. *Non-linear feature extraction:* Here, we represent using $\mathcal{H} = \{h_1, \ldots, h_l\}$, such that

$$h_i = t(f_1, f_2, \ldots, f_L),$$

where t is a non-linear function of the features. For example, if $\mathcal{F} = \{f_1, f_2\}$, then $h_1 = af_1 + bf_2 + cf_1 f_2$ is one such non-linear combination; it is non-linear because we have a term of the form $f_1 f_2$ in h_1.

Autoencoder is a popular, state-of-the-art, non-linear feature extraction tool. Here, a neural network which has an encoder and a decoder is used. The middle layer has l neurons so that the l outputs from the middle layer give an $l(< L)$-dimensional representation of the L-dimensional pattern that is input to the autoencoder. Note that the encoder encodes or represents the L-dimensional pattern in the l-dimensional space while the decoder decodes or converts the l-dimensional pattern into the L-dimensional space. Note that it is called autoencoder because the same L-dimensional pattern is associated with the input and output layers.

1.5.4   Model Selection

Selection of the model to be used to train an ML system depends upon the nature of the data and knowledge of the application domain. For some applications, only a subset of the ML models can be used. For example, if some features are numerical and others are categorical, then classifiers based on perceptrons and support vector machine (SVM) are not suitable as they compute the dot product between vectors and dot products do not make sense when some values in the corresponding vectors are non-numerical. On the other hand, Bayesian models and decision tree–based models are ideally suited to deal with such data as they depend upon the frequency of occurrence of values.

1.5.5   Model Learning

This step depends on the size and type of the training data. In practice, a subset of the labelled data is used as training data for learning the model and another subset is used for model validation or model evaluation. Some of the ML models are highly transparent while others are opaque or black box models. For example, decision tree–based models are ideally suited to provide transparency; this is because in a decision tree, at each internal or decision node, branching is carried out based on the value assumed by a feature. For example, if the height of an object is larger than 5 feet, it is likely to be an adult and not a child; such easy-to-understand rules are abstracted by decision trees. Neural networks are typically opaque as the outputs of intermediate/hidden layer neurons may not offer transparency.

1.5.6   Model Evaluation

This step is also called **model validation**. This step requires specifically earmarked data called **validation data**. It is possible that the ML model works well on the training data; then we say that the model is well trained. However, it may not work well on the validation data. In such a case, we say that the ML model overfits the training data. In order to overcome overfitting, we typically use the validation data to tune the ML model so that it works well on both the training and validation data sets.

1.5.7   Model Prediction

This step deals with testing the model that is learnt and validated. It is used for prediction because both classification and regression tasks are predictive tasks. This step employs the test data set earmarked for the purpose. In the real world, the model is used for prediction as new patterns keep coming in. Imagine an ML model built for medical diagnosis. It is like a doctor who predicts and makes a diagnosis when a new patient comes in.

1.5.8   Model Explanation

This step is important to explain to an expert or a manager why a decision was taken by the ML model. This will help in explicit or implicit feedback from the user to further improve the model. Explanation had an important role earlier in expert systems and other AI systems. However, explanation has become very important in the era of DL. This is because DL systems typically employ neural networks that are relatively opaque. So, their functioning cannot be easily explained at a level of detail that can be appreciated by the domain expert/user. Such opaque behaviour has created the need for explainable AI.

1.6   SEARCH AND LEARNING

Search is a very basic and fundamental operation in both ML and AI. Search had a special role in conventional AI where it was successfully used in problem solving, theorem proving, planning and knowledge-based systems.

Further, search plays an important role in several computer science applications. Some of them are as follows:

- Exact search is popular in databases for answering queries, in operating systems for operations like *grep*, and in looking for entries in symbol tables.
- In ML, search is important in learning a classification model, a proximity measure for clustering and classification, and the appropriate model for regression.
- Inference is search in logic and probability. In linear algebra, matrix factorization is search. In optimization, we use a regularizer to simplify the search in finding a solution. In information theory, we search for purity (low entropy).

So, several activities including optimization, inference and matrix factorization that are essential for ML are all based on search. Learning itself is search. We will examine how search aids learning of each ML model in the respective chapters.

1.7   EXPLANATION OFFERED BY THE MODEL

Conventional AI systems were logic-based or rule-based systems. So, the corresponding reasoning systems naturally exhibited transparency and, as a consequence, explainability. Both forward and backward reasoning was possible. In fact, the same knowledge base, based on experts' input, was used in both diagnosis and in teaching because of this flexibility. Specifically, the knowledge base used by the MYCIN expert system was used in tutoring medical students using another expert system called GUIDON.

However, there were some problems associated with conventional AI systems:

- There was no general framework for building AI systems. Acquiring knowledge, using additional heuristics and dealing with exceptions led to adhocism; experience in building one AI system did not simplify the building of another AI system.
- Acquiring knowledge was a great challenge. Different experts typically differed in their conclusions, leading to inconsistencies. Conventional logic-based systems found it difficult to deal with such inconsistent knowledge.

There has been a gradual shift from using knowledge to using data in building AI systems. Current-day AI systems, which are mostly based on DL, are by and large data dependent. They can learn representations automatically. They employ variants of multi-layer neural networks and backpropagation algorithms in training models.

Some difficulties associated with DL systems are:

- They are data dependent. Their performance improves as the size of the data set increases. So, they need larger data sets. Fortunately, it is not difficult to provide large data sets in most of the current applications.
- Learning in DL systems involves a simple change of weights in the neural network to optimize the objective function. This is done with the help of backpropagation, which is a gradient-descent algorithm and which can get stuck with a locally optimal solution. Combining this with large data sets may possibly lead to overfitting. This is typically avoided by using a variety of simplifications in the form of regularizers and other heuristics.
- A major difficulty is that DL systems are black box systems and lack explanation capability. This problem is currently attracting the attention of AI researchers.

We will be discussing how each of the ML models is equipped with explanation capability in the respective chapters.

1.8   DATA SETS USED

In this book, we make use of two data sets to conduct experiments and present results in various chapters. These are:

- **Data Sets for Classification**

  1. *MNIST Handwritten Digits Data Set:*
     - There are 10 classes (corresponding to digits 0, 1, ..., 9) and each digit is viewed as an image of size 28×28 ($= 784$) pixels; each pixel having values in the range 0 to 255.
     - There are around 6000 digits as training patterns and around 1000 test patterns in each class and the class label is also provided for each of the digits.
     - For more details, visit `http://yann.lecun.com/exdb/mnist/`

  2. *Fashion MNIST Data Set:*
     - It is a data set of Zalando's article images, consisting of a training set of 60,000 examples and a test set of 10,000 examples.

- Each example is a 28×28 greyscale image, associated with a label from 10 classes.
- It is intended to serve as a possible replacement for the original MNIST data set for benchmarking ML models.
- It has the same image size and structure of training and testing splits as the MNIST data.
- For more details, visit `https://www.kaggle.com/datasets/zalando-research/fashionmnist`

3. *Olivetti Face Data Set:*

   - It consists of 10 different images each of 40 distinct subjects. For some subjects, the images were taken at different times, varying the lighting, facial expressions (open / closed eyes, smiling / not smiling) and facial details (glasses / no glasses).
   - All the images were taken against a dark homogeneous background with the subjects in an upright, frontal position (with tolerance for some side movement).
   - Each image is of size $64 \times 64 = 4096$.
   - It is available on the scikit-learn platform.
   - For more details, visit `https://ai.stanford.edu/~marinka/nimfa/nimfa.examples.orl\_images.html`

4. *Wisconsin Breast Cancer Data Set:*

   - It consists of 569 patterns and each is a 30-dimensional vector.
   - There are two classes *Benign* and *Malignant*. The number of patterns from *Benign* is 357 and the number of *Malignant* class patterns is 212.
   - It is available on the scikit-learn platform.
   - For more details, visit
     `https://scikit-learn.org/stable/modules/generated/sklearn.datasets.load\_breast\_cancer.html`

- **Data Sets for Regression**

1. *Boston Housing Data Set:*

   - It has 506 patterns.
   - Each pattern is a 13-dimensional vector.
   - It is available on the scikit-learn platform.
   - For more details, visit
     `https://scikit-learn.org/0.15/modules/generated/sklearn.datasets.load\_boston.html`

2. *Airline Passengers Data Set:*

   - This data set provides monthly totals of US airline passengers from 1949 to 1960.
   - This data set is taken from an inbuilt data set of Kaggle called AirPassengers.
   - For more details, visit
     `https://www.kaggle.com/datasets/chirag19/air-passengers`

3. *Australian Weather Data Set:*

   - It provides various weather record details for cities in Australia.
   - The features include location, min and max temperature, etc.
   - For more details, visit `https://www.kaggle.com/datasets/arunavakrchakraborty/australia-weather-data`

SUMMARY

Machine learning (ML) is an important topic and has affected research practices in both science and engineering significantly. The important steps in building an ML system are:

- Data acquisition that is application domain dependent.
- Feature engineering that involves both data preprocessing and representation.
- Selecting a model based on the type of data and the knowledge of the domain.
- Learning the model based on the training data.
- Evaluating and tuning the learnt model based on validation data.
- Providing explanation capability so that the model is transparent to the user/expert.

EXERCISES

1. You are given that 9×17 is 153 and 4×17 is 68. From this data, you need to learn the value of 13×17. Which learning paradigm is useful here? Specify any assumptions you need to make.

2. You are given the following statements:

   a. The sum of two even numbers is even.
   b. 12 is an even number.
   c. 22 is an even number.

   What can you deduce from the above statements?

3. Consider the following statements:

   a. If x is an even number, then $x + 1$ is odd and $x + 2$ is even.
   b. 34 is an even number.

   Which learning paradigm is used to learn that 37 is odd and 38 is even?

4. Consider the following reasoning:

   a. If x is odd, then $x + 1$ is even.
   b. 22 is even.

   You have learnt from the above that 21 is odd. Which learning paradigm is used? Specify any assumptions to be made.

5. Consider the following attributes. Find out whether they are nominal, ordinal or numerical features. Give a reason for your choice.

   a. Telephone number
   b. Feature that takes values from {ball, bat, wicket, umpire, batsman, bowler}
   c. Temperature
   d. Weight
   e. Feature that takes values from {short, medium height, tall}

6. Let x_i and x_j be two l-dimensional unit norm vectors; that is, $\| x_i \| = 1$ and $\| x_j \| = 1$. Derive a relation between the Euclidean distance $d(x_i, x_j)$ and cosine of the angle between x_i and x_j.

7. Consider the data set:

$$(1, 1, 1), (1, 1, 2), (1, 1, 3), (1, 2, 2), (1, 1, -), (6, 6, 10).$$

Predict the missing value in the pattern vector $(1, 1, -)$ using KNNs with $K = 3$. What happens when $K = 5$?

8. Consider the data in Q7 above. Use this data to predict the missing value in the pattern vector $(100, -, 100)$ with $K = 1$. Is there any problem with $(100, -, 100)$? What can you say about this pattern?

9. Consider the following 4 two-dimensional vectors. Let

$$X_1 = (1, 100000), X_2 = (2, 100000), X_3 = (1, 200000), X_4 = (2, 200000)$$

   a. How do we use range scaling scheme in this example.
   b. Normalize the data using standard scaler.

10. Show that feature selection is a special case of feature extraction.

Bibliography

- Sastry, PS. "Introduction to Statistical Pattern Recognition" in *Pattern Recognition*, National Programme on Technology Enhanced Learning. https://nptel.ac.in/courses/117108048

- Murphy, KP. 2012. *Machine Learning: A Probabilistic Perspective*, The MIT Press: Cambridge, England.

- Murty, MN and Biswas, A. 2019. *Centrality and Diversity in Search: Roles in A.I., Machine Learning, Social Networks, and Pattern Recognition*, Springer Briefs in Intelligent Systems, Springer.

- Aggarwal, M and Murty MN. 2021. *Machine Learning in Social Networks: Embedding Nodes, Edges, Communities, and Graphs*, Springer Nature.

- Bishop, CM. 2005. *Neural Networks for Pattern Recognition*, Oxford University Press.

- Murty MN and Susheela Devi, V. 2015. *Introduction to Pattern Recognition and Machine Learning*, World Scientific Publishing Co. Pte. Ltd.: Singapore.

- Murty MN and Susheela Devi, V. 2011. *Pattern Recognition: An Algorithmic Approach*, Springer Science and Business Media.

- Ranga Suri, NNR, Murty MN and Athithan G. 2019. *Outlier Detection: Techniques and Applications - A Data Mining Perspective*, Intelligent Systems Reference Library, Springer.

Colour Plate 1

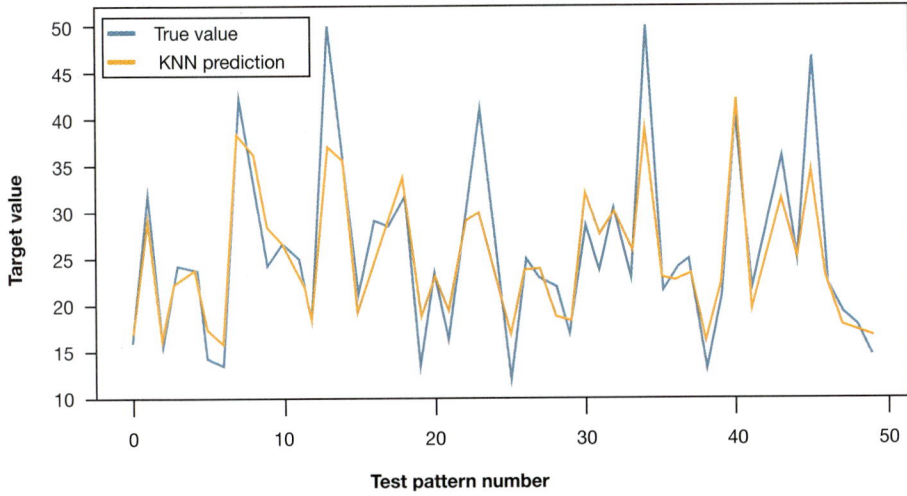

FIG. 2.7 KNN regression: results on the Boston Housing data

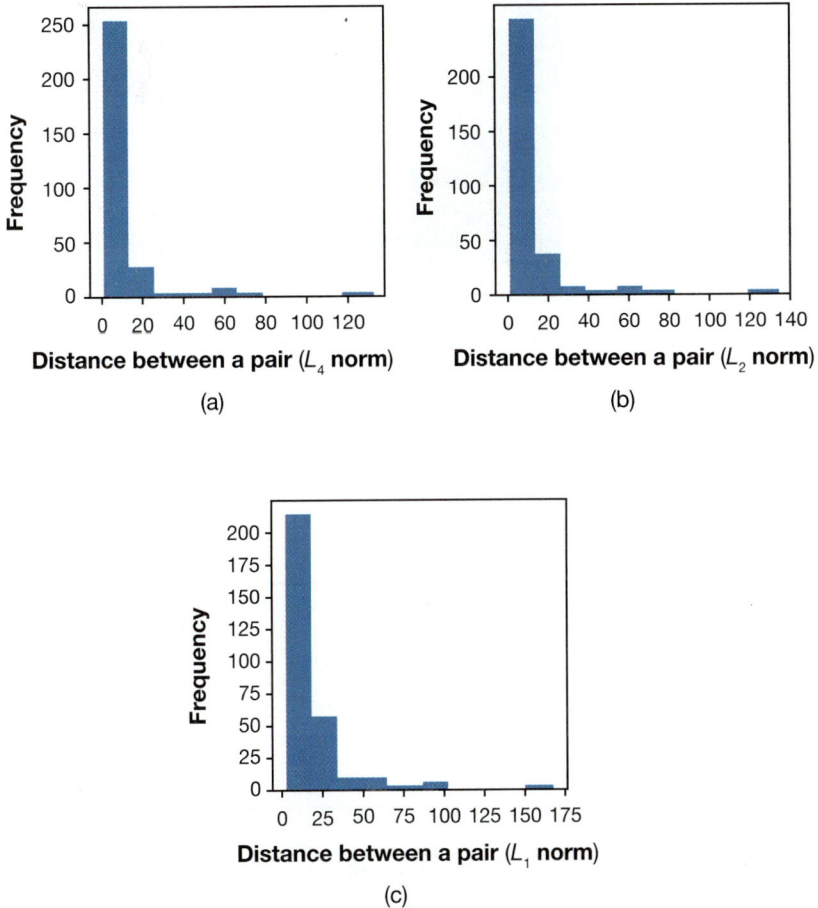

(a)

(b)

(c)

FIG. 2.9 Concentration of distances using different norms

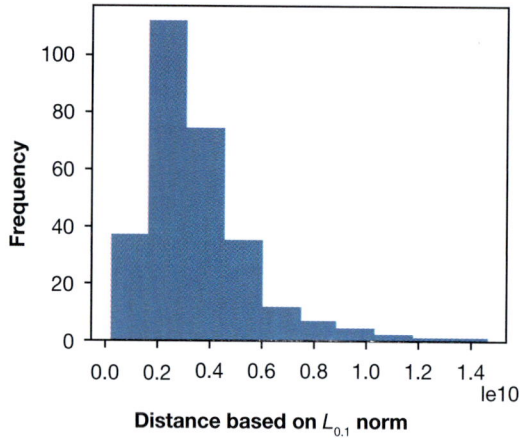

(a)

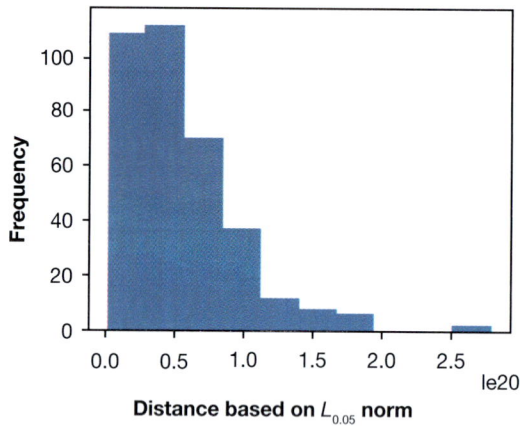

(b)

FIG. 2.10 Concentration of distances using different norms

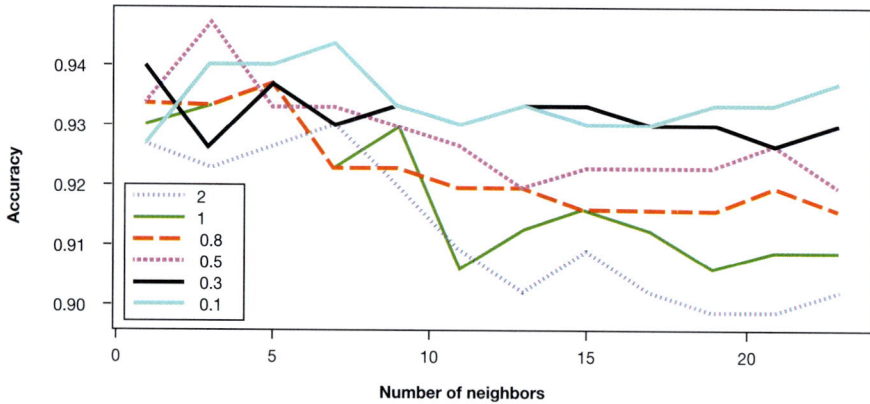

FIG. 2.11 Accuracy of KNN using different norms

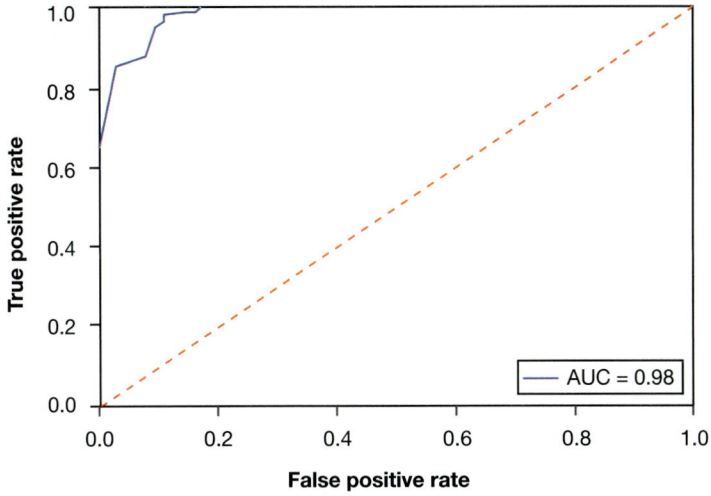

Fig. 2.12 ROC curve for Breast Cancer data using KNN

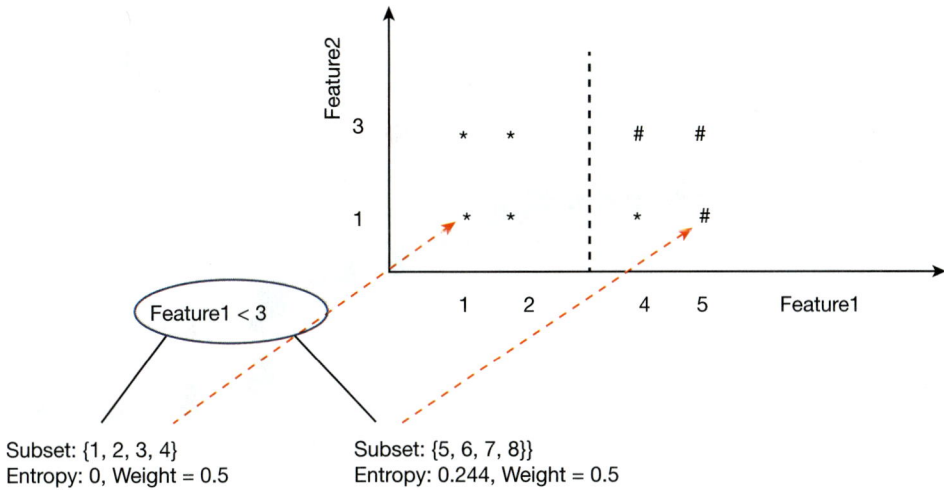

Fig. 3.2 The process of splitting at the root

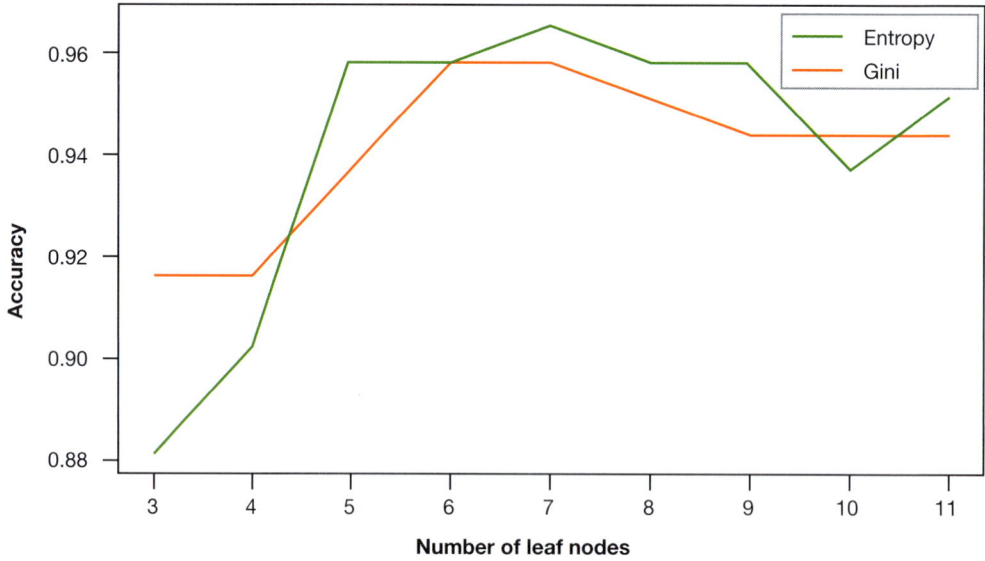

FIG. 3.11 Accuracies of decision trees using Gini index and entropy

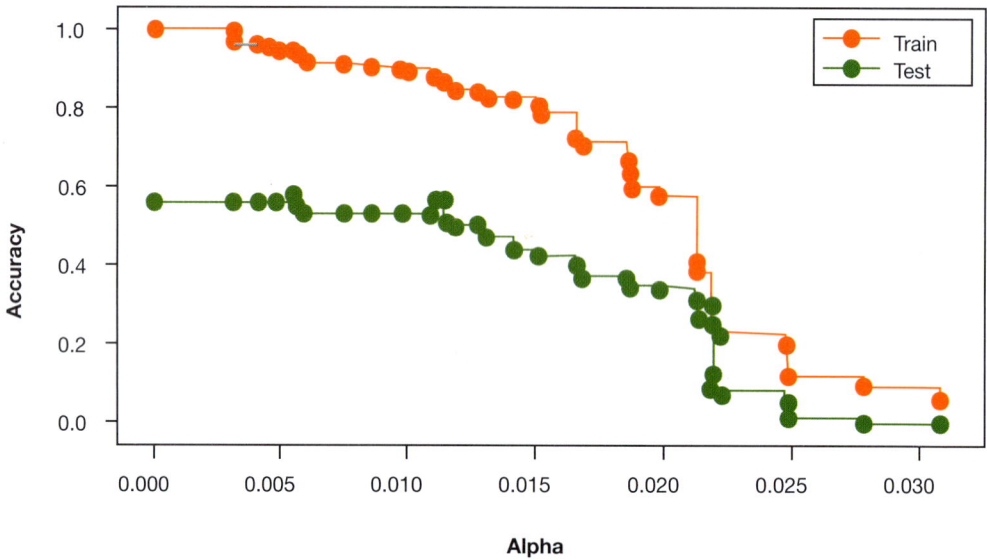

FIG. 3.14 Alpha versus accuracy on the Olivetti Face data set

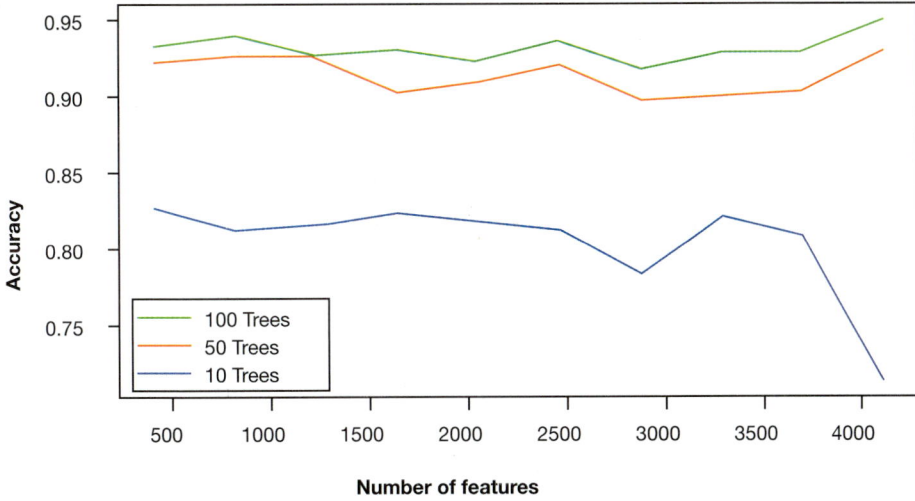

FIG. 3.16 Number of features versus accuracy using a random forest classifier on the Olivetti Face data set

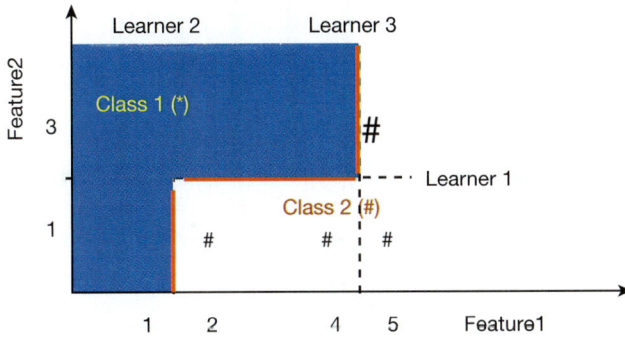

FIG. 3.20 Illustration of the boundary between the two classes using AdaBoost

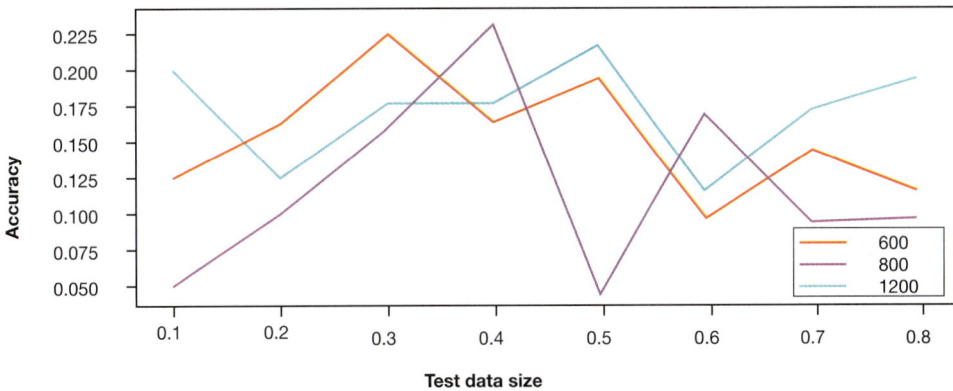

FIG. 3.23 Results of the AdaBoost classifier on the Olivetti Face data set with different number of weak learners (600, 800, 1200) and different test data sizes

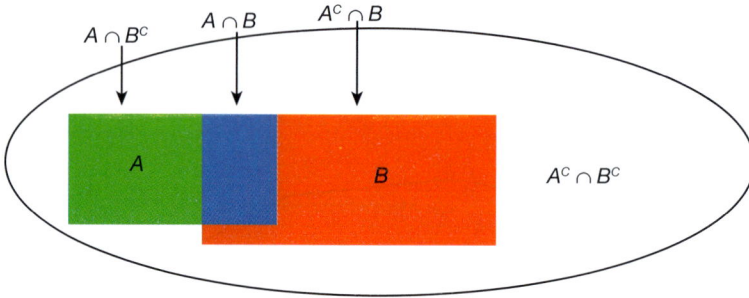

FIG. 4.1 Some events related to two given events A and B

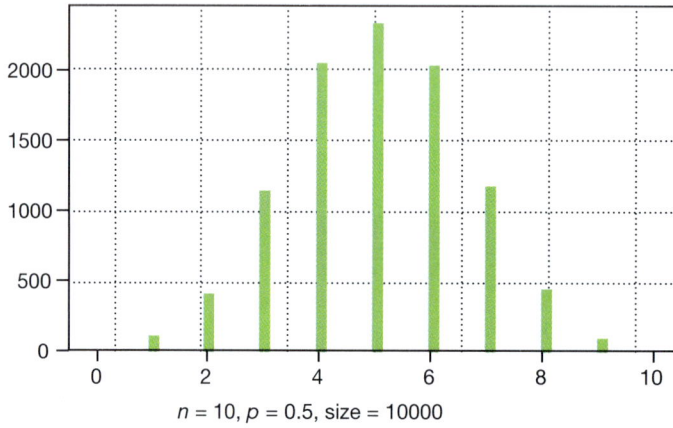

$n = 10$, $p = 0.5$, size = 10000

(a)

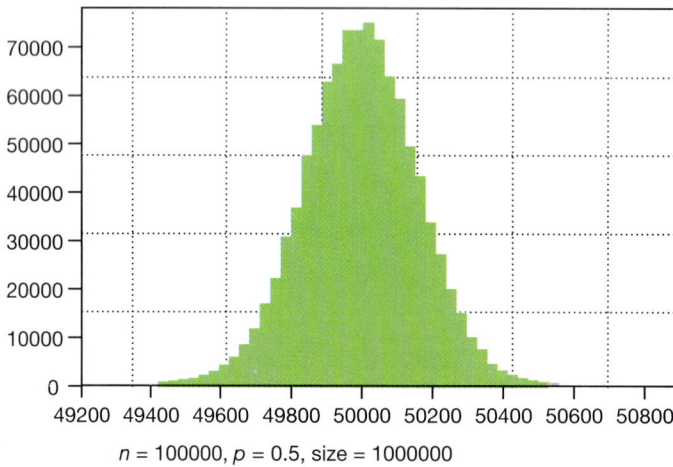

$n = 100000$, $p = 0.5$, size = 1000000

(b)

FIG. 4.4 Probability mass function of the binomial RV

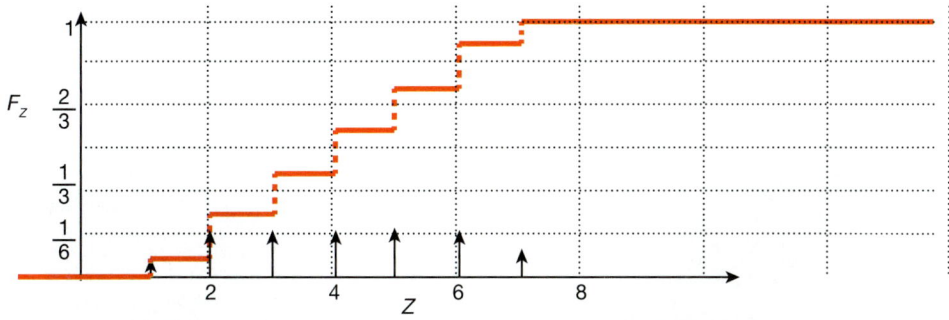

FIG. 4.5 An example of the cumulative distribution function of an RV

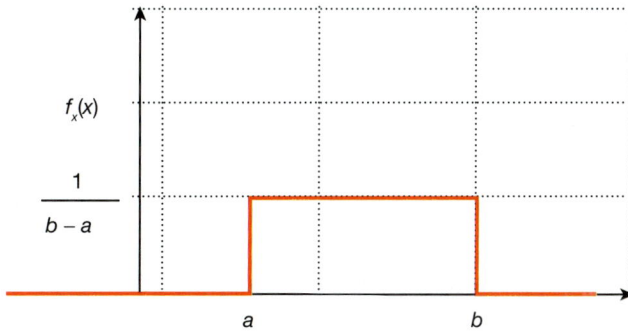

FIG. 4.6 Probability density function of a uniform RV X

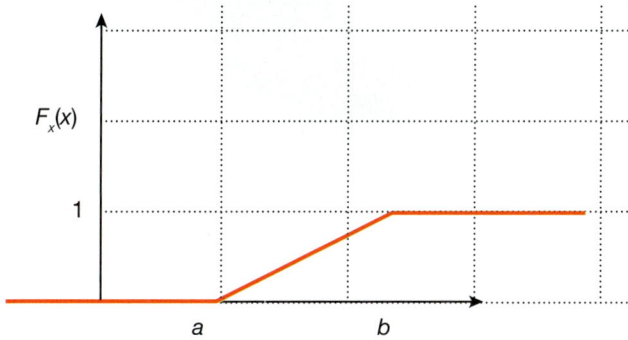

FIG. 4.7 Cumulative distribution function of the uniform RV X

Covariance between x_1 and $x_2 = -0.66$

Covariance between x_1 and $x_2 = 0$

$$\Sigma = \begin{bmatrix} 1 & -0.66 \\ -0.66 & 1 \end{bmatrix}$$

(a)

$$\begin{bmatrix} 1 & 0 \\ 0 & 1 \end{bmatrix}$$

(b)

Covariance between x_1 and $x_2 = 0.66$

$$\begin{bmatrix} 1 & 0.66 \\ 0.66 & 1 \end{bmatrix}$$

(c)

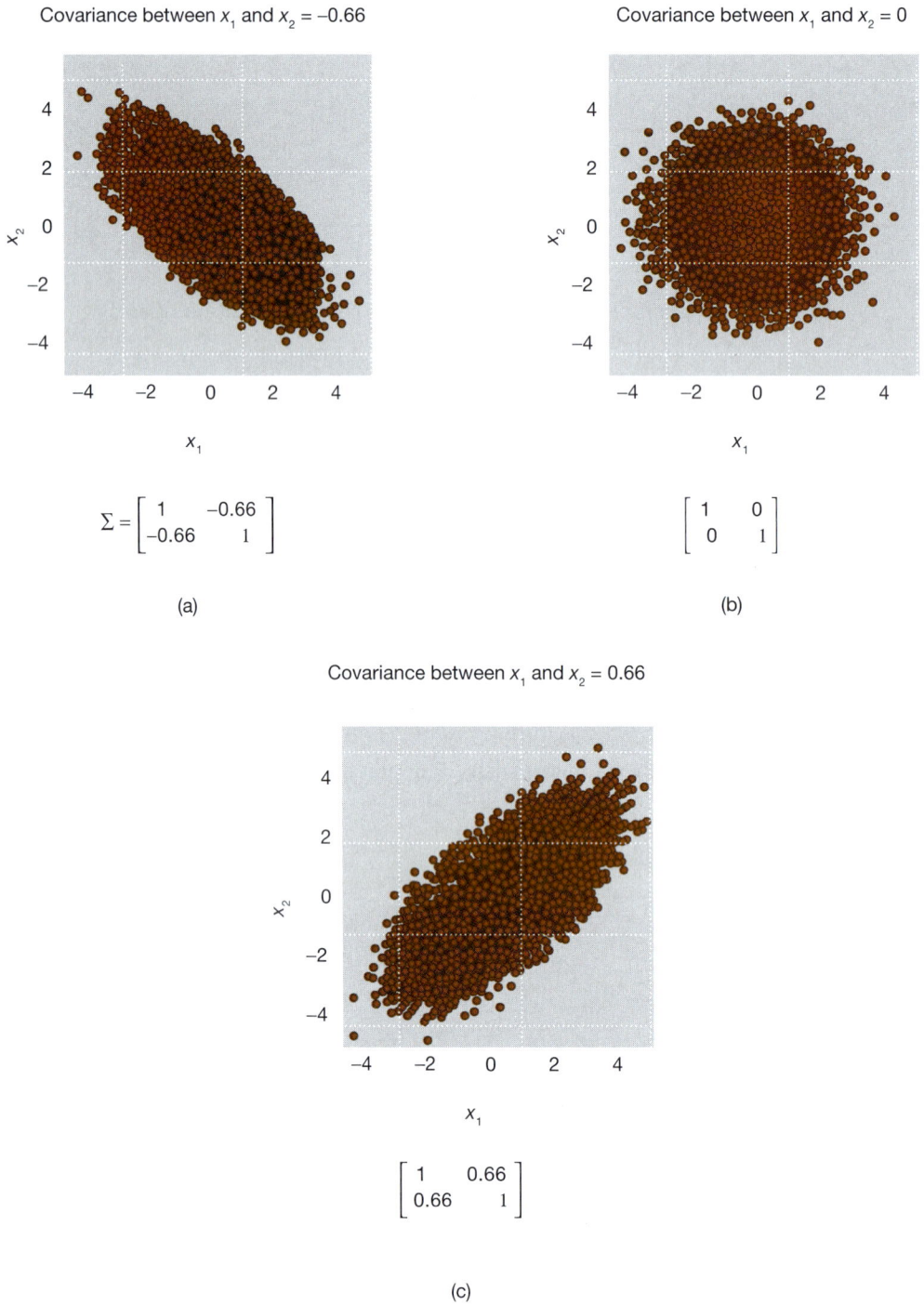

FIG. 4.10 Normally distributed two-dimensional points

CHAPTER 2

Nearest Neighbor-Based Models

Learning Objectives

At the end of this chapter, you will be able to:

- Explain the proximity measures of classification
- Give a brief overview of distance measures
- Describe the ML models that use the nearest neighbor concept

2.1 INTRODUCTION TO PROXIMITY MEASURES

Proximity measures are used to quantify the degree of similarity or dissimilarity between two or more pattern vectors. These pattern vectors can represent documents, images or even entire audio or video files. Proximity measures are often used by machine learning (ML) algorithms to compare and classify or group or make predictions using these patterns. There are many types of proximity measures and the popular ones are:

- **Euclidean distance:** This the most popular distance as it is intuitively appealing. It measures the straight-line distance between two points in a multi-dimensional space. So, it is also called *as the crow flies* distance.
- **Cosine similarity:** This measures the cosine of the angle between two vectors and is often used in text analysis to compute similarity between a pair of documents.
- **Jaccard similarity:** This measures the ratio of the cardinalities of intersection over union of two sets and is often used in recommendation systems to compare users' preferences.
- **Hamming distance:** This measures the number of positions at which two binary strings differ and is often used in error correction codes.

In the following section, we will discuss various distance measures.

2.2 DISTANCE MEASURES

A distance measure is used to find the dissimilarity between patterns represented as vectors. Patterns which are more similar should be closer. The distance function (d) could be a metric or a non-metric. The most popularly used family of distance metrics is called the **Minkowski metric**.

A **metric** is a type of measure that possesses three key attributes: positive reflexivity, symmetry and triangular inequality:

- *Positive reflexivity:* $d(x, x) = 0$
- *Symmetry:* $d(x, y) = d(y, x)$
- *Triangular inequality:* $d(x, y) \leq d(x, z) + d(z, y)$, where x, y, z are any three patterns.

The following sections describe different types of dissimilarity/similarity (proximity) measures between pattern vectors.

2.2.1   Minkowski Distance

The Minkowski metric is a commonly used family of distance metrics which can be expressed as

$$d^r(p, q) = \left(\sum_{k=1}^{L} (|p_k - q_k|^r) \right)^{\frac{1}{r}},$$

where r is a parameter that determines the type of metric being used and p and q are l-dimensional vectors. Some variations based on selecting the value of r are:

- L_∞ *norm:* Here, $r = \infty$ and $d(p, q) = maximum_k(|p(k) - q(k)|)$, $k \in \{1, \ldots, L\}$.
- L_2 *norm:* In this case, $r = 2$ and $d(p, q) = (\sum_{k=1}^{L} (|p(k) - q(k)|^2))^{\frac{1}{2}}$ is the Euclidean distance; this is the most popular variation.
- L_1 *norm:* In this case, $r = 1$ and $d(p, q) = (\sum_{k=1}^{L} (|p(k) - q(k)|))$ is the city-block distance.
- *Fractional norm:* It is possible that r is a fraction. In such a case, the resulting distance is called fractional norm. It is not a metric as it violates the triangle inequality.

The importance of different norms will be examined while explaining the nearest neighbor classifiers. It is important to ensure that all features used in the distance measure have the same range of values, as attributes with larger ranges may gain undue advantage. **Normalisation** of feature values can help to ensure that they are in the same range.

The **Mahalanobis distance** is another popular distance measure that is used in classification, and it is computed using the covariance matrix. The squared Mahalanobis distance is given by

$$d^2(x, y) = (x - y)^t \Sigma^{-1} (x - y)$$

EXAMPLE 1:   If $x = (5, 2, 4)$ and $y = (3, 4, 2)$, the Euclidean distance between them is $d(x, y) = \sqrt{(5 - 3)^2 + (2 - 4)^2 + (4 - 2)^2}$, which equals 3.46.

2.2.2   Weighted Distance Measure

To assign greater importance to certain attributes, a weight can be applied to their values in the weighted distance metric. This metric takes the form

$$d(x, y) = \left(\sum_{k=1}^{L} w_k \times (x(k) - y(k))^r \right)^{\frac{1}{r}},$$

where w_k represents the weight associated with the kth dimension or feature.

EXAMPLE 2: If $x = (5, 2, 4)$ and $y = (3, 4, 2)$, with weights assigned as $w_1 = 0.3$, $w_2 = 0.5$ and $w_3 = 0.2$, then $d(x, y)$ is calculated as $0.3 \times (5 - 3)^2 + 0.5 \times (2 - 4)^2 + 0.2 \times (4 - 2)^2 = 4$.

The weights determine the significance of each feature, with the second feature being more important than the first, and the third feature being the least significant in this example.

2.2.3   Non-Metric Similarity Functions

This category includes similarity functions that do not obey the triangular inequality or symmetry. They are commonly used for image or string data and they are resistant to outliers or extremely noisy data. The squared Euclidean distance is an example of a non-metric, but it provides the same ranking as the Euclidean distance metric.

One example of a non-metric similarity function is the k-median distance between two vectors. Given $x = (x(1), x(2), \ldots, x(l))$ and $y = (y(1), y(2), \ldots, y(l))$, the formula for the k-median distance is

$$d(x, y) = k\text{-median}\{|x(1) - y(1)|, \ldots, |x(n) - y(n)|\},$$

where the k-median operator returns the kth value of the ordered difference vector.

Another similarity measure is

$$S(x, y) = \frac{x^t y}{\|x\| \|y\|},$$

which corresponds to the cosine of the angle between the vectors x and y. It is symmetric because $cos(\theta) = cos(-\theta)$ and it represents the similarity between x and y. A possible distance function corresponding to the cosine similarity, $S(x, y)$ is

$$d(x, y) = 1 - S(x, y)$$

Note that $d(x, y)$ is symmetric as $S(x, y)$ is symmetric. However, $d(x, y)$ violates the triangular inequality. Example 3 demonstrates how it violates the inequality.

EXAMPLE 3: Let x, y and z be three vectors in a two-dimensional space, where the angle between x and z is 45 and the angle between z and y is 45. Here, $d(x, y) = 1 - cos(\pi/2) = 1$, while $d(x, z) + d(z, y) = 1 - cos(\pi/4) + 1 - cos(\pi/4) = 2 - \frac{2}{\sqrt{(2)}} = 0.586$. So, $d(x, z) + d(z, y) < d(x, y)$, violating the triangle inequality.

2.2.4   Levenshtein Distance

The Levenshtein distance, also known as edit distance, is a measure of the distance between two strings. It is determined by calculating the minimum number of mutations needed to transform string $s1$ into string $s2$, where a mutation can be one of three operations: changing a letter, inserting a letter or deleting a letter. The edit distance can be defined using the following recurrence relation:

- $d(\text{" "}, \text{" "}) = 0$, (*two empty strings match*)
- $d(s, \text{" "}) = d(\text{" "}, s) = \|s\|$, (*distance from an empty string*)
- $d(s1 + ch1, s2 + ch2) = min\ [d(s1, s2) + \{\text{if } ch1 = ch2 \text{ then } 0 \text{ else } 1\}, d(s1 + ch1, s2) + 1, d(s1, s2 + ch2) + 1]$

If the last characters of the two strings are identical, they can be matched without penalty, resulting in an edit distance of $d(s1, s2)$. If they are different, $ch1$ can be changed into $ch2$ with an overall cost of $d(s1, s2) + 1$, or $ch1$ can be deleted and $s1$ edited into $s2 + ch2$, resulting in a cost of $d(s1, s2 + ch2) + 1$, or $ch1$ can be inserted into $s2$ resulting in a cost of $d(s1 + ch1, s2) + 1$. The minimum value of these possibilities gives the edit distance. For instance, if we consider the strings "CAT" and "RAT", the edit distance between them is 1 because only one letter needs to be changed. However, if we consider the strings "CAT" and "LAN", the edit distance between them is 2 because multiple changes are required to transform one into the other.

2.2.5   Mutual Neighborhood Distance (MND)

In this case, the function used to measure the similarity between two patterns, x and y, is defined as $S(x, y) = f(x, y, \epsilon)$, where ϵ denotes the context, that is, the surrounding points. In this context, all other data points are labelled in increasing order of some distance measure, starting with the nearest neighbor as 1 and ending with the farthest point as $N - 1$. The label of x with respect to y is denoted by NN (x, y), and the mutual neighborhood distance (MND) is calculated as MND $(x, y) = $ NN $(x, y) + $ NN (y, x). MND is symmetric, with NN (x, x) set to 0; it is also reflexive. However, it does not satisfy the triangle inequality and is not a metric.

Example 4 explains how the context, the collection of points in the vicinity, changes the MND between points p, q and r.

EXAMPLE 4:   Consider Fig. 2.1.

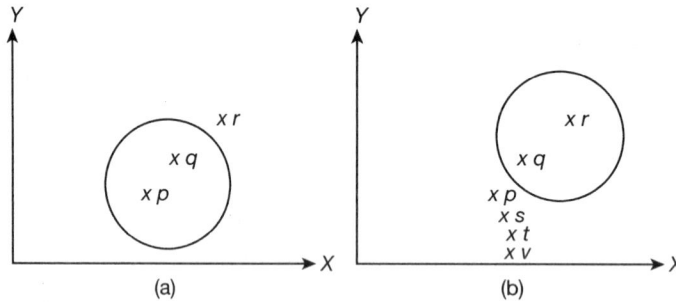

FIG. 2.1 Mutual neighborhood distance illustration

The ranking of the points p, q and r can be represented as shown in Table 2.1 and their mutual nearest neighbor distances are shown in Table 2.2.

TABLE 2.1 Relative positional ranking

	1	2
p	q	r
q	p	r
r	q	p

TABLE 2.2 Mutual distances

$\text{MND}(p, q) = 2$
$\text{MND}(q, r) = 3$
$\text{MND}(p, r) = 4$

Corresponding to Fig. 2.1 (b), the ranking of the points p, q, r, s, t and v can be represented as shown in Table 2.3 and their mutual nearest neighbor distances are shown in Table 2.4. It can be seen that in the first case, the least MND is between p and q, whereas in the second case, it is between q and r. This occurs by changing the context or introducing more points in the vicinity of one of the points.

TABLE 2.3 Relative positional ranking

	1	2	3	4	5
p	s	t	v	q	r
q	p	r	s	t	v
r	q	p	s	t	v

TABLE 2.4 Mutual distances

$\text{MND}(p, q) = 5$
$\text{MND}(q, r) = 3$
$\text{MND}(p, r) = 7$

2.2.6 Proximity Between Binary Patterns

We can view l-dimensional binary patterns as binary strings of length l. Let p and q be two l-bit binary strings. Some of the popular proximity measures on such binary patterns are:

- **Hamming Distance (HD):** If $p(i) = q(i)$, we say that p and q match on their i^{th} bit, else $(p(i) \neq q(i))$, and p and q mismatch on the ith bit. Hamming distance is the number of mismatching bits of the l-bit locations.

  EXAMPLE 5: Consider the two 10-bit patterns p and q given by
  $p = 1000001000$
  $q = 0000001001$
  The Hamming distance is 2 as they mismatch in bit positions 1 and 10.

- **Simple Matching Coefficient (SMC):** Let us define the following:
  M_{01} is the number of bits where p is 0 and q is 1
  M_{10} is the number of bits where p is 1 and q is 0
  M_{00} is the number of bits where p is 0 and q is 0
  M_{11} is the number of bits where p is 1 and q is 1
  Now we define SMC as follows:

  $$\text{SMC}(p, q) = \frac{M_{11} + M_{00}}{M_{00} + M_{01} + M_{10} + M_{11}}$$

- **Jaccard Coefficient (JC):** It is defined as

  $$\text{JC}(p, q) = \frac{M_{11}}{M_{01} + M_{10} + M_{11}}$$

  EXAMPLE 6: Consider again
  $p = 1000001000$
  $q = 0000001001$
  Here, $M_{11} = 1$, $M_{00} = 7$, $M_{01} = M_{10} = 1$. So,
  $\text{SMC}(p, q) = \frac{1+7}{7+1+1+1} = \frac{8}{10} = 0.8$ and
  $\text{JC}(p, q) = \frac{1}{1+1+1} = \frac{1}{3} = 0.33$

2.3 DIFFERENT CLASSIFICATION ALGORITHMS BASED ON THE DISTANCE MEASURES

Many classification algorithms (classifiers) inherently depend upon some distance between a pair of patterns. We discuss them in this section.

2.3.1   Nearest Neighbor Classifier (NNC)

Here, a test pattern x is classified based on its nearest neighbor (NN) in the training data. Specifically, let

$$\mathcal{X} = \{(x_1, l_1), (x_2, l_2), \ldots, (x_n, l_n)\}$$

be the labelled training data set of n patterns, where each element is a tuple; the first component of the tuple is the pattern vector and the second component is its class label.

Let each pattern be a vector in some l-dimensional space. Here, x_i, $i = 1, 2, \ldots, n$ is the ith training pattern and l_i is its class label. So, if there are p classes with their labels coming from the set

$$\mathcal{L} = \{C_1, C_2, \ldots, C_p\},$$

then $l_i \in \mathcal{L}$, for $i = 1, 2, \ldots, n$. Now the nearest neighbor of the test pattern x is given by

$$\text{NN}(x) = \arg \min_{x_j \in \mathcal{X}} d(x, x_j),$$

where x_j is the jth training pattern and $d(x, x_j)$ is the distance between x and x_j. Intuitively, $\text{NN}(x)$ is in the proximity of x; so $\text{NN}(x)$ is at a minimum distance from x and is maximally similar to x. The NN rule assigns x the class label of $\text{NN}(x)$.

EXAMPLE 7:   Let the training set consist of the following two-dimensional patterns with associated labels:

TABLE 2.5 Example data set

$x_1 = (0.7, 0.7)$, $l_1 = 1$;	$x_2 = (0.8, 0.8)$, $l_2 = 1$;	$x_3 = (1.1, 0.7)$, $l_3 = 1$
$x_4 = (0.7, 1.1)$, $l_4 = 1$;	$x_5 = (1.1, 1.1)$, $l_5 = 1$;	$x_6 = (3.0, 2.0)$, $l_6 = 2$
$x_7 = (3.7, 2.7)$, $l_7 = 2$;	$x_8 = (4.1, 2.7)$, $l_8 = 2$;	$x_9 = (3.7, 3.1)$, $l_9 = 2$
$x_{10} = (4.1, 3.1)$, $l_{10} = 2$;	$x_{11} = (4.3, 2.7)$, $l_{11} = 2$;	$x_{12} = (4.3, 3.1)$, $l_{12} = 2$
$x_{13} = (3.1, 0.3)$, $l_{13} = 3$;	$x_{14} = (3.1, 0.6)$, $l_{14} = 3$;	$x_{15} = (3.7, 0.4)$, $l_{15} = 3$
$x_{16} = (3.4, 0.6)$, $l_{16} = 3$;	$x_{17} = (3.9, 0.9)$, $l_{17} = 3$;	$x_{18} = (3.9, 0.6)$, $l_{18} = 3$

For the ith pattern x_i, the class label is l_i and $l_i \in \{1, 2, 3\}$ for $i = 1, 2, \ldots, 18$. This can be seen in Fig. 2.2. Here '▲' corresponds to Class 1, '+' corresponds to Class 2 and '*' corresponds to Class 3. Now if there is a test pattern $T = (2.1, 0.7)$, it is necessary to find the distance from T to all the training patterns.

Let the distance between a training pattern x and T be the Euclidean distance

$$d(x, T) = \sqrt{(x(1) - T(1))^2 + (x(2) - T(2))^2}$$

The distance from a point T to every point in the training set can be computed using the above formula. For $T = (2.1, 0.7)$, the distance to x_1 is

$$d(x_1, T) = \sqrt{(0.7 - 2.1)^2 + (0.7 - 0.7)^2} = 1.4$$

We find, after calculating the distance from all the 18 training points to T, that the closest neighbor of T is x_3, which has a distance of 1.0 from T and x_3, which belongs to Class 1. Hence, T is classified as belonging to Class 1.

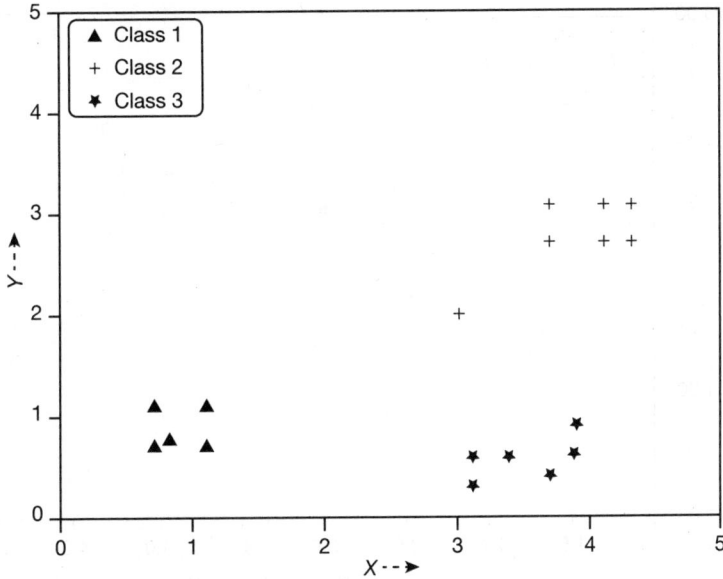

FIG. 2.2 Example data set in graphical form

2.3.2 *K*-Nearest Neighbor Classifier

In the k-nearest neighbor (KNN) algorithm, we find the k nearest neighbors of a test pattern x from the training data $\mathcal{X}$, and then assign the majority class label among the k neighbors to x. To determine the k nearest neighbors of x, it is necessary to calculate the distance between x and each of the n training patterns in the l-dimensional space. The distance metric used depends on the specific problem at hand, and can be Euclidean distance, Manhattan distance or cosine distance, among others. The class label of x is then determined based on the majority class label among its k nearest neighbors.

Assuming that Fig. 2.2 is a visual representation of the KNN algorithm being applied to a test pattern T, if the value of k is set to 5, the five nearest neighbors of T are x_3, x_{14}, x_{13}, x_5 and x_{16}. Among these five patterns, the majority class is Class 3. By using this method of selecting the majority class label among the k nearest neighbors, the errors in classification can be reduced, especially when the training patterns are noisy. While the closest pattern to the test pattern may belong to a different class, considering the number of neighbors and the majority class label increases the likelihood of correct classification. Figure 2.3 visually demonstrates this phenomenon.

It indicates that the test point T is closest to point 5, which is an outlier in Class 1 and is represented as 'x'. If the KNN algorithm is used, the point T will be classified as part of Class 2, which is represented by '$+$'. The selection of k is a crucial aspect of this algorithm. In the case of large data sets, k can be increased to decrease the error. The value of k can be determined through experimentation by keeping aside a subset of the training data as the validation data and classifying patterns from the validation set using different values of k using the training patterns to compute the neighbors. The value of k can be selected based on the lowest error observed in classification.

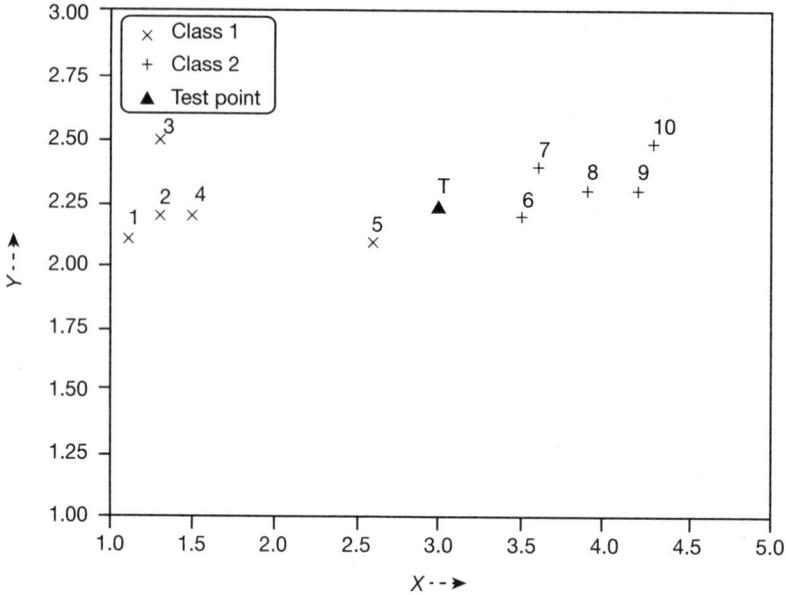

FIG. 2.3 Classification of T using the KNN classifier

EXAMPLE 8: Based on Fig. 2.2, if the pattern T is $(2.1, 0.7)$, its nearest neighbor is x_3, and it would be classified as part of Class 1 if the nearest neighbor algorithm is employed. However, if the five nearest neighbors are considered, they are x_3 and x_5, both belonging to Class 1, and x_{14}, x_{13} and x_{16}, belonging to Class 3. Following the majority class rule, the pattern T would be classified as part of Class 3.

2.3.3 Weighted K-Nearest Neighbor (WKNN) Algorithm

This algorithm is similar to the KNN algorithm, as it also considers the k nearest neighbors. However, this algorithm takes into account the distance of each of the k neighbors from the test point by weighting them accordingly. It is also known as the distance-weighted k-nearest neighbor algorithm. Each neighbor is associated with a weight w, which is determined by the following formula:

$$w_j = \begin{cases} \frac{(dk-dj)}{(dk-d1)} & if \quad (d_k \neq d_1) \\ 0 & if \quad (d_k = d_1) \end{cases}$$

Here, j represents the neighbor's index in the list of k nearest neighbors, while d_k and d_j are the distances between the test point and the kth neighbor and the jth neighbor, respectively. The value of w_j ranges from 1 for the nearest neighbor to 0 for the kth (most distant) neighbor. Using these weights, the WKNN algorithm assigns the test pattern to the class with the highest total weight of its representative neighbors among the k nearest neighbors. Unlike the traditional KNN algorithm, the WKNN algorithm employs a weighted majority rule, which takes into account the effect of outlier patterns in classification.

EXAMPLE 9:   Consider $T = (2.1, 0.7)$ in Fig. 2.2. For the five nearest points, the distances from T are

$d(T, x_3) = 1.0$; $d(T, x_{14}) = 1.01$; $d(T, x_{13}) = 1.08$; $d(T, x_5) = 1.08$; $d(T, x_{16}) = 1.30$

The weight values will be

$w_3 = 1.0$

$w_{14} = \frac{(1.30-1.01)}{(1.30-1.00)} = 0.97$

$w_{13} = \frac{(1.30-1.08)}{(1.30-1.00)} = 0.73$

$w_5 = \frac{(1.30-1.08)}{(1.30-1.00)} = 0.73$

$w_{16} = 0$

Summing up for each selected class, Class 1 to which x_3 and x_5 belong sums to 1.73, and Class 3 to which x_{14}, x_{13} and x_{16} belong sums to 1.7. Therefore, the point T belongs to Class 1.

It is possible that the KNN and WKNN algorithms assign the same pattern a different class label. This is also illustrated in the following example.

EXAMPLE 10:   In Fig. 2.2, when $T = (1.9, 2.4)$, the five nearest patterns are x_6, x_5, x_4, x_7 and x_3. The distances from T to these patterns are

$d(T, x_6) = 1.17$; $d(T, x_5) = 1.53$; $d(T, x_4) = 1.77$; $d(T, x_7) = 1.83$; $d(T, x_3) = 1.88$

The weight values are

$w_6 = 1$

$w_5 = \frac{(1.88-1.53)}{(1.88-1.17)} = 0.493$

$w_4 = \frac{(1.88-1.77)}{(1.88-1.17)} = 0.155$

$w_7 = \frac{(1.88-1.83)}{(1.88-1.17)} = 0.07$

$w_3 = 0$

Summing up for each class, Class 2 to which x_6 and x_7 belong sums to 1.07 and Class 1 to which x_5, x_4 and x_3 belong sums to 0.648, and therefore, T is classified as belonging to Class 2 by WKNN. Note that the same pattern is classified as belonging to Class 1 when we use the KNN algorithm with $k = 5$.

2.3.4   Radius Distance Nearest Neighbor Algorithm

This algorithm is an alternative to the KNN algorithm that considers all the neighbors within a specified distance r of the point of interest. This algorithm can be described as follows:

1.   Given a point T, identify the subset of data points that fall within the radius r centred at T, denoted by

$$Br(T) = \{x_i \in \mathcal{X} \text{ s.t. } \|T - Xi\| \leq r\}$$

2.   If $Br(T)$ is empty, output the majority class of the entire data set.
3.   If $Br(T)$ is not empty, output the majority class of the data points within $Br(T)$.

This algorithm is useful for identifying outliers, as any pattern that does not have similarity with the patterns within the chosen radius can be identified as an outlier. The choice of the value of radius r is critical as it can affect the performance of the algorithm.

By considering all neighbors within the specified radius, this algorithm can be more effective than the traditional KNN algorithm, especially when the nearest neighbor is too far away to be relevant.

EXAMPLE 11: For the example shown in Fig. 2.2, from point $T = (2.1, 0.7)$, the patterns which are located within a radius of 1.45 are x_1, x_2, x_3, x_5, x_{13}, x_{14} and x_{16}. The majority of these patterns belong to Class 1. T is therefore assigned to Class 1.

2.3.5 Tree-Based Nearest Neighbor Algorithm

This section explores how to find nearest neighbors in the transaction databases. Transaction databases store data collected from various sources such as businesses, scientific experiments, physical systems monitoring or supermarket transactions. Also known as market basket data, these databases contain the transactions made by each customer, with each transaction consisting of items bought by the customer. The transactions can differ in size, and the objective of analysing this data is to establish a relationship between certain items in the transactions. This process, known as **association rule mining**, aims to identify the occurrence of one item based on the occurrence of other items. To simplify and expedite the process, only frequently occurring items are considered, with a minimum support value chosen to exclude items occurring less than the minimum support. One of the tree data structures used to represent the processed transactional database is the frequent pattern (FP) tree.

To construct the FP tree:

1. The first step is to determine the frequency of each item in the transaction database. Frequency of an item is the number of transactions in which it occurs in the transaction database. Consider only those items whose frequency is greater than or equal to the user-defined minimum support; sort them in descending order of frequency.
2. Each entry in the transaction database is then arranged in the same order of frequency of items from largest to smallest, ignoring the infrequent items.
3. The root of the FP tree is created and labelled *null*. The first transaction in the database is used to construct the first branch of the FP tree based on the ordered sequence.
4. The second transaction is added to the tree using the same order. The common prefix between the second and first transaction is added to the existing path, increasing the count by one. For the remaining part of the transaction, new nodes are created. This process is repeated for the entire database.

The following example shows how the FP tree can be constructed from a transaction database, which is in the form of a table of size 4×4. Consider a 4×4 square to represent digits, with each square serving as a pixel. The squares are assigned a positional value, as illustrated in Table 2.6. For instance, the digit 0 can be represented using the 4×4 square depicted in Table 2.7 and denoted by the positional values 1, 2, 3, 4, 5, 8, 9, 12, 13, 14, 15 and 16. Table 2.8 displays how the digits 0, 1, 4, 6 and 7 can be represented using this method.

TABLE 2.6 Positional representation of a pixel 4 × 4 table

1	2	3	4
5	6	7	8
9	10	11	12
13	14	15	16

TABLE 2.7 Representation of '0' in a 4 × 4 table

1	1	1	1
1			1
1			1
1	1	1	1

The frequency of each item can be determined while scanning the transaction database presented in Table 2.8. Sorting the items from the highest to the lowest frequency results in the following list: (12:5), (1:4), (16:4), (4:3), (5:3), (8:3), (9:3) and (15:3). With the minimum support set to 3, only items with a frequency of 3 or more are included in this list. It is important to note that any ties that may arise are settled arbitrarily.

TABLE 2.8 Transaction database

Digit	Transaction (Positional information of a digit)
0	1, 2, 3, 4, 5, 8, 9, 12, 13, 14, 15, 16
1	4, 8, 12, 16
4	1, 5, 7, 9, 10, 11, 12, 15
6	1, 5, 9, 10, 11, 12, 13, 14, 15, 16
7	1, 2, 3, 4, 8, 12, 16

The transaction database in Table 2.9 shows the transactions with the items ordered based on their frequencies. With support value set to 3, only items with a frequency of three or more are retained in this table. The items with a support of two or less are removed.

TABLE 2.9 Transaction database with transactions ordered according to frequency of items

Digit	Transaction (After removing non-frequent items)
0	12, 1, 16, 4, 5, 8, 9, 15
1	12, 16, 4, 8
4	12, 1, 5, 9, 15
6	12, 1, 6, 5, 9, 15
7	12, 1, 16, 4, 8

Using this ordered database, an FP tree is constructed, as shown in Fig. 2.4.

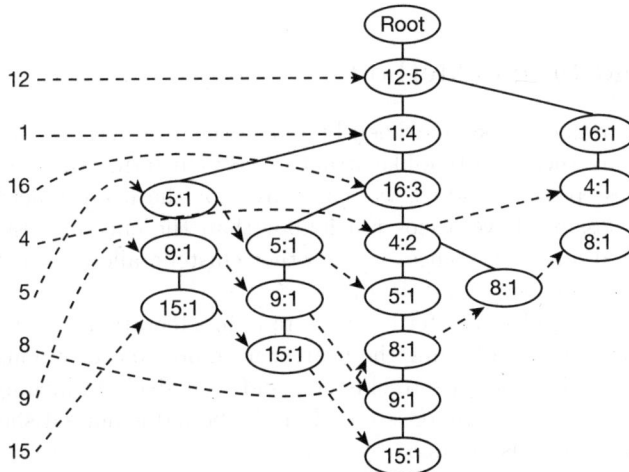

FIG. 2.4 FP tree for the transaction database in Table 2.8

The root node points to the starting item of the transactions, which in this case is 12. Each transaction is processed sequentially, and the corresponding nodes are added to the tree. The count of each item is incremented along the path that already exists, and new nodes are created for the remaining items in the transaction.

For example, in the first transaction, the path goes from the root node to 12, then to 1, 16, 4, 5, 8, 9 and 15. Each of these nodes has a count of 1. In the second transaction, since the next item after 12 is not 1, a new branch is created from 12 to 16, and the counts of 12, 16, 4 and 8 are incremented by 1. This process continues for each transaction until the entire database has been processed.

The resulting FP tree is a compressed representation of the frequent items in the transaction database. Each node in the tree corresponds to an item, and the count of each item is stored at the corresponding node. The header node for each item points to all the nodes in the tree that contain that item.

FP Tree-Based Nearest Neighbor: To identify the nearest neighbor of a test pattern in a transaction database, the FP tree can be utilised with the following approach:

1. Firstly, remove items in the test pattern that are below the minimum support threshold. Then, arrange the remaining items in the pattern according to their order in the FP tree.
2. Next, search the tree from the root node for the branch containing the first item in the test pattern. If this branch exists, continue to search for the next item in the pattern, and so on.
3. In the event that an item does not exist, examine all the branches from that point and choose the one with the maximum number of common items with the test pattern. This will determine the nearest neighbor.

EXAMPLE 12:   Suppose we have an FP tree as shown in Fig. 2.4 for the transaction data set in Table 2.8. Let us consider a test pattern with features 1, 2, 3, 4, 6, 7, 8, 12 and 16. Removing items below the minimum support threshold, we are left with 1, 4, 8, 12, 16. By arranging these items in the order in which they appear in the FP tree, we get 12, 1, 16, 4, 8. Starting from the root node of the FP tree (12), we can compare the remaining items in the test pattern. It is observed that the test pattern has the maximum number of items in common with digit 7. Therefore, it can be classified as belonging to digit 7.

2.3.6   Branch and Bound Method

The branch and bound method seeks to efficiently find the nearest neighbor by taking advantage of an ordered data structure such as a tree-like structure. By clustering the data into representative groups with the smallest possible radius, we can search for the nearest neighbor while avoiding branches that cannot possibly have a closer neighbor than the current best value found. Lower bounds can be computed for the distances in the other clusters, allowing us to eliminate clusters that cannot possibly contain the nearest neighbor.

This recursive method involves clustering the data points hierarchically into subsets until there are clusters of one point, and then finding the nearest neighbor to a new point T. This is done by computing lower bond b_j with reference to cluster j and recursively branching to the cluster with the smallest b_j until the nearest neighbor is found or the bound is not satisfied.

Note that b_j for a cluster j is obtained by

$$b_j = d(T, \mu_j) - r_j,$$

where μ_j is the centre of the cluster j and r_j is its radius.

The branch and bound method has shown significant improvements over the standard nearest neighbor algorithm on an average.

For the patterns shown in Table 2.5, the branch and bound method is demonstrated using Fig. 2.5. The first step involves clustering the points into subsets. In this example, the points of Class 1 form Cluster 1, the points of Class 2 form Cluster 2, and the points of Class 3 form Cluster 3. Further sub-clusters are formed by clustering Cluster 1 into Clusters 1a and 1b, Cluster 2 into Clusters 2a and 2b, and Cluster 3 into Clusters 3a and 3b. At the next level, each point is taken as a sub-cluster. Figure 2.5 also shows the centres and radii of the clusters created for the patterns shown in Table 2.5.

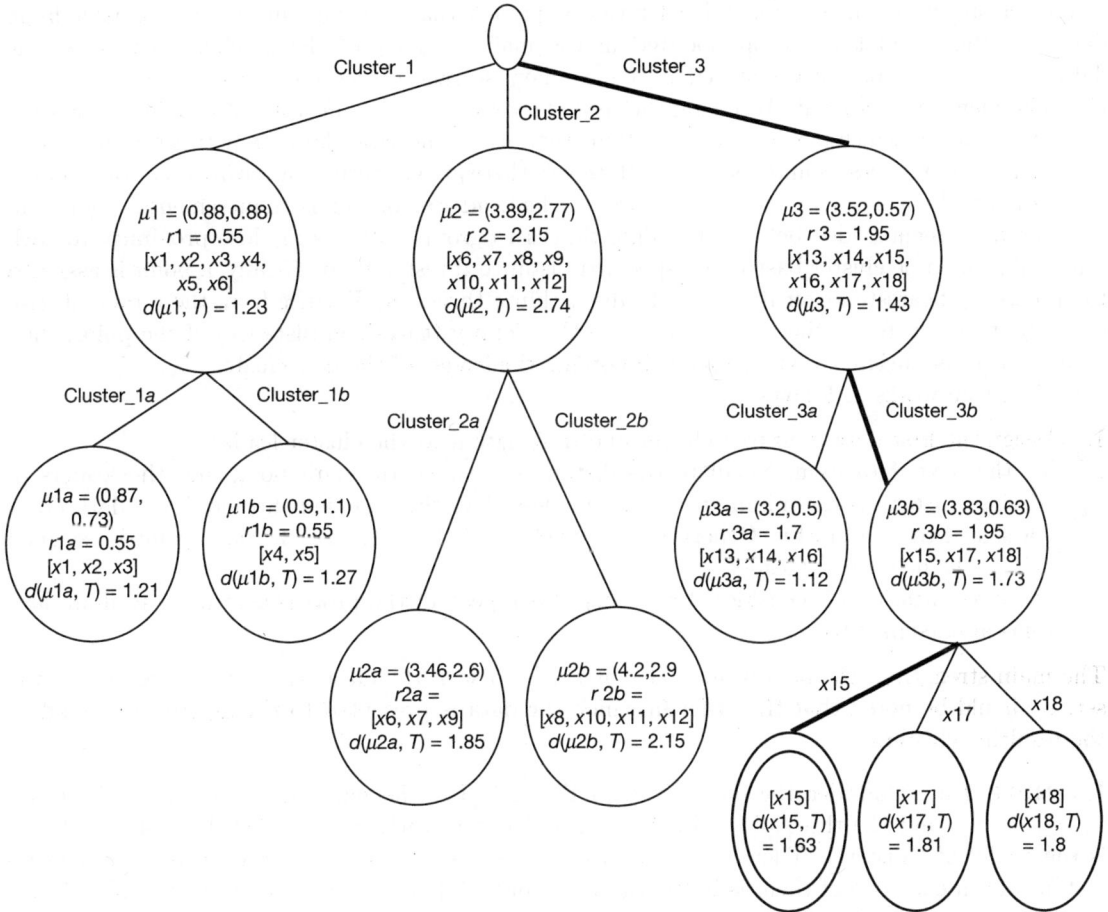

FIG. 2.5 Working of branch and bound algorithm for Table 2.5

EXAMPLE 13:   To find the nearest neighbor of a new point $T = (2.1, 0.7)$, the lower bounds b_j for each cluster j are computed.

For Cluster 1, $b_1 = d(T, \mu_1) - r_1 = 0.68$. For Cluster 2, $b_2 = 0.59$. For Cluster 3, $b_3 = 0.52$.

Since b_3 is the smallest, the sub-clusters of Cluster 3 are searched. The centre of Cluster 3a is (3.2, 0.5) with a radius of 1.7, and the centre of Cluster 3b is (3.83, 0.63) with a radius of 1.95. This gives rise to $b_{3a} = 0.58$ and $b_{3b} = 0.22$. The second sub-cluster of Cluster 3 is searched and point x_{15} is found to be the closest point to T.

The bound d is calculated as the distance from T to point x_{15}, which is 1.63. Since b_1 and b_2 are greater than d, Clusters 1 and 2 need not be searched. Therefore, x_{15} is declared as the nearest neighbor of T.

2.3.7 Leader Clustering

Leader clustering is an incremental clustering approach that is commonly used to cluster large data sets that cannot be accommodated in the main memory of the machine processing the data. As a result, the data is stored in a secondary storage device and must be transferred to the main memory as needed. Accessing the secondary storage is significantly slower than accessing the main memory, which means that algorithms that access the same data many times require more secondary storage accesses and disk scans. However, there are clustering algorithms that only access the data once, known as incremental algorithms. The Leader algorithm is an incremental algorithm.

The fundamental idea behind this algorithm is to group patterns in close proximity to each other into the same cluster based on a specified distance threshold. Specifically, a point is assigned to an existing nearest cluster if the point falls within a threshold distance from the representative (leader) of the cluster; if there is no cluster in the vicinity (threshold distance) of the point, then a new cluster is initiated with the point becoming the leader of the new cluster.
The algorithm works as follows:

1. Assign the first data item to a cluster and designate it as the cluster leader.
2. For the next data item, calculate the distances between the data point and the leaders of existing clusters. If the minimum distance is less than the specified threshold, the data point is assigned to that cluster. Otherwise, a new cluster is formed, and the data point is assigned to it as the new cluster leader.
3. The next data item is considered and Step 2 is repeated; the process continues until all data items are assigned to clusters.

The main strength of the algorithm is that it needs to scan the data set only once to cluster the set. It should be noted that the order in which the data is presented to the algorithm can affect the resulting clusters.

EXAMPLE 14:   Consider the data given in Fig. 2.5. Let the data be processed in the order $x_1, x_2, \ldots, x_{18}$ and the threshold T be set to 1.5. To start with, x_1 is assigned to Cluster 1 and is the leader of Cluster 1. Then x_2, x_3, x_4 and x_5 are assigned to Cluster 1 since the Euclidean distance from each one of them is below the threshold 1.5 (that is, 0.1, 0.4, 0.4, 0.6, respectively) from x_1.

The Euclidean distance from x_1 to x_6 is 2.6 (which is more than the threshold) and hence forms a new cluster, Cluster 2, for which x_6 is the Cluster leader; x_7, x_8 and x_9 are assigned to Cluster 2, since the Euclidean distance from them to x_6 is below the threshold (that is, 1.0, 1.3, and 1.3, respectively, from x_6). The Euclidean distance from x_{10} is more than 1.5 with respect to x_1 and x_6 (which are, respectively, 4.2 and 1.6). So, a new cluster, Cluster 3 is formed with x_{10} as the cluster leader; x_{11} and x_{12} are assigned to Cluster 3, since the Euclidean distance from x_{10} to them is below the threshold (that is, 0.4 and 0.2, respectively, from x_{10}).

The Euclidean distance from x_{13} is more than 1.5 with respect to x_1, x_6 and x_{10} (which are, respectively, 2.4, 1.7 and 2.9). So, a new cluster, Cluster 4 is formed with x_{13} as the cluster leader; x_{14}, x_{15}, x_{16}, x_{17} and x_{18} are assigned to Cluster 4, since the Euclidean distance from x_{13} to them is below the threshold (that is, 0.3, 0.6, 0.4, 1.0 and 0.9, respectively, from x_{13}).

Figure 2.6 shows the clusters formed using the Leader algorithm with a threshold value of 1.5.

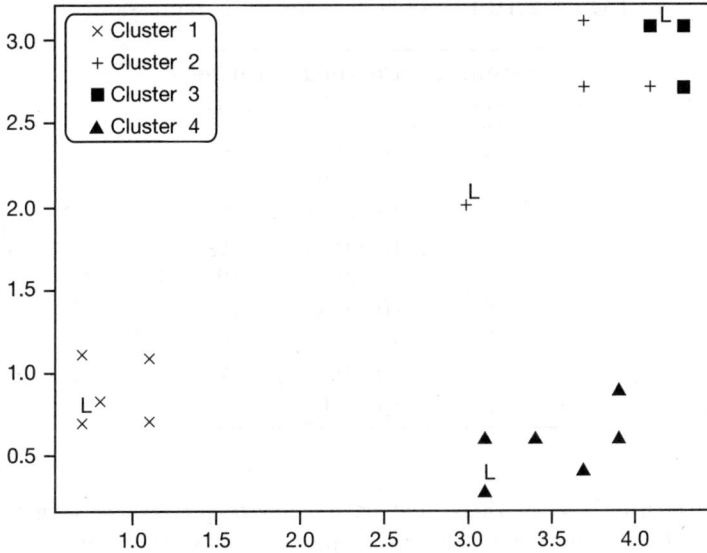

FIG. 2.6 Clusters formed using the Leader algorithm for the data set in Table 2.5 with leaders shown as 'L'

2.4   KNN REGRESSION

It is possible to use KNN for regression also. So, in this case we are given a set $\mathcal{X}$ of n labelled examples, where

$$\mathcal{X} = \{(x_1, y_1), (x_2, y_2), \ldots, (x_n, y_n)\}$$

Here x_i, $i = 1, 2, \ldots, n$ is a data vector and y_i is a scalar. It is possible for y_i also to be a vector in some applications. However, we restrict our attention to scalar y_is. The regression model needs to use $\mathcal{X}$ to find the value of y for a new vector x. In the case of regression based on KNN, we perform the following:

1.   Find the k nearest neighbors of x from n data vectors. Let them be $x^1, x^2, \ldots, x^k$.
2.   Consider the y values associated with these x^is. Let them be $y^1, y^2, \ldots, y^k$.

3. Take the average of these y^is and declare this average value to be the predicted value of y associated with x. So, the predicted value of y, call it $\hat{y}$ is,

$$\hat{y} = \frac{1}{k}(y^1 + y^2 + \cdots + y^k)$$

We will illustrate it using the following example.

EXAMPLE 15: Consider the data shown in Table 2.10.

TABLE 2.10 Example data for KNN regression

Number (i)	Pattern (x_i)	Target (y_i)
1	(0.2,0.4)	8
2	(0.4,0.2)	8
3	(0.6,0.4)	12
4	(0.8,0.6)	16
5	(1.0,0.7)	19
6	(0.8,0.4)	14
7	(0.6,0.2)	10
8	(0.5,0.5)	12
9	(0.2,0.6)	10

Let the new pattern be $x = (0.3, 0.4)$. Let us see how to predict the target value for x using KNN regression. Let $k = 3$; the 3 NNs of x from the patterns in the table are $(0.2, 0.4)$, $(0.4, 0.2)$ and $(0.5, 0.5)$. The corresponding target values observed in the table are 8, 8 and 12, respectively. The average of these values is $\frac{8+8+12}{3} = 9.33$. So, the predicted target value for $x = (0.3, 0.4)$ is 9.33.

Further, we are given that the underlying model is a linear model such that $y_i = f(x_i) = ax_i(1) + bx_i(2) + c$, where $x_i(1)$ is the first component and $x_i(2)$ is the second component of the vector x_i.

If we use the first three entries in the table and solve for a, b and c, we get $a = b = 10$ and $c = 2$. Note that the remaining 6 rows satisfy this linear model. Using the model, the target value for $x = (0.3, 0.4)$ is 9. The value predicted by KNN regression is 9.33. The squared error is $(9 - 9.33)^2 = 0.1111$ that is obtained by using the predicted value (9.33) and the value given by ground truth (9).

We will now examine the role of KNN regression on a real-world data set.

We consider the Boston Housing data set, it contains information collected by the U.S. Census Service concerning housing in the Boston area. It has 13 features including crime rate in the town and the number of rooms in the dwelling. The target is the median value of owner-occupied homes in $1000s. We consider the first 250 pattern vectors in the set for this experiment. We use 200 patterns for training and the remaining 50 patterns for testing. The true target values and the predicted target values of the 50 test patterns are shown in Fig. 2.7.

The corresponding MSE values for different values of k are plotted in Fig. 2.8 where MSE is the average of the squared errors across the collection of patterns.

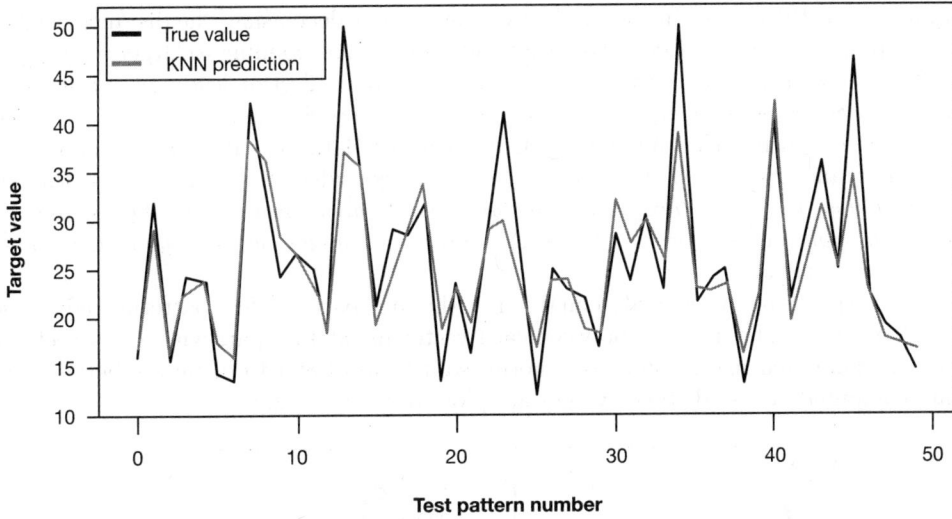

FIG. 2.7 KNN regression: results on the Boston Housing data (for colour figure, please see Colour Plate 1)

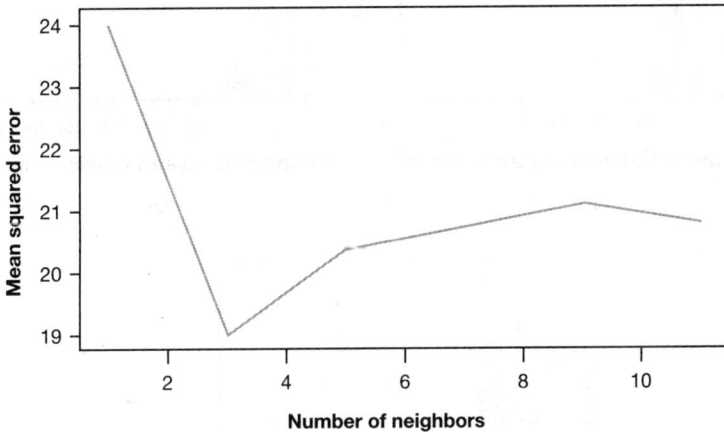

FIG. 2.8 KNN regression: MSE values on the Boston Housing data

2.5   CONCENTRATION EFFECT AND FRACTIONAL NORMS

A major difficulty encountered while using some of the popular distance measures like the Euclidean distance is that the distance values, between various pairs of points, may not show much dynamic range. Consider the following example.

EXAMPLE 16:   Let $p = (4, 2)$ and $q = (2, 4)$ be two points in a two-dimensional space. Values of the distance using some popular distance norms are:

- L_∞ norm or the Max norm: $Max(|4 - 2|, |2 - 4|) = 2$.
- L_2 norm or Euclidean distance: $2\sqrt{2}$
- L_1 norm or City-Block distance: $|4 - 2| + |2 - 4| = 4$.

Observe that as the r value in the Minkowski norm keeps decreasing, the distance between the pair (p, q) keeps increasing. So, Euclidean norm and norms corresponding to larger r tend to assign smaller values to distances between various pairs of points in a collection leading to **concentration effect**; distance values between pairs of points come close to each other.

We illustrate it using the Wisconsin Breast Cancer data set. It is a two-class data set with *Malignant* and *Benign* as the class labels. We consider the following 11 features for illustration: *radius_mean, texture_mean, texture_worst, area_mean, smoothness_mean, compactness_mean, compactness_worst, concavity_mean, perimeter_worst, symmetry_worst* and *fractal_dimension_worst*.

The data is split into two roughly equal parts; the first part has 284 patterns and the second part has 285 patterns. The distance between each vector in the first part with every vector in the second part is computed using the chosen norm; so a total of 80,940 distances between pattern vectors are computed. These distance values are plotted in Figure 2.9.

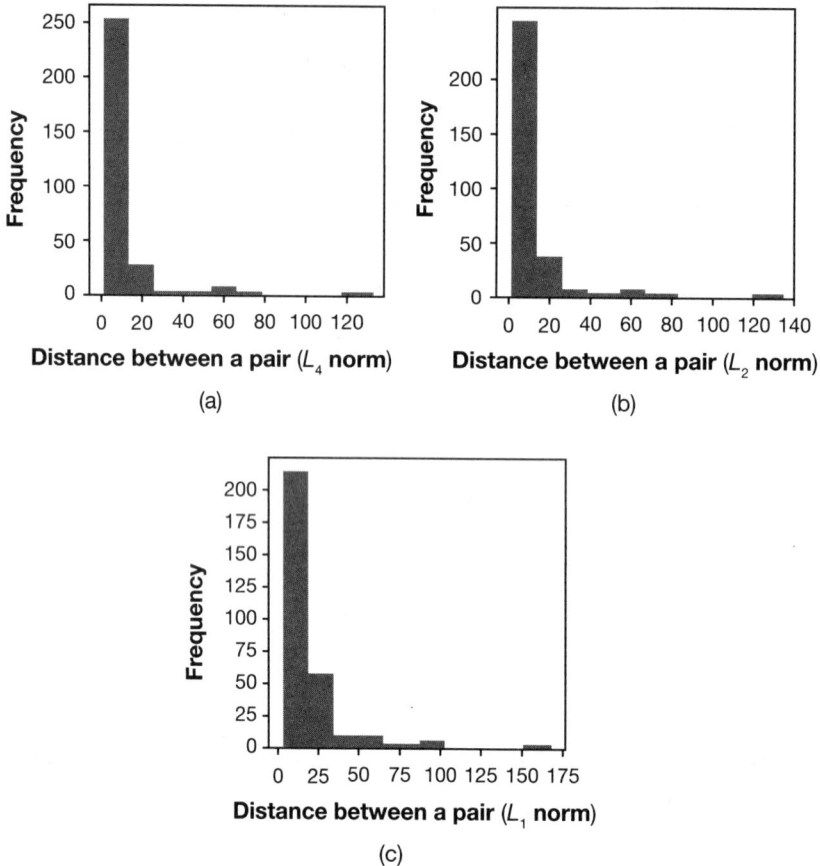

(a)

(b)

(c)

FIG. 2.9 Concentration of distances using different norms (for colour figure, please see Colour Plate 1)

Here, the X-axis shows the distance value and the Y-axis shows the frequency of the value. In Fig. 2.9 (a), the distance frequency values for L_4 norm are shown. In (b), the values correspond to the Euclidean norm and in (c), values for city-block distance are shown. Note that there is higher concentration of distances in (a), followed by (b) and the least concentration is seen when the L_1 norm is used for distance computation as shown in (c). It is shown in the literature that this concentration effect will be worse as the dimensionality of the data vectors increases.

This behaviour prompted researchers to use fractional norms, where r is a fraction, to increase the dynamic range of the values or decrease the concentration effect. We illustrate it with an example.

EXAMPLE 17: We have $p = (4,2)$ and $q = (2,4)$. The fractional norms give the distances as shown below:

- $L_{0.5}$ norm: $(\sqrt{|4-2|} + \sqrt{|2-4|})^{\frac{1}{0.5}} = (2\sqrt{2})^2 = 8.$
- $L_{0.25}$ norm: $(2^{0.25} + 2^{0.25})^4 = 32.$
- $L_{0.1}$ norm : $(2^{0.1} + 2^{0.1})^{10} = 2048.$

Note the increase in the distance value between the pairs as the value of r decreases from 0.5 to 0.1.

We can examine the impact of these fractional norms on the concentration effect by reconsidering the Wisconsin Breast Cancer data. The results are shown in Fig. 2.10 for the $L_{0.1}$ and $L_{0.05}$ norms. Note the difference between the dynamic ranges using the $L_{0.1}$ norm in Fig. 2.10 (a) and the $L_{0.05}$ norm in Fig. 2.10 (b); specifically note the scale factors that are 1e10 and 1e20. The dynamic ranges in this figure are significantly larger than those in Fig. 2.9. Further, the concentration effect reduces as the value of r decreases from ∞ to 0.05.

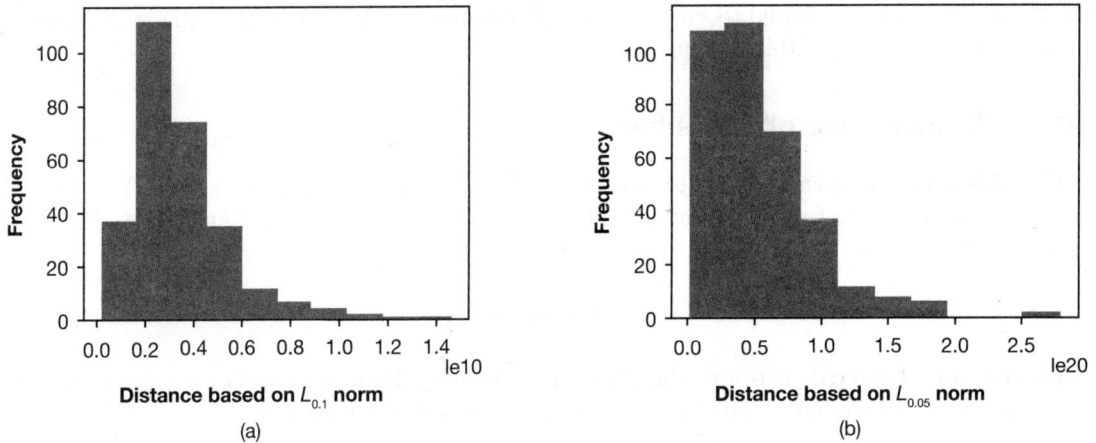

FIG. 2.10 Concentration of distances using different norms (for colour figure, please see Colour Plate 1)

An important observation is that in the process of improving the dynamic range of distance values, the fractional norms can improve the classification performance. This is illustrated with the help of accuracy results using different norms on the Wisconsin Breast Cancer data, as shown in Fig. 2.11. Here, accuracy is defined as the ratio of the number of correctly classified test patterns to the total number of test patterns. Observe that the fractional norm-based distances give better accuracy of KNN. The $L_{0.1}$ norm-based KNN is the best, followed by the $L_{0.3}$ and $L_{0.5}$ norms.

The Euclidean distance-based KNN seems to be the worst performer. Further, because of the enhanced dynamic range, the fractional norms perform better as the value of k is increased.

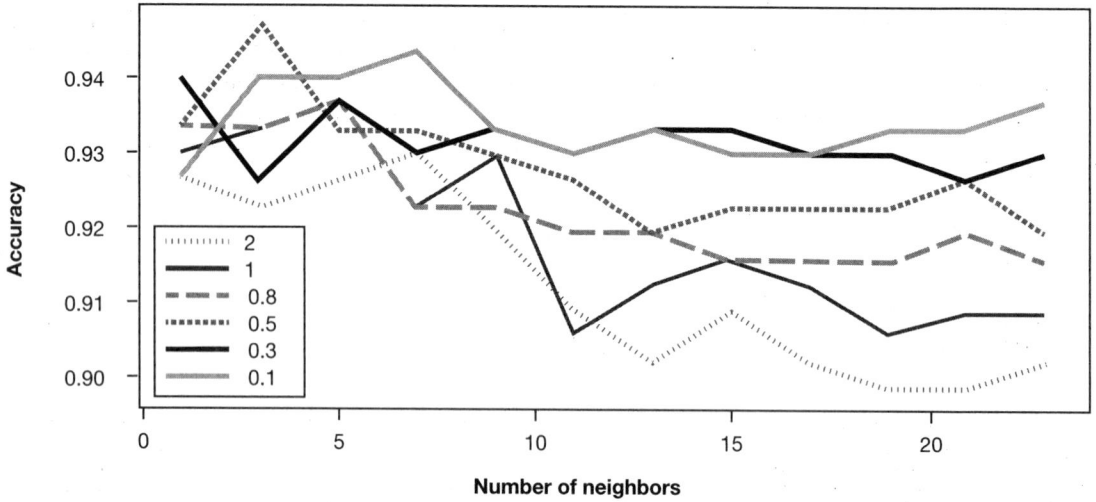

FIG. **2.11** Accuracy of KNN using different norms (for colour figure, please see Colour Plate 1)

2.6 PERFORMANCE MEASURES

There are several measures used to evaluate the performance of ML models. We will consider some of the popular measures in this section.

2.6.1 Performance of Classifiers

- **Classification Accuracy:** Let n be the total number of patterns that are classified by a classification algorithm. Let n_c be the number of correctly classified patterns. Then the classification accuracy is

$$Classification\ accuracy = \frac{n_c}{n}$$

- **Confusion Matrix:** It captures the results in a compact form. This data can be compacted further to better analyse the results. We illustrate this with an example.

EXAMPLE 18: Let there be three classes C_1, C_2 and C_3. Let there be 200, 100 and 50 patterns from classes C_1, C_2 and C_3, respectively. Let a classifier classify these patterns into three classes, as shown in Table 2.11.

TABLE 2.11 Confusion matrix for a three-class problem

True/Predicted	C_1	C_2	C_3
C_1	180	15	5
C_2	5	85	10
C_3	3	2	45

The first row of the table shows how the 200 patterns from C_1 are classified: 180 of them are assigned to C_1, 15 are assigned to C_2 and 5 are assigned C_3. Similarly row 2 accounts for 100 patterns from C_2 and row 3 shows the assignment of 50 patterns from C_3.

Note that classification accuracy is obtained by considering the correctly classified patterns. Note that 180 patterns from C_1, 85 patterns from C_2 and 45 patterns from C_3 are correctly classified, as indicated by the diagonal entries in the matrix. They add up to 310. So, classification accuracy is $\frac{310}{350} \approx 0.8857$. The percentage accuracy is obtained by multiplying this number by 100. So, the percentage accuracy is 88.57%.

It is possible to compact the confusion table by looking at the entries with respect to C_1. The compact table of size 2×2 is shown in Table 2.12.

TABLE 2.12 Compact confusion matrix for C_1

True/Predicted	C_1	$\bar{C}_1$
C_1	180	20
$\bar{C}_1$	8	142

This compact table gives us useful information that could be used in calculating other evaluation measures. Now we are concerned with C_1 and $\bar{C}_1$ or not C_1 (classes other than C_1). The first row of this table shows that 180 patterns of C_1 are correctly classified. This entry corresponds to the correctly classified number of C_1 patterns to Class C_1. So, these are called true positives or TP. TP = 180.

The second column in the first row has 20 patterns. These patterns are from C_1 that are not classified as belonging to C_1; they are assigned to $\bar{C}_1$. So, they are called false negatives or FN. Here, FN = 20.

The first entry in the second row shows that 8 patterns from the other classes are assigned to Class C_1. So, they are called false positives or FP. So, FP = 8. The second entry in the second row corresponds to 142 patterns from $\bar{C}_1$ that are assigned to $\bar{C}_1$. These are called true negatives or TN. So, TN = 182.

We can use the confusion matrix entries, specifically TP, TN, FP and FN, to define additional evaluation measures. They are:

- *Precision*: It is the ratio of TP to the total number of predicted positives (TP + FP). So, precision in this example is $\frac{180}{188} \approx 0.9574$.
- *Recall*: It is the ratio of TP to the total number of positives in the training data (TP + FN). So, in this example, it is $\frac{180}{200} = 0.9$.
- *F1 Score*: It is the harmonic mean of precision and recall. More specifically, it is $\frac{2}{\frac{1}{Precision} + \frac{1}{Recall}}$. In this example, F1 Score ≈ 0.9278.

We can have one more popular measure called **area under the ROC (receiver operating characteristic) curve** based on some additional quantities derived from the table:

- True positive rate or TPR is given by

$$TPR = \frac{TP}{TP + FN}$$

- False positive rate or FPR is given by

$$FPR = \frac{FP}{FP + TN}$$

- Area under the curve (AUC) is obtained by plotting a graph between FPR on the X-axis and TPR on the Y-axis. This is called the receiver operating characteristic (ROC). We illustrate this with the help of the following example.

EXAMPLE 19: Consider the entries in Table 2.12. The values are TP = 180, FP = 8, FN = 20 and TN = 182. So, TPR = $\frac{180}{180+20}$ = 0.9 and FPR = $\frac{8}{190}$ = 0.042. So, $(0.042, 0.9)$ give us a point for the ROC plot.

We obtain multiple points by computing probabilities and thresholding them as explained next. Let there be two classes with labels 1 and 0. Let the training patterns from each class be given by

Class label 1: (1,1), (2,2), (3,3), (4,4), (5,5)

Class label 0: (4,1), (5,2), (6,3), (7,4), (8,5)

Let the validation patterns be given by

Class label 1: (2,1), (1,3)

Class label 0: (8,4), (7,5)

If we find $k = 5$ nearest neighbors for each validation pattern from the training set, we obtain details as shown in Table 2.13. Note that the table has 7 columns. The first column corresponds to the validation pattern considered. There are 4 rows in the table corresponding to the 4 validation patterns. The second column corresponds to the class label of these validation patterns. Note that the first two are from Class 1 and the remaining two are from Class 2.

TABLE 2.13 Computing probabilities and thresholding

Pattern	Class	5 NNs	Probability	Th>0.3	Th>0.5	Th>0.7
(2,1)	1	(1,1),(2,2),(4,1),(3,3),(5,2)	0.6	1	1	0
(1,3)	1	(2,2),(1,1),(3,3),(4,4),(4,1)	0.8	1	1	1
(8,4)	0	(8,5),(7,4),(6,3),(5,5),(5,2)	0.2	0	0	0
(7,5)	0	(8,5),(7,4),(5,5),(6,3),(4,4)	0.4	1	0	0

The third column lists ($k =$) 5 NNs, out of the 10 training patterns, of the validation pattern given in column 1. The fourth column gives the probability that the validation pattern belongs to Class 1. This probability is the ratio of the number of neighbors, out of 5, from Class 1 to the total number of NNs considered, that is, 5.

For example, for the validation pattern (2,1), three patterns, namely, (1,1), (2,2), (3,3) are from Class 1. So, the probability is $\frac{3}{5} = 0.6$. Similarly, probabilities are calculated for the remaining three validation patterns.

In columns 5 to 7, we use thresholds (Th) to convert the probability into 1 or 0. For example in column 5, if the probability is more than (Th =) 0.3, we put a 1, otherwise, a 0. Similarly, we get a 1, 0, 1 in rows 2, 3 and 4 based on the threshold (Th) value of 0.3. Entries in columns 6 and 7 are obtained as in column 5 but with thresholds 0.5 and 0.7, respectively.

For each of the thresholdings, we have a possible assignment of the four validation patterns. By comparing with the true class label given in column 2 of the table, we get an (FPR, TPR) pair as follows:

- For Th > 0.3, we get TP = 2, FN = 0, FP = 1, TN = 1. So, FPR = $\frac{1}{1+1}$ = 0.5 and TPR = $\frac{2}{2+0}$ = 1.
- For Th > 0.5, we get TP = 2, FN = 0, FP = 0, TN = 2. So, FPR = $\frac{0}{0+2}$ = 0 and TPR = $\frac{2}{2+0}$ = 1.
- For Th > 0.7, we get TP = 1, FN = 1, FP = 0, TN = 2. So, FPR = $\frac{0}{0+2}$ = 0 and TPR = $\frac{1}{1+1}$ = 0.5.

So, by thresholding, we obtain different (FPR, TPR) pairs that are used in plotting the ROC curve. The standard rule is to convert the class decisions into probabilities and then threshold them to get FPR and TPR pairs.

Note that schemes for computing probabilities can differ when other classifiers are used. However, a similar thresholding scheme can be used to obtain FPR and TPR pairs once the probabilities are available.

2.6.2 Performance of the Regression Algorithms

- **Mean squared error (MSE):** It is the most popular metric used for regression. It is defined by

$$\text{MSE} = \frac{1}{n}\sum_{i=1}^{n}(y_i - \hat{y}_i)^2,$$

where n is the number of patterns, y_i is the target value for the ith pattern and $\hat{y}_i$ is the value predicted by the regression model.
- **Mean absolute error (MAE):** It is the average of the difference between the target and the predicted values. It is given by

$$\text{MAE} = \frac{1}{n}\sum_{i=1}^{n}|y_i - \hat{y}_i|$$

2.7 AREA UNDER THE ROC CURVE FOR THE BREAST CANCER DATA SET

We consider the Breast Cancer data set for this experiment.

- The total number of patterns is 569 and each is a 30-dimensional vector.
- There are two classes: *Benign* and *Malignant*.
- We use KNN for classification with the value of $k = 15$.
- We use 171 patterns for validation and 398 patterns for training.
- We used the sklearn platform for conducting the experiment.
- We show the ROC curve in Fig. 2.12.
- Note the broken line in the figure. This indicates the possible lower bound.
- Ideally the grey line, the ROC curve, should go vertical on the Y-axis from 0 to 1 and then go horizontal at $Y = 1$, parallel to the X-axis from $X = 0$ to $X = 1$.

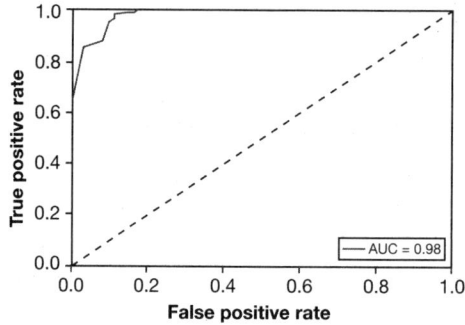

FIG. 2.12 ROC curve for Breast Cancer data using KNN (for colour figure, please see Colour Plate 1)

- So, in the ideal case, the area under the ROC should be 1.
- The ROC curve is almost ideal, but for the top-left corner.
- Note that the area under the curve, AUC, is 0.98.

SUMMARY

- We have considered neighborhood-based ML models in this chapter. They include:
  - KNN classifier; it is the NNC when $k = 1$. Evaluation of its performance is based on classification accuracy.
  - Regression or function value estimation using KNN. Evaluation of the KNN-based estimation scheme is by using the MSE.
  - Clustering based on spherical neighborhood using a threshold value as the radius of the sphere. In the Leader clustering algorithm, each of the resulting clusters is represented by its leader.
- We have identified the role of proximity measures in neighborhood models. We have explored several proximity measures that are popularly used.
- We have detailed a tree-based NNC and a branch and bound-based NNC for efficiency.
- We have studied the concentration problem that results as a consequence of using the popular integral norms including the Euclidean distance. We have discussed how fractional norms can overcome this problem by expanding the dynamic range of the distance values.
- We have shown experimental results on the Wisconsin Breast Cancer data set for classification and the Boston Housing data set for regression.
- We have detailed the performance evaluation measures that can be used to analyse various ML models. Classification accuracy and MSE are the most popular for classification and regression, respectively.
- When there is class imbalance, that is, a class has a smaller number of patterns and the other class has a very large number of patterns, then accuracy is not the right choice. Other measures like F1 score need to be used.

- **Explanation Capability:** These neighborhood ML models are based on

  - The intuitively appealing fact that similar things exist in close proximity of each other.
  - NNC is a simple and transparent classifier as the decision to assign a class label to a test pattern is based on the class label of its nearest pattern. It is not difficult for experts in an application domain to appreciate and understand how it works.
  - Even KNN and regression based on KNN is very transparent for the respective domain experts to understand the decision-making process based on extended neighborhood.
  - The Leader clustering algorithm is the simplest clustering algorithm. It is order dependent; it gives different clusterings and sets of leaders for different orders in which the data is presented to the algorithm. However, it is transparent and easy to appreciate as the leaders are good representatives for clusters.

EXERCISES

Use the Euclidean distance for computing the distance unless specifically stated otherwise.

1. Consider two l-bit binary strings p and q. How are $HD(p, q)$ and $SMC(p, q)$ related?
2. Derive the condition under which $SMC(p, q) > JC(p, q)$ for l-bit strings p and q.
3. Let $p = (4, 2, 5)$ and $q = (2, 1, -7)$. Find the distance between these two patterns using

   a. Euclidean distance
   b. City-block distance
   c. Max norm (L_∞)
   d. Fractional norm $(r = 0.25)$

4. Consider the points $p = (3, 0), q = (0, 3)$ and $s = (0, 0)$. Consider the fractional norm with $m = 0.5$. Is it a metric?
5. Give a two-dimensional example having 2 classes, where NNC is better than KNN with $k = 3$. This example illustrates that NNC can perform better than KNN $(k > 1)$ in some cases.
6. Consider the set of two-dimensional patterns: $((1, 1.5), 1), ((1, 2.5), 1), ((1, 3.5), 1), ((2, 1.5), 1), ((2, 2.5), 1), ((2, 3.5), 1), ((2, 4),1), ((2.5, 2.5), 1),((3.5, 1.5), 1), ((3.5, 2.5), 1), ((3.5, 3.5), 2), ((3.5, 4.5), 2), ((4.5, 1.5), 2), ((4.5, 2.5), 2), ((4.5, 3.5), 2),((5, 4.5), 2), ((5, 5.5),2),((6, 3.5), 2), ((6, 4.5), 2), ((6, 5.5), 2)$, where each pattern is represented by feature 1, feature 2 and the class.

   a. If a test pattern T is at $(2.6, 5.5)$, find the class of T using the nearest neighbor algorithm.
   b. Find the class of T using the KNN algorithm, where k is 3.
   c. Find the class of T using the WKNN algorithm, where k is 3.
   d. Find the class of T using the radius distance nearest neighbor algorithm with the radius $= 2.5$.

7. Consider two sets of two-dimensional points from two classes:
   Class 1: (3.5,3), (4.5,1), (1,5), (2,6), (3.5,4), (4.5,3)
   Class 2: (2,3.5), (2.5,1), (3.5,2), (3,2), (2.5,2), (2,1)
   Consider a variant defined using the centroid data set. Let Centroid1 be the mean of the six vectors in Class 1. Similarly Centroid2 is the mean of the six vectors in Class 2. Note that

the mean vector, Centroid, is defined on a collection of n vectors $X_1, X_2, \ldots, X_n$ as

$$\text{Centroid} = \frac{1}{n} \sum_{i=1}^{n} X_i$$

Now the centroids are used as the training data, that is, each of the new classes have only the centroid patterns $NewClass_1 : Centroid1$; $NewClass_2 : Centroid2$. Using this training data, classify the 12 patterns given using NNC.

a. This classifier is called the minimal distance classifier (MDC). What is the advantage of using MDC?

b. What is the classification accuracy of MDC in this problem on the 12 patterns?

8. Consider the two-dimensional vectors from two classes. Plot the decision boundaries for NNC and MDC. These decision boundaries are piece-wise linear and they are obtained by considering pairs of points, one from Class 1 and the other from Class 2. One side of the decision boundary will have Class 1 patterns and the other side will have Class 2 patterns.

Class 1	Class 2
(1,0)	(−2,0)
(−1,0)	(0,1)
(0,0)	(2,0)

9. Consider the data given in the following table.

Pattern	Vector (x)	Class label	Function $f(x)$
1	1 1	C_1	3
2	1 2	C_1	2
3	2 2	C_1	5
4	3 2	C_1	8
5	4 2	C_2	11
6	4 3	C_2	10
7	4 4	C_2	9
8	5 3	C_2	13

a. Use the Leader clustering algorithm on the 8 two-dimensional vectors using a threshold value of 3. Each vector has two components. So, you need not consider the last two columns for this problem. Give the leaders and the clusters obtained.

b. For this problem, we need to consider all the columns in the data other than the class labels.

   i.   Given that $f((x_1, x_2)) = ax_1 + bx_2 + c$, obtain the values of a, b and c using the first three patterns and the respective f values. Verify whether the remaining 5 patterns also satisfy this equation or not.

   ii. Consider the pattern (3, 1).

   A. Obtain its f value using $f((3,1)) = 3a + b + c$ where a, b and c are obtained in part (i) of the problem.

   B. Find its ($k =$) 3 nearest neighbors from the 8 patterns in the table.

   C. Let $\hat{f}((3,1))$ be the average of the f values of these 3 nearest neighbors. Obtain the value of squared error for (3, 1) by using $(f((3,1)) - \hat{f}((3,1)))^2$.

10. Cluster the 12 patterns given in Q7 using the Leader algorithm. Use a threshold of 1.5 units to cluster.

   a. Cluster the patterns in each class separately using the order in which the patterns are given. Let S_1^L be the set of leaders obtained. Give S_1^L.
   b. Cluster all the 12 patterns without using the class label. Let S_2^L be the set of leaders obtained. Assume that each leader belongs to a class if the cluster represented by the leader has a majority of patterns from that class. What is S_2^L?
   c. Is there a difference between S_1^L and S_2^L?

11. Consider the data in Q9(a) and reduce the data set size as follows. Take a training pattern x. Find out 3 $(k = 3)$ nearest neighbors of x; break any ties in favour of the class to which x belongs. If all the k neighbors are from the same class as that of x, then mark x. Perform this over all training patterns and then drop all the marked patterns and report the reduced set.

12. Consider the following process called bootstrapping:
   Consider a pattern x from the data given in Q9(a) and find 2 nearest neighbors x^1, x^2 from the same class as x; compute x' by the mean of the $k+1$ (3) vectors, that is, x and its 2 NNs. Repeat this process on all the training patterns in the collection. Replace the training data points with the x values obtained. What is the effect of this bootstrapping process? What happens if x values are added to the training set instead of replacing it?

13. Consider a 4×4 square to represent a digit, with each square assigned a positional value as illustrated in Table 2.6. The table given below depicts the digits represented using this method.

Transaction database

Digit	Transaction (Positional information of a digit)
0	1, 2, 3, 4, 5, 8, 9, 12, 13, 14, 15, 16
1	4, 8, 12, 16
7	1, 2, 3, 4, 8, 12, 16
9	1, 2, 3, 4, 5, 8, 9, 10, 11, 12, 16
4	1, 5, 7, 9, 10, 11, 12, 15

   Construct the FP tree for this table with minimum support $= 3$.

14. Consider the FP tree constructed in Q13, identify the label (digit) for the test pattern, $T = 1, 5, 9, 10, 11, 12, 15$.

15. Consider the data given in Table 2.12. Use $k = 3$ to compute the (FPR, TPR) pairs for the thresholds Th > 0.3, Th > 0.5 and Th > 0.7 on the probabilities obtained.

PRACTICAL EXERCISES

1. Randomly generate 100 values of x in the range [0,1]. Let them be $x_1, x_2, \ldots, x_{100}$. Perform the following based on the data set generated.

   a. Label the first 50 points $\{x_1, \ldots, x_{50}\}$ as follows: if $(x_i \leq 0.5)$, then $x_i \in Class_1$, else $x_i \in Class_2$.
   b. Classification.

      i. Classify the remaining points, $x_{51}, \ldots, x_{100}$ using KNN. Perform this for $k = 1, 2, 3, 4, 5, 20, 30$.

    ii. Classify the remaining points, $x_{51}, \ldots, x_{100}$ using WKNN. Perform this for $k = 1, 2, 3, 4, 5, 20, 30$.

    iii. Classify the remaining points, $x_{51}, \ldots, x_{100}$ using radius-based NNC. Perform this for $k = 1, 2, 3, 4, 5, 20, 30$.

c. Compute the classification accuracy in all three cases and report. [Note: Classification accuracy $= \frac{nc}{n}$, where $n = 50$, $nc =$ number correctly classified]

2. Cluster the entire set of 100 points (as mentioned in Q1) using Leader clustering. Choose different values for threshold (T) and carry out the clustering.

3. Let the clustering obtained using some threshold, T_i be $\mathcal{C}_i = \{Cluster_1, Cluster_2, \ldots, Cluster_{il}\}$. Computer the purity value for each clustering, which is given by

$$Purity(\mathcal{C}_i) = \sum_{j=1}^{il} maximum(|Cluster_j \cap Cluster_1|, |Cluster_j$$
$$\cap Cluster_2|, \ldots, |Cluster_j \cap Cluster_{il}|)$$

4. Use the Digits data set available under sklearn: `https://scikit-learn.org/stable/modules/generated/sklearn.datasets.load\_digits.html`
Consider 10% of the data for training (179 samples). Each pattern is an 8×8-sized character where each value is an integer in the range 0 to 16. Convert it into binary form by replacing a value below 8 by 0 and other values (≥ 8) by 1.

a. Use these 179 patterns with labels and the remaining without labels for this subtask. Use KNN and label the patterns without labels. Obtain the % classification accuracy. Perform this task with k values from the set $\{1, 3, 5, 10, 20\}$.

b. Obtain the frequent itemsets for these 179 patterns using FP-growth by viewing each binary pattern as a transaction of 64 items. Repeat this task with different *minsup* values from $\{0.1, 0.3, 0.5, 0.7\}$.

Bibliography

- Andoni, A and Indyk, P. 2008. Near-Optimal Hashing Algorithms for Approximate Nearest Neighbor in High Dimensions, *Communications of the ACM*, 51(1): 117–122.

- Beyer, K, Goldstein, J, Ramakrishnan, R and Shaft, U. 1998. When is "Nearest Neighbour" Meaningful?, *Proceedings of ICDT*, 217–235.

- Flexer, A and Schnitzer, D. 2015. Choosing l^p norms in high-dimensional spaces based on Hub Analysis, *Neurocomputing*, 169: 281–287.

- Murty, MN and Susheela Devi, V. 2015. *Introduction to Pattern Recognition and Machine Learning*, World Scientific Publishing Co. Pte. Ltd.: Singapore.

- Duda, RO, Hart, PE and Stork, DG. 2007. *Pattern Classification*, New York: John Wiley & Sons.

- Han, JW, Kamber, M and Pei, J. 2012. *Data Mining Concepts and Techniques, Third Edition*, Morgan Kaufmann Publishers, Waltham.

CHAPTER 3

Models Based on Decision Trees

At the end of this chapter, you will be able to:

- Describe decision trees and impurity measures
- Explain how decision trees are used in classification and regression
- Explain the bias–variance trade-off
- Describe predictive models based on decision trees
- Explain how AdaBoost, random forest and gradient boosting are used in prediction

3.1 INTRODUCTION TO DECISION TREES

Decision tree is a popular structure in ML. It is built by splitting the data set associated with a node into subsets that typically have less entropy or better purity. Splitting is carried out using the values of a selected feature; this feature-based decision-making at each of the internal nodes offers transparency and is ideally suited for explanation and for easier understanding by application domain experts.

Some properties of the decision tree (DT) building algorithm are:

- **It is greedy**: The best feature, in terms of purity in splitting, among the given set of features is invariably used at the root node of the DT. So, the feature used at each of the child nodes will depend on the feature selected at the parent node. Such a sequential greedy process may fail to be optimal on the whole.
- **It can be expensive**: Selection of a feature depends both on the number of features and the number of patterns. So, tests conducted on the feature values at each node need to be simple. Even simple tests can become computationally demanding when the dimensionality is large.
- **It can overfit**: The depth of the DT can increase as more splits are considered sequentially. Such a deep DT can overfit the training data and perform poorly on the validation/test data. A solution to this problem is provided by building the DT using the training data without worrying about depth and then pruning the more recent subtrees based on its performance on a validation data set. This process is called **pruning** and it helps in solving the overfitting problem.

The first two issues mentioned above are solved using a combination of classifiers. Some of the popular combinational ML models are:

- **Random Forest**: It is a collection of decision trees. The model predicts by combining the predictions of the trees in the collection.
- **AdaBoost**: It combines multiple weak learners to realize a good learning algorithm; a weak learner is a learner that provides more than 50% accuracy. One can use a variety of ML models as possible weak learners; however, simple DT-based weak learners are popular. Here, the combinational model employs weighted majority voting to realize the overall outcome.
- **Gradient Boost**: It is based on a sequence of ML models where each model is built to correct the error in the prediction of the previous model. Here also, it is possible to use any ML model in the sequence, but again, DT models are more popular. In this case, learnt models are used one after the other to predict the overall outcome.

These state-of-the-art models have become popular because they can deal with large-scale high-dimensional data sets that are common in current-day applications. We examine all these DT-based models in this chapter. The notion of overfitting is linked to a very fundamental concept in ML called the bias–variance trade-off. Simply stated, simple models tend to have higher bias while complex models tend to have higher variance.

3.2 DECISION TREES FOR CLASSIFICATION

A decision tree is a tree with a root node and each child of the root node forms the root node of the subtree below it. We associate the entire training data set with the root node. The process of building the decision tree depends upon splitting the data set associated with a node.

A popular approach is to split, based on a feature, the set of patterns associated with a node so that the resulting subsets are least impure. These subsets are associated with the children nodes of the node. Impurity of the split at a node is captured using an impurity measure; the most popular one is **Shannon's entropy**. It is given by

$$-\sum_{i=1}^{C} p_i log(p_i),$$

where p_i is the probability of class i, $i = 1, 2, \cdots, C$ assuming that there are C classes. These probabilities are typically estimated based on the fraction of elements from each class among the set of patterns associated with the node.

EXAMPLE 1: Consider a two-class problem, that is, $C = 2$. Consider the values of p_1 and p_2 corresponding to the 5 different cases shown in Table 3.1. The number of elements in the set associated with the node in each case is assumed to be 100.

Here, we use the fact that the limit of $plog(p)$ is 0 as $p \to 0$ in cases 1 and 5 in the table where the entropy is 0. Some interesting observations from the table are:

- Entropy is non-negative; it is greater than or equal to zero. This is because $p_i \geq 0$ and $log(p_i) \leq 0$ for all i. So, $-log(p_i) \geq 0$ for all i. As a consequence, if there are C classes, then

$$-\sum_{i=1}^{C} p_i log(p_i) \geq 0$$

So, the minimum possible value is 0.

TABLE 3.1 Five different probability distributions and their entropy values

Case	Class 1 (p_1)	Class 2 (p_2)	Entropy
1	100 $(p_1 = 1.0)$	0 $(p_2 = 0.0)$	0.0
2	75 $(p_1 = 0.75)$	25 $(p_2 = 0.25)$	0.811
3	50 $(p_1 = 0.5)$	50 $(p_2 = 0.5)$	1.0
4	25 $(p_1 = 0.25)$	75 $(p_2 = 0.75)$	0.811
5	0 $(p_1 = 0.0)$	100 $(p_2 = 1.0)$	0.0

- In a two-class scenario, entropy is symmetric, that is

$$entropy(p_1, p_2) = entropy(p_2, p_1)$$

This is because

$$entropy(p_1, p_2) = -p_1 log(p_1) - p_2 log(p_2) = -p_2 log(p_2) - p_1 log(p_1) = entropy(p_2, p_1)$$

- In a two-class problem, entropy is maximum when $p_1 = p_2 = 0.5$. This can be explained by computing the partial derivatives of entropy and equating them to 0. The partial derivative of $entropy(p_1, p_2) = -p_1 log(p_1) - p_2 log(p_2)$ with respect to p_1 is $\frac{\partial entropy(p_1, p_2)}{\partial p_1} = -1 - log(p_1)$. Similarly, the partial derivative with respect to p_2 is $-1 - log(p_2)$. If we equate these two quantities to 0, then we get

$$-1 - log(p_1) = -1 - log(p_2) = 0 \Rightarrow log(p_1) = log(p_2) \Rightarrow p_1 = p_2$$

However, $p_1 + p_2 = 1$ and if $p_1 = p_2$, then $p_1 = p_2 = 0.5$.
- This corresponds to the maximum as $\frac{\partial^2 entropy(p_1, p_2)}{\partial^2 p_1}$ equals $\frac{-1}{p_1}$ and is negative as $p_1 \geq 0$; $\frac{\partial^2 entropy(p_1, p_2)}{\partial p_1 \partial p_2} = \frac{\partial^2 entropy(p_1, p_2)}{\partial p_2 \partial p_1} = 0$ and $\frac{\partial^2 entropy(p_1, p_2)}{\partial^2 p_2}$ equals $\frac{-1}{p_2}$ and is negative as $p_2 \geq 0$. The Hessian matrix, H, is

$$H = \begin{bmatrix} -2 & 0 \\ 0 & -2 \end{bmatrix}$$

when $p_1 = p_2 = 0.5$ as $-\frac{1}{p_1} = -\frac{1}{p_2} = -2$.
- A matrix $A_{C \times c}$ is **negative definite** if $x^t A x < 0$ for any real vector x of size $C \times 1$. So, in this case, the Hessian matrix, H, is negative definite as

$$x^t H x = -2x(1)^2 - 2x(2)^2$$

- So, the value of $p_1 = p_2 = 0.5$ corresponds to the maximum.

Entropy depends on the probability distribution, the values of p_1 and p_2 in a two-class case. However, for the sake of brevity, we use entropy without arguments when the context is clear.

EXAMPLE 2:   We will consider the example data set shown in Table 3.2 to examine the splitting process. We show the collection of these two-dimensional patterns in Fig. 3.1. The two classes are represented using $*$ (Class 1) and $\#$ (Class 2).

TABLE 3.2 A data set used to illustrate splitting

Pattern	Feature1	Feature2	Class
1	1	1	1
2	1	3	1
3	2	1	1
4	2	3	1
5	4	1	1
6	4	3	2
7	5	1	2
8	5	3	2

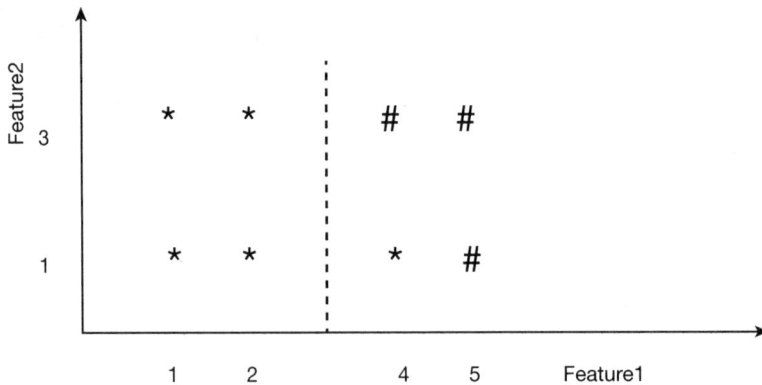

FIG. 3.1 Two-dimensional patterns of the data set given in Table 3.2

The entire collection of 8 two-dimensional patterns is associated with the root node. We split this set into subsets based on test or a splitting criterion that employs a single feature. We have only two features, *feature*1 and *feature*2, to select from.

If we use *feature*1, then one possible split is based on *feature*1 < 3 or not. The corresponding subsets are {1,2,3,4} and {5,6,7,8}. We compute the entropy of each subset and weigh it with the ratio of the size of the subset to the size of the set of its parent. The first subset is pure, that is, all the elements are from the same class (1). The weights are equal and are $\frac{4}{8} = \frac{1}{2}$ for each subset.

We use the proportions of classes as estimates for probabilities. In this example, $C = 2$. So, $p_1 = 1$ and $p_2 = 0$ for the first subset. Hence, entropy is

$$\frac{4}{4}log(1) + 0log0 = 0$$

So, for 4 out of 8 patterns the entropy is 0. Similarly the entropy for the second subset is

$$-\frac{1}{4}log\left(\frac{1}{4}\right) - \frac{3}{4}log\left(\frac{3}{4}\right) = \frac{1}{4}log(4) + \frac{3}{4}log\left(\frac{4}{3}\right) = 0.244219$$

We obtain the weighted value of the entropy of the split by using the weighted combination of the entropy values of the two subsets; weights are proportions of the number of elements in each subset. Here, each of the subsets has 4 elements out of the total of 8 elements present at their parent. So, the weights are equal to $\frac{4}{8} = \frac{1}{2}$. So, the weighted value of entropy impurity is

$$\frac{1}{2} \times 0 + \frac{1}{2} \times 0.244219 = 0.12211$$

So, total weighted entropy associated with the split is 0.12211.

It is possible to split the set based on other values of *feature*1. One possible split is based on *feature*1 $\leq$ 4. The split will result in two subsets, the first set having 6 out of 8 patterns and the second set having the remaining 2 elements from the same class (2). So the entropy is 0.19567 and if we weigh it in proportion $\frac{6}{8}$, we get 0.146757; this is the weighted impurity associated with the split. This is larger than the entropy obtained for the earlier split, which was 0.12211. So, the best (minimum) value of entropy of splitting based on *feature*1 is 0.12211.

The other possibility is to consider splitting of *feature*2; there is only one possible split in this case. Using the test *feature*2 $\leq$ 1 or not, we can split the data into two sets: {1,3,5,7} and {2,4,6,8}. The entropy of the first set is 0.244219 with a weight of $\frac{1}{2}$ (4 out of 8 patterns). The second set has entropy of 0.301 and the weight is $\frac{1}{2}$ (4 out of 8 patterns). So, the weighted impurity is

$$\frac{1}{2}(0.244219 + 0.301) = 0.2726$$

Observe that this value is larger than the 0.12211 obtained using *feature*1. So, *feature*1 is used at the root as it leads to purer subsets. This is depicted in Fig. 3.2.

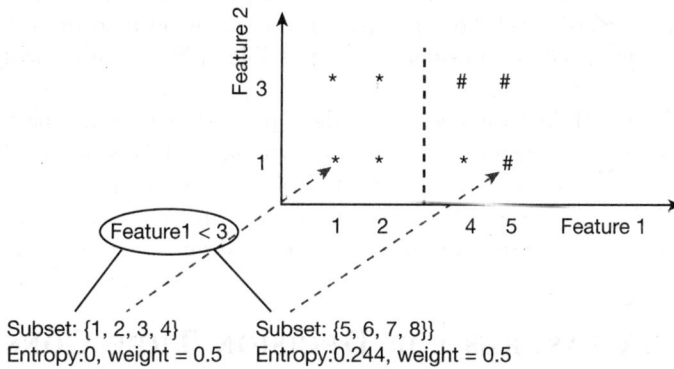

FIG. 3.2 The process of splitting at the root (for colour figure, please see Colour Plate 1)

Note that there could be several (infinite) ways to indicate the same split. For example, *feature*1 $>$ 1, *feature*1 $>$ 1.5 or even *feature*1 $\leq$ 2 will all lead to the same split in terms of the resulting subsets, and so, all of them are equivalent.

The left child of the root is pure, that is, all the elements are from the same class, Class 1; so, there is no need to split it further as all the patterns are from the same class. However, the right child corresponding to the second subset is not pure. So, we need to split the set at the right child further. This process is recursive and goes on till some stopping criterion is satisfied. The final decision tree is shown in Fig. 3.3.

We depict leaf or terminal nodes using rectangles and non-terminal or internal nodes using ellipses in Fig. 3.3. Each non-terminal node is also called a decision node as there is a branching below the node based on the outcome of a test (satisfying a condition) associated with the non-terminal node.

These tests are typically binary and are used to split the set of patterns into two subsets. Note the difference in the entries used at the root node in Figs 3.1 and 3.3; in Fig. 3.1, the test is $Feature1 < 3$ and in Fig. 3.3, it is $Feature1 < 4$. Note that both yield the same subsets after splitting. This is done intentionally, to alert the reader that different tests could be equivalent. For example, $Feature1 \leq 2$ is another equivalent test that yields the same split.

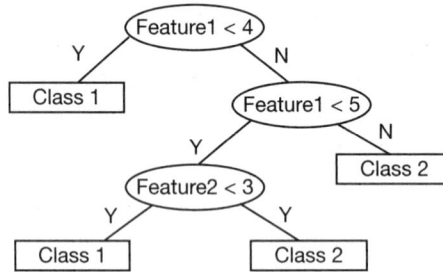

FIG. 3.3 Decision tree for the data set

The process of constructing a decision tree (DT) may be abstracted as follows:

1. Associate the entire training data set with the root and select the best feature and optimal splitting value for splitting. If the split leads to p subsets, then the root will have p children, each associated with a subset. If there are n l-dimensional patterns, then every one of the l features needs to be considered. For each feature, n values need to be sorted; for n patterns, in the worst case, we need to consider $n - 1$ possible splits for each feature. This requires $\mathcal{O}(n \log n)$ time.

2. Consider each of the p children nodes. If a node is pure, that is, it has patterns from only one class, then terminate by declaring the node as a terminal node; else select the best feature and optimal split position and keep splitting recursively, that is, keep splitting the nodes till the termination condition is satisfied. If there are m levels in the tree, the time required will be $\mathcal{O}(mkn \log n)$, where k is the maximum number of children.

3.3   IMPURITY MEASURES FOR DECISION TREE CONSTRUCTION

Even though Shannon's entropy is popular for capturing the impurity associated with a split, there are other impurity measures. Here, by p_i, we mean the probability of the ith class. Some of the important ones are:

- **Gini Index:** The impurity at a node for a C class problem is defined as

$$1 - \sum_{i=1}^{C} p_i^2$$

- **Misclassification Impurity:** It is defined as

$$1 - \max_{j} \ p_j$$

Consider a node at which the associated set has 80% of patterns from Class 1, 15% from Class 2 and 5% from Class 3. So, $p_1 = 0.8$, $p_2 = 0.15$, $p_3 = 0.05$. Then, the misclassification impurity of the node is $1 - 0.8 = 0.2$. Note that it is the sum of the two smaller probabilities p_2 and p_3.

Note that entropy, Gini index and misclassification impurity are all impurity measures. They can be used to find the impurity of a node. We can also use them in computing the impurity associated with a split by considering the weighted impurity used in Example 2. In the same example, we used entropy as the impurity measure.

EXAMPLE 3:   We illustrate the Gini impurity measure and some of the important properties of the decision tree classifier (DTC) using the example data set shown in Table 3.3. There are three independent features: *Outlook*, *Temperature* in degrees Fahrenheit, and percentage *Humidity*. For the classes, there are two possibilities: P standing for Play and NP standing for No Play.

TABLE 3.3 A data set used to illustrate properties of decision trees

Pattern	Outlook	Temp(F)	Humidity	Class
1	Sunny	65	91	NP
2	Sunny	78	70	P
3	Overcast	85	82	P
4	Overcast	62	71	P
5	Rainy	70	65	NP
6	Rainy	79	62	P

If we use *Outlook* for splitting, based on its value, we will obtain three subsets: $\{1,2\}$, $\{3,4\}$ and $\{5,6\}$. The second subset is pure $p_1 = 1$ and $p_2 = 0$ and the Gini index value for it is 0 ($1 - 1^2 - 0^2 = 0$). In the cases of the first and third subsets, $p_1 = p_2 = 0.5$. So, the Gini index value for each of them is $1 - (0.5)^2 - (0.5)^2 = 0.5$ and each subset has 2 out of 6 elements. So, the weights are $\frac{2}{6}$ each.

So, the Gini impurity value for the whole split is $\frac{2}{6} \times 0.5 + \frac{2}{6} \times 0 + \frac{2}{6} \times 0.5 = \frac{1}{3}$. So, using *Outlook* for splitting, we get $\frac{1}{3}$ as the impurity value.

If we use *Humidity* for splitting and use the test *Humidity* < 91, then we get two subsets, $\{1\}$ and $\{2,3,4,5,6\}$. Note that this is an optimal split using *Humidity* which will be considered in the exercises at the end of the chapter. The first subset $\{1\}$ is pure and has 0 Gini index value. For the second set, $p_1 = \frac{4}{5}$ and $p_2 = \frac{1}{5}$. So, the Gini index value is $1 - (\frac{4}{5})^2 - (\frac{1}{5})^2 = \frac{8}{25}$ and the weight is $\frac{5}{6}$ as 5 out of 6 patterns are present in the second subset.

So, the weighted value is

$$\frac{1}{6} \times 0 + \frac{5}{6} \times \frac{8}{25} = \frac{4}{15}$$

If we perform the split based on *Temperature* > 70, then the two subsets we get are $\{2, 3, 6\}$ and $\{1,4,5\}$. Note that the first subset is pure and its Gini index value is 0. For the second subset, $p_1 = \frac{1}{3}$ and $p_2 = \frac{2}{3}$ and the Gini value is $1 - \frac{4}{9} - \frac{1}{9} = \frac{4}{9}$. The weight for this subset is $\frac{3}{6}$ and the weighted value of impurity is $\frac{2}{9}$. This value is the smallest among all the three features considered. So, we choose *Temperature* at the root.

By using *Temperature* > 70 at the root, we obtain a pure subset {2,3,6} and the second subset {1,4,5} needs to be split further. Without getting into details, we have two options for splitting the set {1,4,5}. One is based on *Temperature* > 62 (Fig. 3.4 (a)) and other is based on *Outlook* (Fig. 3.4 (b)); both of them lead to the same impurity value of 0.

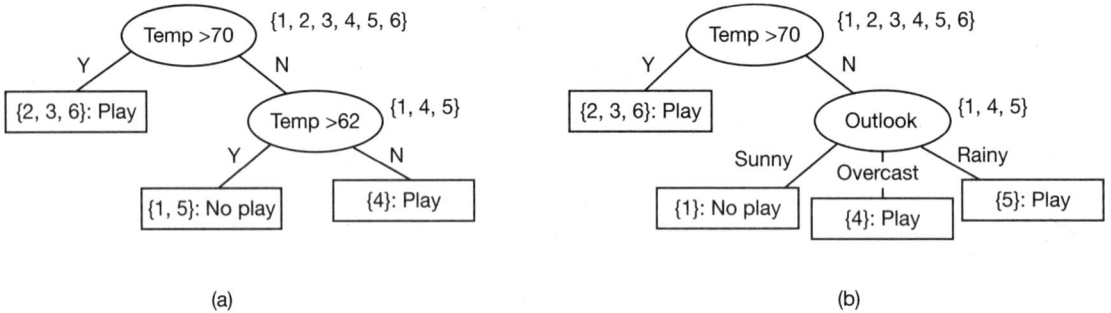

(a)   (b)

FIG. 3.4 Decision trees for the data set

We skipped some of the details involved in building the decision tree. They will be considered in some exercise problems at the end of the chapter.

3.4   PROPERTIES OF THE DECISION TREE CLASSIFIER (DTC)

We have seen how to build decision trees from the given training data. We will now examine some important details in building decision trees and using them in classification.

1. **Splitting rule:** We have considered a simple test for splitting.

   a. Split is based on the values of a single feature. In Fig. 3.3, we have considered tests of the form *Feature*1 < 4. This is called an **axis-parallel split** and is depicted in Fig. 3.2 using a broken line that is parallel to the axis of *Feature*2. It is possible to use more complex rules for splitting. For example, a split rule of the form $20 \times Feature1 + 10 \times Feature2 > 100$ will split the data set shown in Table 3.2 into two subsets: {1,2,3,4,5} and {6,7,8}, leading to minimum impurity in these subsets. So, one test is adequate to build the decision tree (has only one node) as shown in Fig. 3.5. Such a split is called an **oblique split**. However, exploring all possible oblique splits is computationally expensive; it will require time that is exponential in the number of features. So, almost all of the practical implementations use axis-parallel splits.

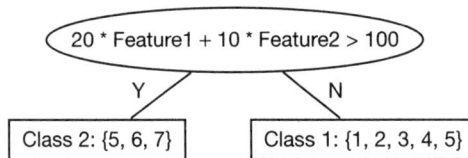

FIG. 3.5 Decision tree for the data set given in Table 3.2

b. We considered binary splits, that is, there will be two children for any node; a set of patterns is split into two subsets. Such binary splits are popular when the feature is numerical. However, for categorical features, it may be convenient to use non-binary splits where the number of children is more than 2. In Example 3, we considered such a categorical feature, *Outlook*, where the set of patterns is split into 3 subsets, as shown in Fig. 3.4.

2. **Criterion for splitting:** In Examples 1 and 2, we used the minimal impurity value to select the best feature for splitting. However, instead of minimizing impurity, it is possible to maximize the purity of a split by using **information gain**.

EXAMPLE 4: Information gain (IG) at a node p in the tree is defined as

$$IG(p) = Entropy(p) - Weighted\ Impurity(children(p))$$

For the data in Example 1, we can compute IG at the root node for each of the three features as follows. The root is associated with the collection of all the 6 patterns, out of which 2 are from class NP and the remaining 4 are from P. So, $p_1 = \frac{4}{6}$ and $p_2 = \frac{2}{6}$. So, entropy at the root is

$$-\frac{2}{6}log\left(\frac{2}{6}\right) - \frac{4}{6}log\left(\frac{4}{6}\right) = 0.918$$

Now IG using each of the three features is:

- *Outlook:* The impurity at the root is 0.918 and the weighted impurity, by using *Outlook* for splitting, is $\frac{2}{6}(-\frac{1}{2}log(\frac{1}{2}) - \frac{1}{2}log(\frac{1}{2})) + \frac{2}{6}(-\frac{1}{2}log(\frac{1}{2}) - \frac{1}{2}log(\frac{1}{2})) = \frac{2}{3} = 0.6666$. So, IG is $0.918 - 0.6666 = 0.2514$.
- *Humidity:* Here, IG is $0.918 - 0.602 = 0.316$.
- *Temperature:* The IG value at the root using this feature is $0.918 - 0.459 = 0.459$.

Among the three features, *Temperature* has the maximum IG value of 0.459. So, we select *Temperature* > 70 for testing at the root. Note that selecting a split based on minimum weighted entropy is the same as selecting a split based on maximum IG at any node.

A problem with information gain is that it favours features that have more values. A correction is proposed in the form of **information gain ratio**. It is defined as

$$\frac{IG}{Weighted\ Impurity}$$

for each attribute at a node.

EXAMPLE 5: So, the IG values for the three features considered in Example 2 are:

- *Outlook:* the ratio is $\frac{0.2514}{0.6666} = 0.1676$
- *Humidity:* the ratio is $\frac{0.316}{0.602} = 0.525$
- *Temperature:* the IG ratio is $\frac{0.459}{0.459} = 1.0$

So, we select *Temperature* based on the IG ratio as it has the maximum value among the three IG ratio values.

3. **Binary or Non-binary:** We have observed that we can split a set of patterns associated with a node into two or more subsets; each subset is associated with a child of the node under consideration. Consider the decision trees shown in Fig. 3.4. The entire set of patterns is associated with the root node. A subset {1,4,5} is associated with a child node of the root.

This subset is further split using *Temperature* > 62 into two subsets in Fig. 3.4 (a). Note that the resulting decision tree is a binary tree. However, in Fig. 3.4 (b), the subset {1,4,5} is split further into three subsets, each having a single element using the feature *Outlook*. So, this is not a binary tree. However, we can create a binary decision tree here by changing the test to *Outlook = Overcast*, as shown in Fig. 3.6.

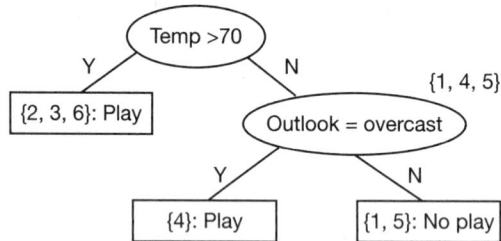

FIG. 3.6 Binary decision tree for the data set given in Fig. 3.4 (b)

4. **Termination condition:** We have seen in the examples that we stop splitting a node once the node is pure or has all the associated patterns from the same class. This is fine in the case of some of the examples we considered. However, in practical applications, it is advisable to permit a small percentage of impurity for termination. Some popular schemes could be:

   a. Terminate when the impurity level is less than, say, one or two percent. That is, out of 100 patterns associated with a node, 99 or 98 are from the same class and the remaining 1 or 2 are from different class/classes.
   b. Use a separate validation data set. Build the decision tree using the training data and check whether the performance of the tree is acceptable on the validation data.

5. **Class labels:** Each terminal node is represented by a rectangular box and it is associated with a class label. If the node is pure, then automatically the class label of the associated patterns is the class label. Consider for example, Fig. 3.6. Here, there are three terminal nodes, one at level 1 and two at level 2. The subset of patterns associated with the class label are depicted in the respective nodes. When an impurity is present, label the node using the class of the majority patterns associated with the node.

6. **Classification:** Given a test pattern, classification involves traversing the tree starting from the root and selecting the edge based on checking for the satisfiability of the condition at each node in the path till a leaf node is encountered. The class label associated with the leaf node is the class label to be assigned to the test pattern.
   For example, consider the patterns (*Outlook = Sunny*) and *Temperature* = 64 and *Humidity* = 72. Using the DT in Fig. 3.6, we take the right branch at the root node as *Temperature* > 70 is violated by the pattern. The next decision node checks for *Outlook = Overcast* and this is also violated by the pattern. So, we take the right branch at this decision node also and encounter a terminal node that is labelled *No Play*. So, the given pattern is classified as *No Play*.

7. **Transparency:** The path from the root to a leaf or terminal node, in a DT, may be viewed as a classification rule. For example, in the binary DT in Fig. 3.6, if we traverse from the root to the rightmost leaf node, we have a rule of the form

   If *Temperature* > 70 and *Outlook* ≠ *Overcast*, then *No Play*

So, a DT can be thought of as an abstraction of several classification rules and each such classification rule could be used to explain the reason behind classifying a pattern. Thus, DT is an excellent structure for explaining the reason behind a prediction.

8. **Handling mixed data types:** Decision tree classifiers are versatile in terms of dealing with both numerical and categorical features characterizing the data. Consider the data set shown in Table 3.3. Here, *Outlook* is categorical and *Temperature* and *Humidity* are numerical features. Using this data set, the decision tree in Fig. 3.6 is constructed. So, DTC can deal with mixed data types.

9. **Eliminating irrelevant features:** The DT automatically abstracts away features that are not relevant for classifying the data. Only relevant features are used. Consider the data set shown in Table 3.3. The feature *Humidity* is not a part of the tree in Fig. 3.6. For the same data set, the DT in Fig. 3.4 (a) employs only *Temperature*; the other two features, *Outlook* and *Humidity*, are irrelevant.

10. **Pruning a decision tree:** In some data sets, the decision tree can keep growing in height and size. Such a tree may overfit the training data; as a consequence, its performance on the testing data may be poor. In order to appreciate this problem, we use the DT shown in Fig. 3.7.

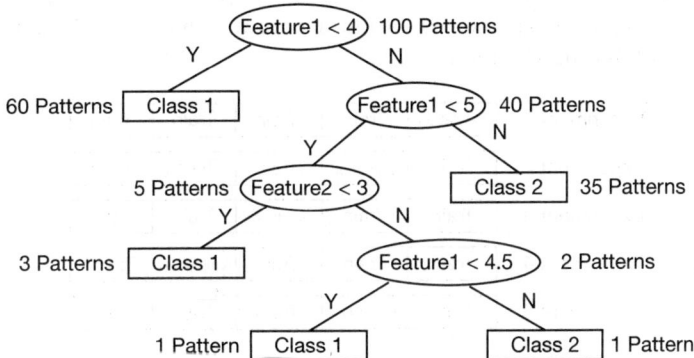

FIG. 3.7 Decision tree that can overfit

- Note that there are 100 training patterns that are associated with the root. Next to each node, non-terminal or terminal, we have indicated the number of patterns associated with the node. Note that there are 64 patterns from Class 1 and 36 patterns from Class 2.
- These are associated with different leaf nodes. The decision tree has 4 levels.
- Observe that split at the third level leading to two terminal nodes at the fourth level explains only two patterns and each resulting leaf node is associated with only one pattern.
- Such a tall tree can lead to overfitting. In some practical applications, the number of levels in the tree can be very large, it can be a few thousands. Some thumb rules say that the number of levels should be around $\log(l)$ where l is the number of features.

Pruning is a technique used to counter this overfitting problem. The solutions offered are:

a. Let the tree grow till the termination condition is satisfied. Now prune the leaf nodes that deal only with less than 3% of the data. In Fig. 3.7, the split based on *Feature1* < 4.5 and the split based on *Feature2* < 3 need not be performed; instead the non-terminal node corresponding to the test *Feature2* < 3 can be converted into a leaf node and associated with Class 1 as 4 out of 5 patterns are from Class 1 (majority decision). This gives us the decision tree shown in Fig. 3.8.

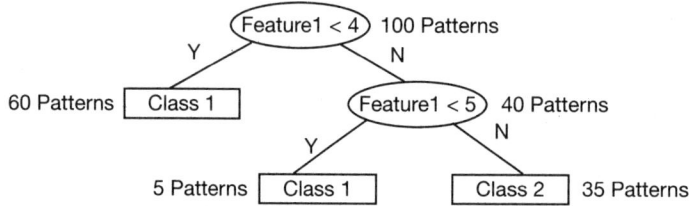

Fig. 3.8 Decision tree after pruning

b. Another possibility is to stop growing the tree beyond some pre-defined number of levels based on the $log(l)$ heuristic or as specified in the termination condition (item d in this list).

c. The other possibility is to build the DT and use k-fold cross-validation to check whether the tree is fine or further splitting is required. Five-fold ($k = 5$) cross-validation may be explained using Fig. 3.9. The training set is split into five equal parts and four parts are used for training and the fifth part for validation. This is repeated five times taking a different part for validation each time. The average validation accuracy over the five experiments is reported as the validation accuracy. This ensures that every part of the data is used in building the model.

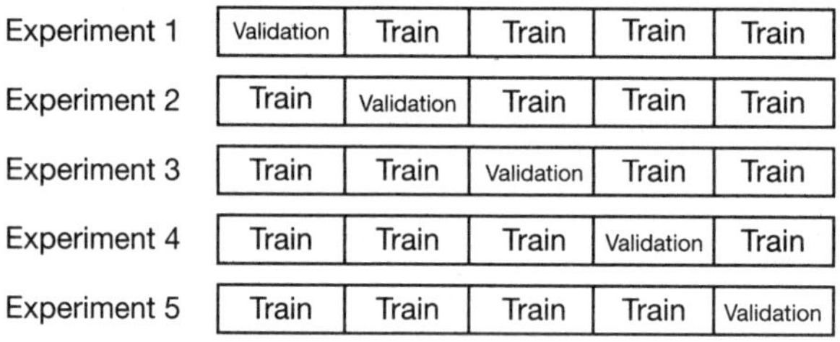

Fig. 3.9 Five-fold cross-validation

d. Another popular scheme is based on **minimal cost-complexity pruning**. It employs a parameter $\alpha \geq 0$; it is called the complexity parameter. The cost-complexity objective

$$C_\alpha(T) = C(T) + \alpha|Term(T)|,$$

where $C(T)$ is the total misclassification rate of the terminal nodes of T, $Term(T)$. The term $|Term(T)|$ is the number of terminal nodes in T. The parameter α tries to strike a balance between the error of the classifier in terms of T and the size of the DT in terms of $|Term(T)|$. We obtain the α that minimizes $C_\alpha(T)$ using the validation set.

3.5   APPLICATIONS IN BREAST CANCER DATA

Decision trees were built for the Wisconsin Breast Cancer data, which has 426 training patterns in a 30-dimensional space. Of the 569 patterns, 426 are used for training and the remaining 143 patterns are used for testing. Each pattern is a 30-dimensional vector.

We depict the decision trees obtained in Fig. 3.10. The DT in Fig. 3.10 (a) is obtained using the Gini index and the one in Fig. 3.10 (b) is obtained using Shannon's entropy. The number of leaf nodes is restricted to 5 in both the cases.

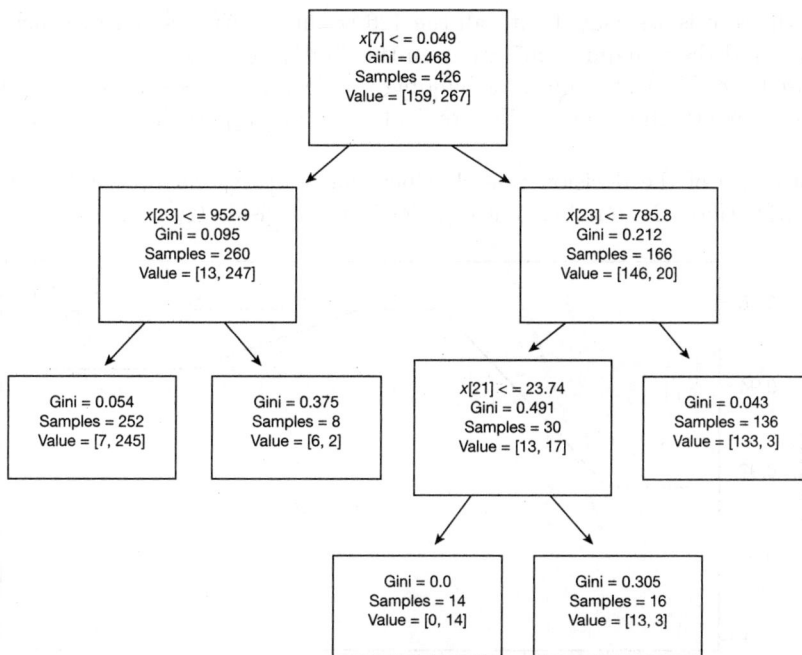

(a)

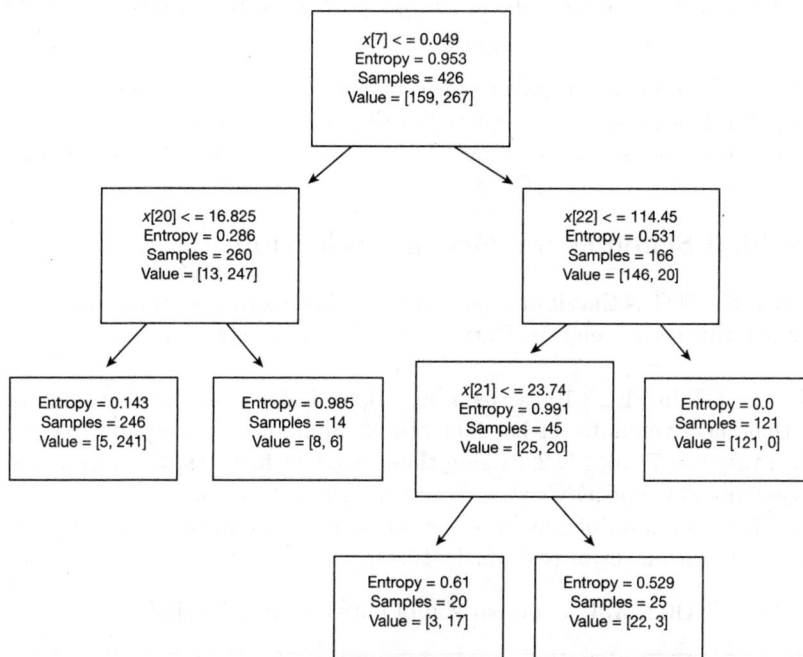

(b)

FIG. 3.10 Decision trees using Gini index and entropy

Note that the root is associated with all the 426 training patterns, out of which 159 are from class *Malignant* and the remaining 267 are from the *Benign* category.

We have used the sklearn platform for building the decision trees shown in Fig. 3.10. We have shown the output directly in the figure. So, non-leaf nodes also appear in rectangular boxes instead of oval boxes.

The performance of the decision tree classifier that employs entropy and Gini impurities is shown in Fig. 3.11. Here, the number of leaf nodes in the decision tree varies from 3 to 11.

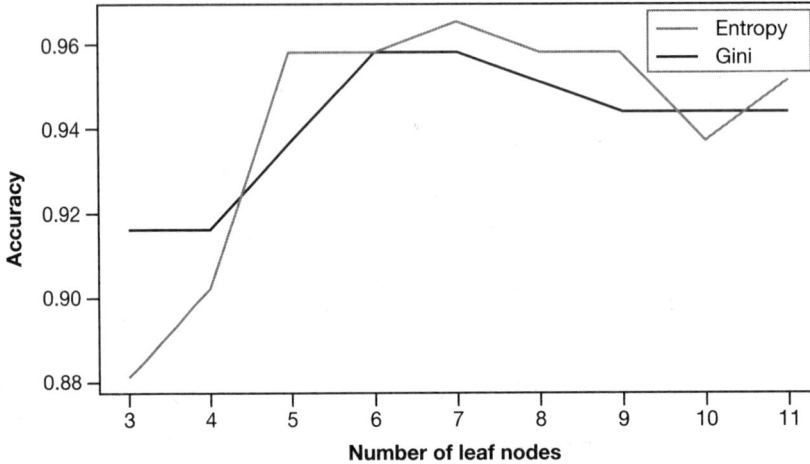

FIG. 3.11 Accuracies of decision trees using Gini index and entropy (for colour figure, please see Colour Plate 1)

Note that the classification accuracy increases to 0.965 in the case of the entropy measure and to 0.958 when the Gini index is used, as depicted in Fig. 3.11. These maximum values are observed when the number of leaf nodes is 7. Accuracy decreases as we increase the number of leaf nodes beyond 7.

3.5.1 Embedded Schemes for Feature Selection

An important property of DT is that it automatically retains relevant features and ignores irrelevant features. We can use this set of relevant features, which is a subset of all the features, in building classifiers.

For example, the DT in Fig. 3.10 (a) uses only three features in decision making; they are 7, 21 and 23; note that the given data set has 30 features. Similarly, the decision tree in Fig. 3.10 (b) uses four features; they are 7, 20, 21, 22. Using these relevant features, the accuracies obtained on the test data based on DTC and KNN ($k = 7$) are shown in Table 3.4.

This entire activity is a kind of feature selection. We call it embedded scheme when the learnt model itself comes out with a set of relevant features.

TABLE 3.4 A comparison of entropy and Gini index

Impurity	Features used	Accuracy of DTC	Accuracy of KNN ($k = 7$)
Entropy	{7,20,21,22}	0.958	0.958
Gini	{7,21,23}	0.937	0.944

We will see some more classifiers that have this property, and hence, they can be used in feature selection.

3.6   REGRESSION BASED ON DECISION TREES

Regression and classification are related predictive tasks. We have seen how the KNN-based scheme handles both classification and regression in the previous chapter. We will now examine the role of decision trees in regression.

EXAMPLE 6:   Let us consider the data set we have used to illustrate KNN regression in the previous chapter; we show it in Table 3.5. There are 9 two-dimensional patterns and the respective target values. We call the features f_1 and f_2.

TABLE 3.5 Data for decision tree-based regression

Number	f_1	f_2	Target (y)
1	0.2	0.4	8
2	0.4	0.2	8
3	0.6	0.4	12
4	0.8	0.6	16
5	1.0	0.7	19
6	0.8	0.4	14
7	0.6	0.2	10
8	0.5	0.5	12
9	0.2	0.6	10

In the case of DTC-based regression, a popular measure for splitting is the squared error (SE). Let a set S of n elements be split into sets S_1 and S_2 having n_1 and n_2 elements, respectively. Further, let the sample means of the target (y) values of the sets S_1 and S_2 be μ_1 and μ_2, respectively. Then the weighted squared error for the split is

$$\frac{n_1}{n} \sum_{x_i \in S_1, i=1}^{n_1} (y_i - \mu_1)^2 + \frac{n_2}{n} \sum_{x_i \in S_2, i=1}^{n_2} (y_i - \mu_2)^2$$

Note that

$$\mu_1 = \frac{1}{n_1} \sum_{x_i \in S_1, i=1}^{n_1} y_i \text{ and } \mu_2 = \frac{1}{n_2} \sum_{x_i \in S_2, i=1}^{n_2} y_i$$

For f_1, we need to consider splits based on $f_1 < 0.4$, $f_1 < 0.5$, $f_1 < 0.6$, $f_1 < 0.8$, $f_1 < 1.0$, $f_1 \leq 1.0$. The respective SEs are given in Table 3.6.

TABLE 3.6 Squared error (SE) values for possible splits

Test	SE of S_1	SE of S_2	Weighted SE
$f_1 < 0.4$	2	82	64.22
$f_1 < 0.5$	2.67	52.93	36.11
$f_1 < 0.6$	11	48.8	32.04
$f_1 < 0.8$	16	12.66	**14.9**
$f_1 < 1.0$	55.5	0	49.33
$f_1 \leq 1.0$	108	0	108

Note that the minimum value of weighted SE, that is, 14.9, is obtained for the split based on the test $f_1 < 0.8$. Subsets S_1 and S_2 for this optimal split are

$$S_1 = \{1, 2, 3, 7, 8, 9\} \text{ and } S_2 = \{4, 5, 6\}$$

Further,

$$\mu_1 = 10 \text{ and } \mu_2 = 16.33$$

It is possible to compute and show that the best split using f_2 is obtained for the test $f_2 < 0.4$; the corresponding weighted SE is 33.55, which is larger than the optimal value of 14.9 obtained using f_1. So, we use the test $f_1 < 0.8$ at the root. Proceeding further in splitting the children nodes, we obtain the DTC shown in Fig. 3.12. Note that for each leaf node, the subset of patterns along with the corresponding average target (y) value is indicated. For example, in the leftmost leaf node, the entry is {1,2,9}:8.66, where the subset indicates the collection of patterns 1, 2 and 9; their target values are 8, 8 and 10, respectively, and their average value is $\frac{8+8+10}{3} = 8.66$.

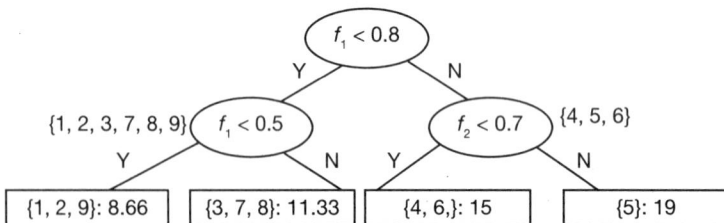

FIG. 3.12 Decision tree for the regression data set

Now if we are given a test pattern of the form $(0.3, 0.4)$, then $f_1 = 0.3$ and $f_2 = 0.4$. So, we test at the root and go to its left child because $f_1 < 0.8$. Now we take the left child of the left child of the root node; we do this because $f_1 < 0.5$. So, we end up in the leftmost leaf node; this node is represented by its average value of 8.66.

We know based on the discussion in Example 15 in Chapter 2 that the true target value of $(0.3, 0.4)$ is 9. So, the squared error is $(9 - 8.66)^2 = \frac{1}{9} = 0.1111$.

Note that for the KNN regression in Example 15 in Chapter 2, we predicted a value of 9.33 for the pattern $(0.3, 0.4)$ which again means a squared error of 0.1111, of course, with a different prediction.

We can abstract the process of building a DT for regression as follows:

1. Associate the entire training data set with the root of the decision tree. Select the best feature and best split point based on the weighted SE value of the target (y) values.

2. Split the set of examples using the selected test (feature and split point) and associate the resulting subsets with the children nodes of the current node.
3. If a child node does not satisfy the termination condition, split it based on best test (feature and split selected) and repeat steps 2 and 3 till there is no need to split.

The termination condition checks for the purity of a node and decides to split if the impurity is above a threshold. Compute the mean value of the target for each leaf node and use it in prediction by associating this value with the leaf node. Given a test pattern, traverse the DT starting from the root and going down a path till a leaf node is encountered. The predicted value of the test pattern is the average target value of the leaf node.

3.7 BIAS–VARIANCE TRADE-OFF

Bias and variance are two important concepts in ML. We illustrate these notions using a simple example.

EXAMPLE 7: Consider the data shown in Table 3.7. There are four examples with their respective target values; consider only the first three columns in the table. Let us select the first 3 rows to train the model and the 4th row for testing. Let us start with a simple linear model; it is a straight-line fit of the form $y = mx + c$, where m and c are the unknowns. The least-square fit for the first three points gives us the following two equations based on the two unknowns:

$$\sum_{i=1}^{3} y_i = 3c + m \sum_{i=1}^{3} x_i \Rightarrow 11 = 3c + 3m$$

and

$$\sum_{i=1}^{3} x_i y_i = c \sum_{i=1}^{3} x_i + m \sum_{i=1}^{3} x_i^2 \Rightarrow 17 = 3c + 5m$$

Solving these equations gives us the values of $m = 3$ and $c = \frac{2}{3}$. Now if we use these values, the estimated value of *target* for the test pattern with $x = \frac{1}{2}$ is $3 \times \frac{1}{2} + \frac{2}{3} = 2.\overset{\bullet}{1}6$. Note that the fourth column in the table corresponds to the straight-line fit given by y_{sl}.

TABLE 3.7 Polynomial fits to illustrate bias and variance

Example	x	Target (y)	y_{sl}	y_{p1}	y_{p2}
1	0	1	$\frac{2}{3}$	1	1
2	1	3	$\frac{11}{3}$	3	3
3	2	7	$\frac{20}{3}$	7	7
4	$\frac{1}{2}$	$\frac{7}{4}$	2.16	1.84	1.87

For this data of three points, we can also fit (overfit) a degree-3 polynomial. In general, we can have infinite polynomials of degree 3 that can fit the three training patterns. We consider two such polynomials that capture all the three training examples, that is, the first 3 rows, perfectly. The polynomials are

$$y_{p1} = 1 + \frac{3}{2}x + \frac{1}{4}x^2 + \frac{1}{4}x^3$$

and

$$y_{p2} = 1 + \frac{5}{3}x + \frac{1}{3}x^3$$

The values predicted using these two polynomials are shown in columns 5 and 6 of the table. Note that y_{p1} and y_{P2} give us the same values as the target y for the training examples. Observe that the 5th and 6th columns are identical and agree with column 3 on the three training examples. However, on the test point (4th row in the table), they give different values: 1.84 and 1.87, respectively.

Now the basic question is which of these models is correct? Note that using three training points, we can fit a degree-2 polynomial uniquely. Starting with a quadratic of the form

$$y = ax^2 + bx + c$$

and plugging in the values of x and y from the first three rows of the table, we get $a = b = c = 1$.

So, the polynomial underlying the data is $1 + x + x^2$; use the 4th row to check that this polynomial exactly satisfies the 4th row also. So, the correct polynomial in this example is of degree-2. When we use the straight-line fit y_{sl}, we are underfitting the data as we get an error even on the training data, as can be seen from columns 3 and 4 of the table. This is a simpler model. Using the least-square fit, we obtain the same linear equation that was seen earlier; so the predicted value of the target is fixed when we use a straight-line fit and the least-square method to estimate the parameters.

On the other hand, we can have infinite higher degree polynomials fitting the first three rows of the table. This is because a degree-3 polynomial fitting the three training examples is not unique; we can draw infinite degree-3 polynomials passing through any three points. Of course, there will be a unique degree-3 polynomial if we use four points.

We have considered two degree-3 polynomials which overfit the training data and can give different estimates for the target, as seen in columns 3, 5 and 6 of the fourth row of the table. Thus, there is non-zero variance among the target values estimated by the two degree-3 polynomial models, even though they successfully fit the training examples correctly.

So, a linear model underfits the training data; it may make mistakes even on the training data and there is no variance in the estimate of the target. This is because we have a single linear model fitting the data points using the least-squares and we predict a unique value, on the test pattern, using such a linear model. However, degree-3 polynomials overfit the training data and each polynomial gives a different target value estimate, leading to variance in the predicted values. So, simple models underfit; they have low variance and high bias as they fail to capture even the training data perfectly.

On the contrary, complex models overfit; they commit no errors on the training data and make different predictions on the test data, leading to low bias and high variance. This phenomenon goes under the name of **bias–variance trade-off**. Making the model simple affects bias and making the model complex affects the variance of the predictions. One needs to learn a model that strikes a balance.

We illustrate this notion further using the data used to construct the decision tree obtained in Fig. 3.12. Consider the decision trees in Fig. 3.13.

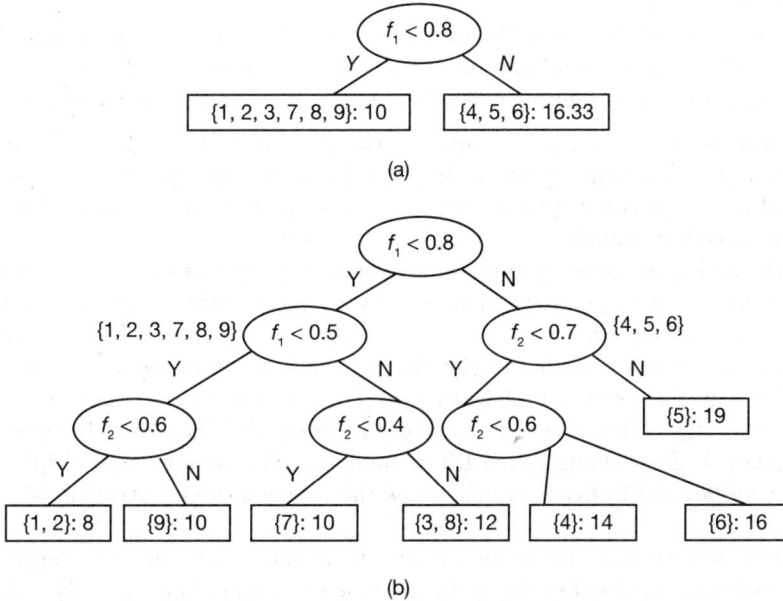

(a)

(b)

FIG. 3.13 Decision trees for the regression data set

The decision tree in Fig. 3.13 (a) is obtained by pruning the decision tree in Fig. 3.12 so as to have the terminal nodes at level 2. On the other hand, the DT in Fig. 3.13 (b) is permitted to have its leaf nodes till level 4. So, the DT in Fig. 3.13 (a) is simpler than the one in Fig. 3.12 and it underfits the training data. Whereas the DT in Fig. 3.13 (b) is more complex than the one in Fig. 3.12 and it overfits the training data.

For the test pattern $(0.3, 0.4)$, we predicted the value to be 9.33 against the correct value of 9, leading to an SE value of 0.1111 using the decision tree in Fig. 3.12. However, the decision tree in Fig. 3.13 (a) predicts the target to be 10, whereas the DT in Fig. 3.13 (b) predicts the value to be 8. So, both of them make a larger error as the DT in Fig. 3.13 (a) is variance-friendly and the one in Fig. 3.13 (b) is bias-friendly; thus both of them make an error on the test data.

Bias–variance is the most important issue in ML model building. There is no clear cut solution to this problem. However, we can tune the parameters of the model, the number of levels and number of nodes in this case, so that the performance of the validation data set is acceptable. Note that pruning DTs is a mechanism to deal with the issue of bias–variance trade-off. It reduces variance.

3.8   RANDOM FORESTS FOR CLASSIFICATION AND REGRESSION

As mentioned earlier in the chapter, DT building can be expensive and the resulting tree may not be optimal because of the greedy nature of the building algorithm; these problems prevail significantly in high-dimensional data sets. So, DTCs are not typically used in applications dealing with high-dimensional data. One popular model used to deal with these problems is the random forest (RF). Its features are as follows:

- It employs a collection of randomly generated DTs in prediction. It is called a random forest as it is a collection of trees.
- Each DT makes a prediction and the overall predicted value is based on majority voting in classification and a suitable combination of the values predicted by the DTs.
- The underlying phenomena may be called bagging or bootstrap aggregation. Here,

   - **Bootstrapping** means sampling multiple sets of training data. In the case of RF, it works as follows. Given n training patterns in some L-dimensional space, m samples, each of size n data points, are selected with replacement; it is possible for the same data point to occur more than once in a sample.
   - In the RF model, we select a random subset of $l(< L)$ features in each sample; so, it is possible that a feature appears in none, one or more samples. So, each sample may be viewed as an $n \times l$ data matrix. So, for m samples, we have m such $n \times l$ data matrices.
   - We construct a DT for each sample. So, each DT is built based on only l features; it is possible that the given set of features is large in size (L is large), but we use only a small number (l) of features for constructing each DT. So, high-dimensional data sets can be handled. Even though each DT is built greedily, because of multiple DTs, different features may be used in decision making at the root nodes and other decision nodes of the m trees.
   - Given a test pattern x, we compute the outputs of the m DTs on x and **aggregate** these m outputs to obtain the final decision. In the case of classification, we take a majority of the m class labels and declare it as the class label of the RF model. In the case of regression, the output of the RF model is the average of all the m DT predictions.

3.8.1   Comparison of DT and RF Models on Olivetti Face Data

We compare the performance of the DT and RF models on the Olivetti Face data set. We show the results of using the DTC, by varying α, on the Olivetti Face data set in Fig. 3.14. Refer to the pruning scheme discussed in Section 3.4 for the use of α. Note that as the value of α increases,

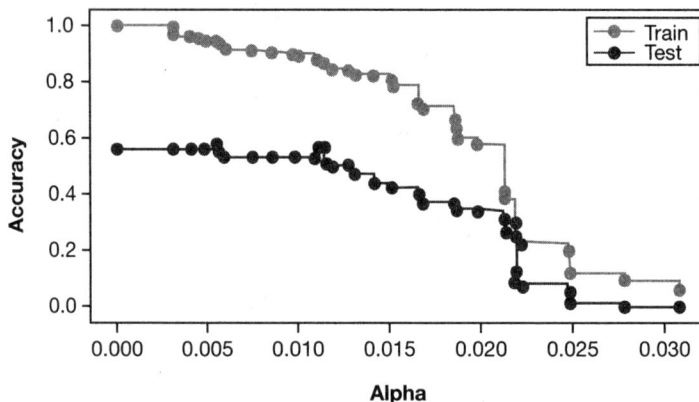

FIG. 3.14 Alpha versus accuracy on the Olivetti Face data set (for colour figure, please see Colour Plate 1)

more importance is given to DTs with a smaller number of terminal nodes, small-sized DTs. So, accuracy decreases as α increases.

Optimal test accuracy is achieved at a value of $\alpha = 0.011$. At this value, the depth of the DT is 26 and the number of nodes is around 80. At this value of α, the accuracy of the DTC is 0.595 using the Gini index and 0.547 using entropy. The number of training images was 360 and the number of test images was 40.

The RF model was used on the data with 10-fold cross-validation using 360 images for training and the remaining 40 for testing. The accuracy is shown in Fig. 3.15. Note that the maximum accuracy of 0.967 is obtained when the number of DTs in the RF is 500.

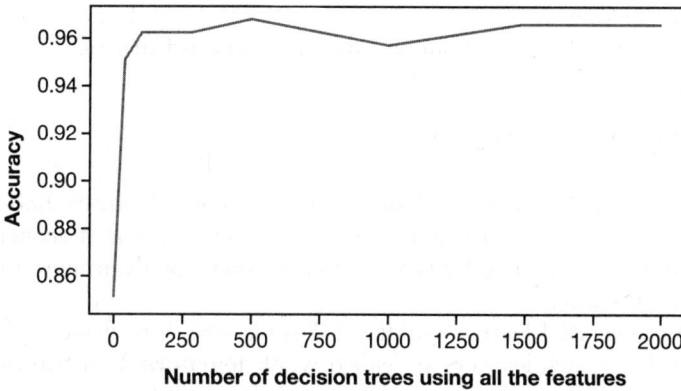

FIG. 3.15 Number of decision trees versus accuracy using a random forest classifier and all the 4096 features

Figure 3.16 shows the accuracies of the RF classifier as the number of features is varied between 10% and 100% in steps of 10% in sampling the data. We present the results based on RFs using 10 DTs, 50 DTs and 100 DTs.

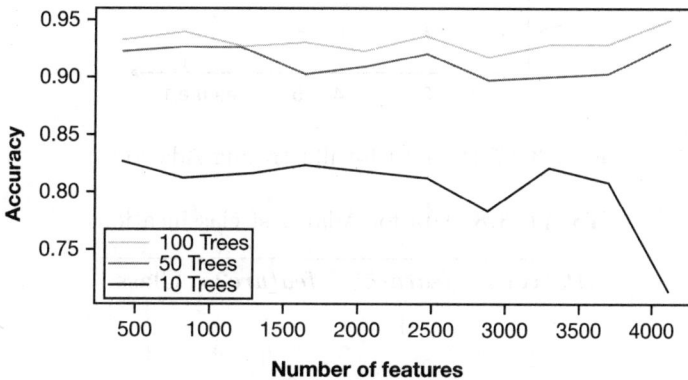

FIG. 3.16 Number of features versus accuracy using a random forest classifier on the Olivetti Face data set (for colour figure, please see Colour Plate 1)

Note that the performance of the RF with 100 DTs is significantly better than the one based on 10 DTs. There is a minor difference between the accuracies of RF based on 50 and 100 DTs.

Some of the important characteristics of the RF model are:

- The decision trees that make up the forest can be built in parallel. The outcome of each DT on a test pattern can be computed in parallel and can be distributed. The aggregation needs some centralised processing.
- Model performance improves if each DT works well. However, the correlation among the DTs should be small.
- It can deal with very large-dimensional data sets.
- The RF model reduces variance by

  1. Using different samples of the data in building the DTs; this way, the correlation between different DTs is reduced.
  2. Each DT is simplified by using only a random subset of features.

3.9   AdaBoost Classifier

Boosting is a process that helps improve prediction performance. Adaptive boosting, AdaBoost for short, is a combinational classifier learning mechanism that works in a sequential manner; data points that are erroneously classified by the previous learner are given more importance (weight) for training the current learner.

Contrary to RF, sequential learner building takes place and we need the learners to perform better than random guess; such learners are called **weak learners**. In a two-class case, we expect the error made by any weak learner to be less than 0.5. We will discuss a two-class problem first.

EXAMPLE 8:   We will illustrate the working of AdaBoost with the help of the data in Fig. 3.17. It is also shown in Table 3.8.

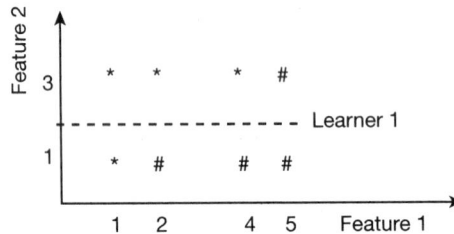

FIG. 3.17 Data set for illustrating AdaBoost

TABLE 3.8 Data for AdaBoost classification

Pattern	*feature*1	*feature*2	Class
1	1	1	*
2	1	3	*
3	2	1	#
4	2	3	*
5	4	1	#
6	4	3	*
7	5	1	#
8	5	3	#

There are two classes labelled '*' (Class 1) and '#' (Class 2) and there are four training patterns from each class in this example data set. Initially, every pattern will have the same weight ($\frac{1}{8}$).

We start with a simple classifier, called the decision stump, a one-node decision tree, as a weak learner. We can use any decision stump that gives less than 50% error; that means the decision stump is a weak learner.

1. We consider a split based on $feature2 > 2$, as shown in Fig. 3.17. From (Learner 1), we get subset1 = {1,3,5,7} and subset2 = {2,4,6,8}. It is convenient for subset1 to be labelled Class 2 (#) and subset2 to be labelled Class 1 (*). This means pattern 1 in the first subset and pattern 8 in the second subset are misclassified; so, the error is $\frac{2}{8} = 0.25$. So, it is a weak learner as the accuracy (0.75) is more than 0.5.

   It is possible to consider labelling the other way, that is, label subset1 as Class 1 (*) and subset2 as Class 2 (#). This gives rise to an error of $\frac{6}{8} = 0.75$. There is no problem. We can always swap labels to get an error ≤ 0.5.

   So, when the error is more than 0.5 in a two-class problem, swap labels to get an error of less than 0.5. However, if the error is exactly equal to 0.5, then swapping the labels will not alter the error. We drop such learners, that make an error of exactly 0.5, and do not consider them in the sequence further.

   Note that Learner 1 gives an error, ϵ_1, of 0.25. We associate a weight with the weak learner, Learner 1 in this case. The weight α_i for the ith learner is given by

   $$\alpha_i = \frac{1}{2} ln \frac{(1 - \epsilon_i)}{\epsilon_i}$$

   For weak learner 1 ($i = 1$), the value of α_1 in this example is

   $$\alpha_1 = \frac{1}{2} ln 3 = 0.55$$

   Note that if the error made by the weak learner is smaller, then the weight associated will be larger.

2. Recall that initially all the patterns are given equal importance of $\frac{1}{8}$ each. Now we update these weights so that erroneously classified patterns receive more importance for the next round. This is achieved by using the following weight update:

   $$w_{t+1}(i) = \frac{w_t(i)}{z} \times e^{\alpha_t}$$

   if the ith data point is misclassified by Learner t and

   $$w_{t+1}(i) = \frac{w_t(i)}{z} \times e^{-\alpha_t}$$

   if the ith data point is correctly classified by Learner t; z is a normalizing quantity to ensure that the sum of the weights for all the patterns is 1.

   So, for patterns 1 and 8, the updated weights are

   $$\frac{1}{8} \times e^{0.55} = 0.2166$$

   and for the rest of the patterns, the weights are

   $$\frac{1}{8} \times e^{-0.55} = 0.072$$

The normalizing factor z is the sum of all the 8 weights, 6 of them having a weight each of 0.072 and the remaining 2 having a weight each of 0.2166. The total is $z = 6 \times 0.072 + 2 \times 0.2166 = 0.8652$. So, the normalized weights for round 2 are $w_2(1) = w_2(8) = \frac{0.2166}{0.8652} = 0.25$ and for the rest of the patterns, the weights are $\frac{0.072}{0.8652} = 0.083$.

Note that the weights of the correctly classified patterns are reduced, from 0.125 to 0.083, and the weights of the incorrectly classified patterns are enhanced, from 0.125 to 0.25.

Now we consider another, the second, weak learner given by $feature1 > 1.5$, as shown in Fig. 3.18.

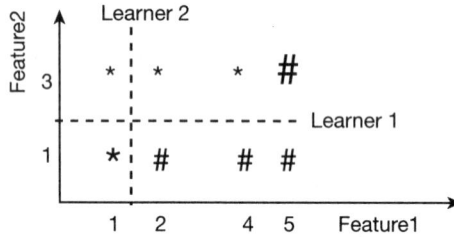

FIG. 3.18 Data set illustrating weak learners

Note that patterns 1 and 8 have a larger weight compared to other patterns as shown in the figure by using bigger sized symbols * and #. Using this test, the two subsets are {1,2} and {3,4,5,6,7,8}. Patterns numbered 4 and 6 are misclassified now. So, the error of this weak learner is obtained by adding the weights for patterns 4 and 6; these weights are 0.083 each. So, error is $\epsilon_2 = 2 \times 0.083 = 0.166$.

So, the weight associated with Learner 2 is $0.5 \, ln\left(\frac{1-0.166}{0.166}\right) = 0.8$. The updated weights are

$$w_3(1) = w_3(8) = 0.15, w_3(4) = w_3(6) = 0.25,$$

and $w_3(2) = w_3(3) = w_3(5) = w_3(7) = 0.05$.

3. Let us consider a split based on $feature1 > 4.5$ for the third weak learner, as shown in Fig. 3.19. Now patterns 3 and 5 are misclassified. So, the error of Learner 3 is $\epsilon_3 = 0.05 + 0.05 = 0.1$. Note that the chosen learners are indeed weak learners as the errors are less than 0.5. The weight for this third learner is $\alpha_3 = 1.1$.

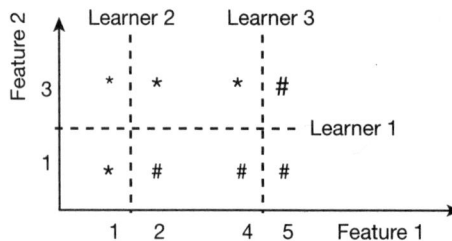

FIG. 3.19 Illustration of weak Learner 3

Now we can test how a combination of these three weak learners can be used to classify patterns.

Consider pattern 8: Learner 1 with weight 0.55 classifies it as a member of Class 1 (*); Learner 2 with weight 0.8 classifies it as a member of Class 2 (#); and Learner 3 with weight 1.1 classifies it as a member of Class 2 (#). So, it is correctly assigned to Class 2 as the sum of the weights for Class 2 is 1.9 > 0.55 (weight for Class 1).

Similarly, if we consider pattern 4, then both Learner 1 and Learner 3 assign it to Class 1 (*) with a total weight of 1.65, whereas Learner 2 assigns it to Class 2 with a weight of 0.8. So, it is correctly classified.

Check that all the eight training patterns are correctly classified. The decision boundary between the two classes is shown using a thick grey piece-wise linear curve that is a combination of three straight line segments in Fig. 3.20. The regions corresponding to Class 1 and Class 2 are clearly separated by the boundary shown by the thick grey line. This is the AdaBoost Classifier for the training data set shown in Table 3.8.

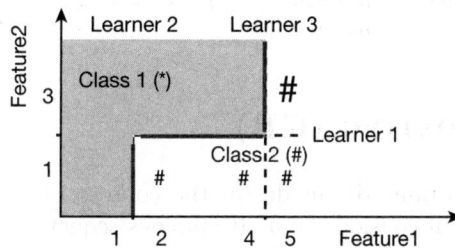

Fig. 3.20 Illustration of the boundary between the two classes using AdaBoost (for colour figure, please see Colour Plate 1)

So, AdaBoost is a model that boosts the performance of the classifier by sequentially adding a weak learner that complements the working of the previous learners. Some important features of AdaBoost are:

- It chooses and combines multiple weak learners in an adaptive manner to achieve a strong learner. The learnt strong learner need not be unique. It depends on the sequence of weak learners used.
- It starts with assigning equal weights to every data point. It selects a weak learner, computes its error (ϵ) and importance (α). It updates the weight associated with each data point using the α value of the current weak learner and uses the updated weights in choosing the next weak learner. The weight update ensures that training patterns misclassified by the current weak learner will have larger weights for the next round to influence the choice and importance of the next weak learner.
- Variance may be viewed as the variation in the prediction when the data set changes. AdaBoost combines multiple models and can hence control variance.
- If m weak learners are used with the error of the ith weak learner on the training data being ϵ_i for $i = 1, 2, \cdots, m$, then it is possible to show that the training error of the model is at most

$$exp\left(-2\sum_{i=1}^{m} \epsilon_i^2\right)$$

3.10 REGRESSION USING DT-BASED MODELS

We have seen in Section 3.6 the role of DTs in regression. It is also possible to use other models like RF and AdaBoost in regression. In the case of RF, training data can be used to learn each DT in the forest. Given a new data point, the function value is predicted by taking the average of the predictions of the DTs in the model.

In the case of regression using AdaBoost, it is the same as the classifier, except for the computation of the error of the weak learner. The error of the tth weak learner may be computed as

$$\epsilon_t = \frac{1}{2} \sum_{i=1}^{n} w_t(i)|y_i - \hat{y}_i|,$$

when the training data has pairs of the form (x_i, y_i), $for\ i = 1, 2, \cdots, n$. So, y_i is the target value and $\hat{y}_i$ is the value predicted by t^{th} weak learner.

3.11 GRADIENT BOOSTING (GB)

Now we will discuss an important ML model in the context of regression. It can be used in classification also, which we will consider later. It employs sequential addition of predictors. The current predictor is built to reduce the error obtained at the previous stage. We will first illustrate its construction and use with the help of an example.

EXAMPLE 9: Consider the two-dimensional regression data shown in Table 3.9. Note that each

TABLE 3.9 Data set to illustrate gradient boosting

$x(1)$	$x(2)$	y
0	1	1
1	1	1.9
2	1	3
3	4	7.1
4	3	7

data point is a two-dimensional vector, x, with its two components being $x(1)$ and $x(2)$ and the target output y is a real number. This is a regression problem where we need to estimate a function f such that $f(x) \approx y$. We go through the following steps to estimate such a function.

1. We need to start with some simple function to estimate. We take function g_1 as $g_1(x) = \mu_y$, where μ_y is the mean (average) of all the y values in the training examples given. So, μ_y for the example data in Table 3.9 is the average of 1, 1.9, 3, 7.1 and 7, which is $\frac{1+1.9+3+7.1+7}{5} = 4$. So, using g_1, every point x is predicted to have the same value of y, that is, μ_y. We calculate the residuals and use them in the prediction of the next predictor g_2.

2. We show in Table 3.10 the residuals based on predictor g_1. We can employ other models for regression but here we use a DT for this regression problem.

TABLE 3.10 Residuals after using $g_1 = 4$

$x(1)$	$x(2)$	y	$y - g_1(x)$ [$y - 4$]
0	1	1	-3
1	1	1.9	-2.1
2	1	3	-1
3	4	7.1	3.1
4	3	7	3

Here, one needs to choose between the two features, $x(1)$ and $x(2)$. The best feature and the split point among $x(1)$ and $x(2)$ for splitting to have the least MSE split is $x(2) < 2$. This leads to two subsets that we can split further to obtain the DT for this regression problem, where for the (x, y) pair, we predict the residual value given by $y - g_1(x)$ as shown in Fig. 3.21.

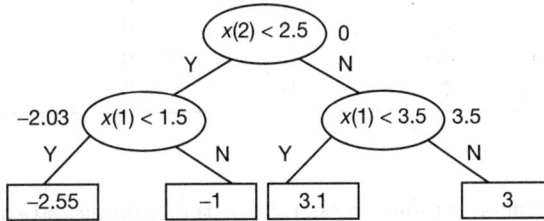

FIG. 3.21 Decision tree to model the residual $y - g_1(x)$

Note that in this DT, we show the average residual value associated with each non-terminal node. The root has an average residual of 0 (average of all the 5 residual values). The best feature and split at the root based on squared error is $x(2) < 2.5$. In a similar manner, further optimal splits are found to be based on $x(1)$. Using this DT, the predictions and the residuals still to be accounted for are shown in Table 3.11.

TABLE 3.11 Residuals after using g_2 in the form of $y - g_1(x) - g_2(x)$

$x(1)$	$x(2)$	y	$y - g_1(x)$	$g_2(x)$	$y - g_1(x) - g_2(x)$
0	1	1	-3	-2.55	-0.45
1	1	1.9	-2.1	-2.55	0.45
2	1	3	-1	-1	0
3	4	7.1	3.1	3.1	0
4	3	7	3	3	0

3. We build another DT for regression to account for the residuals based on $y - g_1(x) - g_2(x)$ as shown in Table 3.11. The resulting DT is shown in Fig. 3.22. We show the residuals resulting after using $g_3(x)$ in Table 3.12.

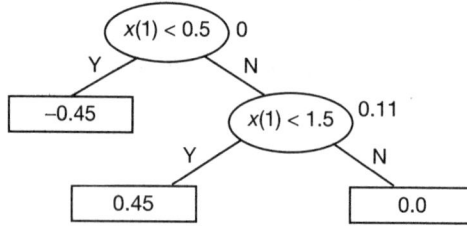

FIG. 3.22 Decision tree to model the residual $y - g_1(x) - g_2(x)$

TABLE 3.12 Residuals after using g_3 in the form of $y - g_1(x) - g_2(x) - g_3(x)$

$x(1)$	$x(2)$	y	$y - g_1(x)$	$g_2(x)$	$y - g_1(x) - g_2(x)$	$g_3(x)$	$y - g_1(x) - g_2(x) - g_3(x)$
0	1	1	-3	-2.55	-0.45	-0.45	0
1	1	1.9	-2.1	-2.55	0.45	0.45	0
2	1	3	-1	-1	0	0	0
3	4	7.1	3.1	3.1	0	0	0
4	3	7	3	3	0	0	0

4. So, using the three learners obtained sequentially, the training data is explained; there is no further residual error after using g_3.

5. **Prediction Using the Model**: We can use this model to predict for any test pattern x, by obtaining $g_1(x)$, $g_2(x)$ and $g_3(x)$ and then declare the predicted value of y, $\hat{y}$, by using the expression

$$\hat{y} = g_1(x) + g_2(x) + g_3(x)$$

Let us say we are given a test pattern $x = (1, 1.9)$; then we have $g_1(x) = 4$, $g_2(x) = -2.55$ and $g_3(x) = 0.45$. So, the predicted value is $4 - 2.55 + 0.45 = 1.9$.

Some important properties of gradient boosting are:

- It is popularly called GBM, gradient boosting machine.
- For regression squared error (SE), loss function is used. Consider the SE function given by

$$\frac{1}{2}(y - g(x))^2$$

We need to obtain optimal $g(x)$ to ensure that the model fits the target value y correctly. By taking the gradient of the SE function with respect to $g(x)$, we get

$$\frac{\partial SE}{\partial g(x)} = -(y - g(x))$$

So, negative of the gradient is given by

$$-\frac{\partial SE}{\partial g(x)} = y - g(x)$$

This means that the residual $y - g(x)$ is the same as the negative of the gradient. Let us recall how the residual predictions help us in predicting the value of y. If we start with some $g_1(x)$, the residual is $y - g_1(x)$ or negative of the gradient. So, if the current estimate for y is $g_1(x)$, then the next estimate of y is obtained by taking

$$y = g_1(x) - \frac{\partial SE}{\partial g(x)} = g_1(x) - (y - g_1(x))$$

So, modelling the residual using the next learner in the sequence is done by using gradient descent on the squared error loss function. So, the whole process of GB may be viewed as a gradient descent process; that is why it is called gradient boosting.

- We have seen that GBM employs a sequence of learners to account for the residuals surfacing at various levels. These learners are additive and can predict the target value for a test pattern.
- For classification, we can use, instead of SE, cross entropy loss, given by

$$-\frac{1}{m} \sum_{i=1}^{n} (y_i log(\hat{y}_i) + (1 - y_i)log(1 - \hat{y}_i))$$

Note that its value is 0 when $y_i = \hat{y}_i$ for $i = 1, 2, \cdots, n$. The GBM follows the same procedure, as in regression, in building the learners sequentially.

3.12 PRACTICAL APPLICATION

We use the AdaBoost classifier on the Olivetti Face data set with different number of weak learners and show the results in Fig. 3.23. Note that the performance of AdaBoost is worse than that of

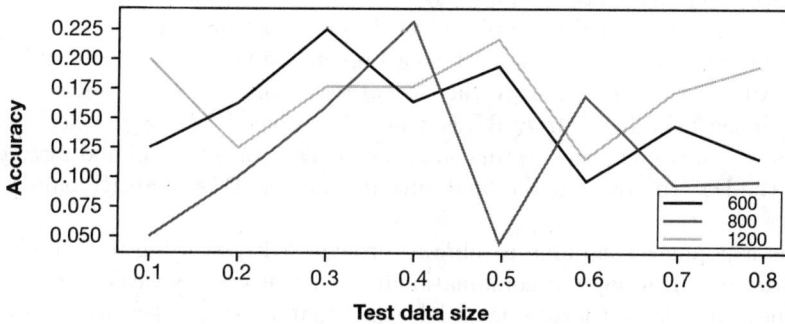

FIG. 3.23 Results of the AdaBoost classifier on the Olivetti Face data set with different number of weak learners (600, 800, 1200) and different test data sizes (for colour figure, please see Colour Plate 1)

even DTC for which the results were shown in Fig. 3.14. This could be because of the large number of features (4096) and the small number of training patterns (<400) in the Olivetti Face data set.

We used the GBM based on XGBoost (extreme gradient boosting algorithm) that is a faster implementation of GBM and is available on the sklearn platform on the Olivetti Face data set with different number of trees (learners) and show the results in Fig. 3.24.

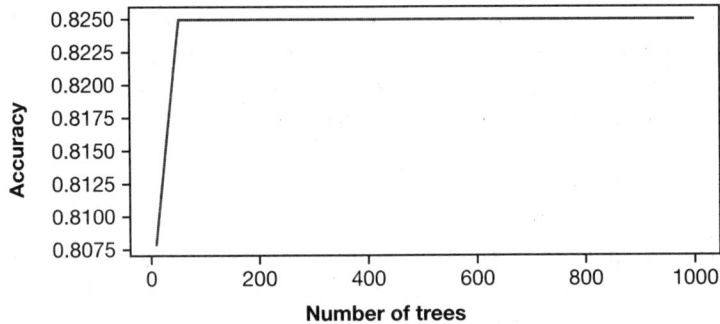

FIG. **3.24** Results of GBM classifier on the Olivetti Face data set with different number of trees

Note that both the boosting algorithms have performed worse than RF on this high-dimensional data.

SUMMARY

In this chapter we examined ML models based on decision trees. Some important issues discussed are:

- DTs are simple, easy to understand structures. They combine multiple rules where the path from the root to each leaf is a classification or a regression rule. So, in terms of explainability, DTs are the best.
- Building DTs beyond the use of axis-parallel splits is computationally expensive as the data set size grows. Even axis-parallel-split–based DTs are not viable as the dimensionality grows because of the greedy algorithm employed in the construction of DTs.
- DTs are built with the help of impurity measures including entropy, Gini index, misclassification impurity and squared error in the case of regression.
- When the dimensionality is large, RF is a popular choice which is a collection of DTs and it combines bootstrapping in selecting samples for DT construction and aggregation of the outputs of the DTs to arrive at the final output. Further, DTs in an RF can be constructed in parallel.
- Boosting is a popular technique to obtain stronger learners using a combination of weak learners. AdaBoost is a popular combinational learner; it creates weak learners sequentially.
- Gradient boosting is another sequential learner that creates a learner to account for the residuals in the estimates made by the previous learner; it is an additive model.
- All these models can be used in both regression and classification.

EXERCISES

1. Show that for a C-class problem, the maximum entropy is $log(C)$.
2. Consider the data set used in Example 19 of Chapter 2. Obtain the weighted entropy if the split on $feature1$ is based on the test $feature1 > 1$ or not.
3. Synthesize a two-dimensional data set having 4 patterns such that the number of split positions to be considered using either of the two features is 4.

4. An impurity measure (variance impurity) for a two-class case is $p_1 p_2$. How is it related to the Gini index for $C = 2$?

5. Consider the data set used in Example 1. We have considered only one split resulting in three children for *Outlook*. However, it is possible to use a binary split (two children) based on *Outlook = Overcast* or not or equivalently *Outlook = Sunny or Rainy*. Then the resulting subsets will be {3,4} and {1,2,5,6}. Compute the Gini index for this binary split. What is the misclassification impurity for this split?

6. Consider the *Temperature* feature for splitting based on the data shown in Table 3.3. How many splits need to be considered for obtaining the best split using this feature. Compute IG for all these splits and show that the split position selected in Example 1, that is *Temperature > 70*, gives the minimum weighted impurity and maximum IG over all these splits.

7. Consider the data set used in Example 1. Consider the split using *Humidity > 65*. Obtain the weighted impurity for this split by using the Gini index and misclassification impurity as the impurity measures.

8. Show that *Humidity < 91* or an equivalent split gives the maximum IG value compared to all other possible splits using *Humidity* for the data in Table 3.3. Use misclassification impurity for the calculation of the weighted impurity.

9. Suppose a weak learner in AdaBoost gives an error of 0.5. What is the α value that characterizes the importance of the weak learner?

10. In Example 6, the final classifier based on three weak learners was shown to correctly classify patterns 4 and 8. Check whether the remaining training patterns, numbered 1, 2, 3, 5, 6 and 7, are correctly classified by the AdaBoost classifier.

11. Use the AdaBoost classifier in Example 6 to classify the following patterns: (3,1) and (3,3).

12. Justify the choice of regions for Class 1 and Class 2 that are shown in Fig. 3.20.

13. Consider the error computation for AdaBoost regression by using

$$\epsilon_t = \frac{1}{2} \sum_{i=1}^{n} w_t(i)|y_i - \hat{y}_i|$$

Show that the same formula was used in classification using AdaBoost, where y_i is the class label assuming a value of either -1 or $+1$ for the two-class problem and $\hat{y}_i$ is the predicted value and is either $+1$ or -1.

14. Check whether the y values for all the x values given in Table 3.9 are properly predicted or not by computing $g_1(x)$, $g_2(x)$ and $g_3(x)$ for each of the 5 patterns, and computing the estimate for y for each x, by summing the corresponding values of $g_1(x)$, $g_2(x)$ and $g_3(x)$.

15. Use the model discussed in Example 7 to predict the value of y when $x = (1.9, 1)$. Estimate $g_1(x)$, $g_2(x)$ and $g_3(x)$ and then predict the corresponding y value.

PRACTICAL EXERCISES

1. Download the Olivetti Face data set. There are 40 classes (corresponding to 40 people), each class having 10 faces of the individual; so there are 400 images in total. Here, each face is viewed as an image of size 64×64 ($= 4096$) pixels; each pixel having values 0 to 255 which are ultimately converted into floating numbers in the range [0,1]. Visit `https://scikit-learn.org/0.19/datasets/olivetti\_faces.html` for more details.

Your Tasks: There are three tasks. For all the tasks, split the data set into train and test parts. Carry out this splitting randomly 10 times and report the average accuracy. You may vary the test and train data set sizes. The tasks are:

a.   **Task 1:** Build a **decision tree** using the training data. Tune the parameters corresponding to pruning the decision tree. Use the best decision tree to **classify the test data set** and obtain the accuracy. Use both Gini and entropy impurities.

b.   **Task 2:** Build a **random forest classifier** using the training data set. Use RF with 50 decision trees. Obtain the classification accuracy on the test data with the number of features as 20%, 40% and 60% of the given set of features.

c.   **Task 3**: Use the **XGBoost classifier** to classify by viewing the entire data set as the training data set. Find out the accuracy on the data set using 50 and 100 trees.

Bibliography

*   Rifkin, RM. 2008. Multiclass Classification, *Lecture Notes*, Spring08, MIT, USA.

*   Witten, IH, Frank, E and Hall, MA. 2011. *Data Mining, Third Edition*, Morgan Kaufmann Publishers, Waltham.

*   Breiman, L. 2001. Random Forests, *Machine Learning*, 45: 5–32.

*   Breiman, L. 2017. *Classification and Regression Trees*, Routledge: New York.

*   Freund, F and Schapire, RE. 1999. A Short Introduction to Boosting, *Journal of Japanese Society for Artificial Intelligence*, 14(5): 771–780.

*   MNIST data set: `https://www.tensorflow.org/datasets/catalog/mnist`

*   ORL Face data set:
    `https://www.kaggle.com/datasets/tavarez/the-orl-database-for-training-and-testing`

*   scikit-learn, *Machine Learning in Python*, `https://scikit-learn.org/stable/`

CHAPTER 4

The Bayes Classifier

Learning Objectives

At the end of this chapter, you will be able to:

- Explain the Bayes classifier
- Describe probability, conditional probability and Bayes' rule
- Define random variables, probability mass function, probability density function, cumulative distribution function, expectation and variance
- Explain optimality of the Bayes classifier
- Describe parametric and non-parametric schemes for density estimation
- Define class conditional independance and the naïve Bayes classifier

4.1 INTRODUCTION TO THE BAYES CLASSIFIER

The Bayes classifier is an optimal classifier. It operates based on the probability structure associated with the domain of application. It employs Bayes' rule to convert the *a priori* probability of a class into posterior probability with the help of probability distributions associated with the class. These posterior probabilities are used to classify the test patterns; the pattern is assigned to the class that has the largest posterior probability. In case the probability structure is not readily available, the training data is used to learn the probability structure.

Some of the important properties of Bayesian classifiers are as follows:

- It minimizes the probability of error associated with classification.
- It can deal with data that employs both categorical and numerical features.
- It is more of a benchmark classifier having sound theoretical properties. However, in most practical applications, the underlying probability structure is not readily available.
- Some additional constraints on the probability structure are applied to make estimation of the probability structure simpler. For example, the **naïve Bayes classifier (NBC)** is one that assumes that features are independent of each other, given that the data points belong to a class.
- Primarily, there are two schemes for estimating the probabilities associated. One of them depends solely on the data; it is called the maximum likelihood estimate or the frequency-based estimate. The other scheme is more general and it combines application domain knowledge with the data in estimation; it is called the **Bayesian estimation or Bayesian learning** of the probability structure.

4.2   PROBABILITY, CONDITIONAL PROBABILITY AND BAYES' RULE

Let us start with a simple scenario where domain knowledge is used in the form of prior probabilities to classify patterns.

EXAMPLE 1:   Let us assume that 10,000 people in a community had undergone the COVID-19 test and 50 of them tested positive, while the remaining 9950 tested negative. In this two-class (binary class) problem, a simple frequency estimate will give us the following values for the probabilities for the two classes:

$$P(positive) = \frac{50}{10000} = 0.005 \text{ and } P(negative) = \frac{9950}{10000} = 0.995$$

These probabilities are called **prior probabilities** as they are obtained using domain knowledge.

Suppose a new person, from the community, who has not undergone the test, needs to be classified. In the absence of any other information from the person, one would try to use the prior probabilities in decision making; so the new person is assigned a COVID-19-negative label as the probability $P(negative)$ is significantly larger than $P(positive)$ (0.995 >> 0.005). So, invariably every other person in the community who has not undergone the test will be classified as COVID-19-negative using this rule of classification, which is influenced by the larger prior probability of being COVID-19-negative.

Such a classification is erroneous if the new person is actually COVID-19-positive. This can occur with a probability of 0.005 ($P(positive)$); so, the *probability of error* is 0.005. Note that every person who is actually COVID-19-negative is correctly classified in this process.

This is not new; in the KNN classifier, if the value of $k = n$, where n is the size of the training data set, and if we have k_1 (out of k) from the COVID-19-positive class and k_2 ($k - k_1$) from the COVID-19-negative class, then the probability estimates are

$$P(positive) = \frac{k_1}{n} \text{ and } P(negative) = \frac{k_2}{n}$$

It is not difficult to see that KNN gives the same result as classification based on prior probabilities.

One can ask whether we can do better if more information is available. In order to answer this question, we need to refresh some basic probability concepts:

- In a random experiment, we have a **sample space**, $\mathcal{S}$, that is, the set of all outcomes. For example, tossing a coin gives us {H,T} as the sample space, where H stands for *head* and T stands for *tail*.
- An **event** is a subset of the sample space. We associate probabilities with events. If A is an event, then its probability $P(A) \in [0,1]$, that is, probability is non-negative and is upper bounded (less than or equal to) 1.
- If A and B are disjoint events ($A \cap B = \phi$), where $A \cap B$ is the intersection of the sets A and B and ϕ is the null set or empty set, then

$$P(A \cup B) = P(A) + P(B),$$

where $A \cup B$ is the union of the sets A and B. This property holds for a countable union of events if they are pairwise disjoint.

EXAMPLE 2:   If a coin is tossed twice, the sample space is {HH,HT,TH,TT}. If $A = \{HH,HT\}$ and $B = \{TT\}$, then $A \cap B = \phi$. Further, $P(A) = \frac{2}{4}$ as A has 2 out of the 4 outcomes, and $P(B) = \frac{1}{4}$. Note that $A \cup B = \{HH, HT, TT\}$; so,

$$P(A \cup B) = \frac{3}{4} = P(A) + P(B) = \frac{2}{4} + \frac{1}{4}$$

However, if $A = \{HH,HT\}$ and $C = \{HT, TT\}$, then $A \cap C = \{HT\} \neq \phi$. So, $P(A \cup C) = \frac{3}{4}$ whereas $P(A) = \frac{2}{4}$ and $P(C) = \frac{2}{4}$. So, here $P(A \cup C) \neq P(A) + P(C)$. It is possible to show that

$$P(A \cup B) = P(A) + P(B) - P(A \cap B)$$

If an event is described as *at least one tail*, then the event is {HT,TH,TT} and the probability of the event is $\frac{3}{4}$, as out of 4 elements of the sample space, 3 elements are favourable to the event.

Given an event A, its compliment A^c is given by the set difference $S - A$. Figure 4.1 shows regions corresponding to various related events, given two events A and B. Note that $A \cap B^c$ is the intersection of events A and B^c. The region for $A \cap B$ is the intersection of events A and B. The event $A^c \cap B$ corresponds to the intersection of the events A^c and B. Finally the remaining region indicates the intersection of the events A^c and B^c.

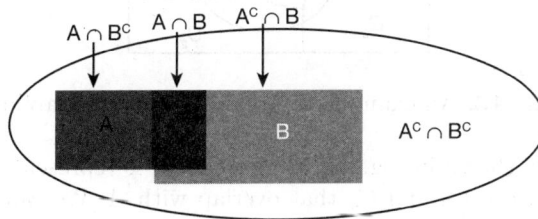

FIG. 4.1 Some events related to two given events A and B (for colour figure, please see Colour Plate 1)

4.2.1   Conditional Probability

We need to update the probability values if new information is given. Consider the following example.

EXAMPLE 3:   Consider tossing a coin twice. We have seen in Example 2 that the probability of at least one tail is $\frac{3}{4}$. The corresponding event is {HT, TH, TT}. If we are given additional information that one of the tosses has resulted in a head, then the sample space is constrained to {HT, HH, TH}. So, under the condition that one toss has resulted in a head, the sample space shrinks. Now for at least one tail, the event in the new sample space is {HT,TH}. So, the probability has reduced from $\frac{3}{4}$ to $\frac{2}{3}$ for the event in the presence of more information. This computation is captured by using the notion of **conditional probability**.

If A and B are two events such that $P(B) \neq 0$, then

$$P(A|B) = \frac{P(A \cap B)}{P(B)}$$

It is the probability of A conditioned on B. It is undefined if $P(B) = 0$. In the current example, if A is the event specified by *at least one tail* and B is the event specified by *one toss that has resulted in a head*, then $B = \{HH, HT, TH\}$; so, $P(B) = \frac{3}{4}$. Note that $A \cap B = \{TH, HT\}$. So,

$$P(A|B) = \frac{\frac{2}{4}}{\frac{3}{4}} = \frac{2}{3}$$

as computed earlier in this example.

We have another important concept called **independent events**. We say that two events A and B are independent if $P(A \cap B) = P(A) \times P(B)$. So, if A and B are independent, then

$$P(A|B) = \frac{P(A \cap B)}{P(B)} = \frac{P(A) \times P(B)}{P(B)} = P(A)$$

4.2.2   Total Probability

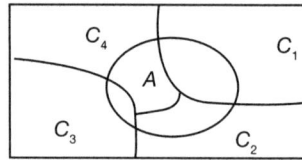

FIG. 4.2 An example to illustrate total probability

Consider the Venn diagram shown in Fig. 4.2. Here, event A is represented by the elliptical region and there are 4 events C_1, C_2, C_3 and C_4 that overlap with A. We can represent A using these overlapping sets by

$$A = (A \cap C_1) \cup (A \cap C_2) \cup (A \cap C_3) \cup (A \cap C_4)$$

The four overlapping sets in the union are disjoint as C_1, C_2, C_3 and C_4 are disjoint. Let

$$B_i = A \cap C_i \text{ for } i = 1, 2, 3, 4$$

So, $P(B_i) = P(A|C_i)P(C_i)$. So,

$$P(A) = P(B_1) + P(B_2) + P(B_3) + P(B_4)$$

Hence,

$$P(A) = P(A|C_1)P(C_1) + P(A|C_2)P(C_2) + P(A|C_3)P(C_3) + P(A|C_4)P(C_4)$$

We know that $P(C_i|A) = \frac{P(A|C_i)P(C_i)}{P(A)}$. So, in general

$$P(C_i|A) = \frac{P(A|C_i)P(C_i)}{P(A|C_1)P(C_1) + P(A|C_2)P(C_2) + P(A|C_3)P(C_3) + P(A|C_4)P(C_4)}$$

4.2.3 Bayes' Rule and Inference

Let us examine Bayes' rule and inference based on conditional probabilities. We consider some of the terms:

- $P(C_i)$: Prior or initial probability of event C_i.
- Model of the world (A) under each C_i: $P(A|C_i)$
- How to infer $P(C_i|A)$, or what is the probability $P(C_i|A)$ or equivalently how the *prior probability* $P(C_i)$ gets updated to the *posterior probability* $P(C_i|A)$. Let us consider an example to appreciate these ideas further.

EXAMPLE 4:   Let C_1 and C_2 be two chests such that C_1 has 20 white balls (WB) and 10 red balls (RB); C_2 has 15 WBs and 15 RBs.

If one of the two chests is picked with equal probability and a ball randomly picked from the chest is WB, what is the probability that it came from C_1? We have the following information:

- Prior: $P(C_1) = P(C_2) = \frac{1}{2}$
- Given: $P(WB|C_1) = \frac{2}{3}$, $P(RB|C_1) = \frac{1}{3}$ and $P(WB|C_2) = P(RB|C_2) = \frac{1}{2}$
- Needed: $P(C_1|WB) = \frac{P(WB|C_1)P(C_1)}{P(WB|C_1)P(C_1)+P(WB|C_2)P(C_2)}$
- $= \frac{(\frac{2}{3})(\frac{1}{2})}{(\frac{2}{3})(\frac{1}{2})+(\frac{1}{2})(\frac{1}{2})} = \frac{4}{7}$

Let us consider one more example.

EXAMPLE 5:   A new COVID-19 test claims to have 90% **true positive rate** (sensitivity) and 98% **true negative rate** (specificity). In a population with a COVID-19 prevalence of $\frac{1}{1000}$ (one out of 1000), what is the chance that a patient who tested positive is truly positive? Let us consider the following:

- Let A be the event that a patient is **truly positive**; So, $P(A) = 0.001$.
- So, A^c $(S - A)$ is the event of being **truly negative**. So, $P(A^c) = 0.999$.
- Let B be the event that the patient **tested positive**.
- We want $P(A|B)$. Let the data of sensitivity and specificity be summarized as shown in Table 4.1.

TABLE 4.1 Relevant probabilities of positive and negative tests

True/Test	Positive	Negative
Positive	0.9	0.1
Negative	0.02	0.98

- The rows in Table 4.1 correspond to being truly positive and truly negative whereas the columns signify tested positive and tested negative.
- Note that the probability of being tested positive when truly positive is 0.9 (true positive rate); so, the probability of being tested negative when truly positive is 0.1.
- Similarly, in the second row, the entry 0.98 is the probability of being tested negative when truly negative (true negative rate) and 0.02 is the probability of being tested positive when the patient is truly negative.

– We need to compute $P(A|B)$; we can use Bayes' rule to obtain $P(A|B)$ given $P(B|A)$, $P(A)$, $P(B|A^c)$ and $P(A^c)$ as follows:

$$P(A|B) = \frac{P(B|A)P(A)}{P(B|A)P(A) + P(B|A^c)P(A^c)}$$

$$= \frac{0.9 \times 0.001}{0.9 \times 0.001 + 0.02 \times 0.999} = 0.0431$$

– So, more than 95% of those testing positive will be actually negative.

We can compute conditional probabilities using the chain rule. The chain rule is typically based on multiple applications of Bayes' rule.

Consider $P(A|B, C)$ for events A, B and C. It may be computed using the following chain rule:

$$P(A|B, C) = \frac{P(A) \times P(B, C|A)}{P(B, C)} = \frac{P(A) \times P(B|A) \times P(C|A, B)}{P(B, C)}$$

Similarly,

$$P(A^c|B, C) = \frac{P(A^c) \times P(B, C|A^c)}{P(B, C)} = \frac{P(A^c) \times P(B|A^c) \times P(C|A^c, B)}{P(B, C)}$$

So, if we compare $P(A|B, C)$ and $P(A^c|B, C)$, we need to consider only the numerators as the denominators of both the quantities are equal to $P(B, C)$. In such cases, we can write

$$P(A|B, C) \propto P(A) \times P(B|A) \times P(C|A, B)$$

and

$$P(A^c|B, C) \propto P(A^c) \times P(B|A^c) \times P(C|A^c, B)$$

4.2.4   Bayes' Rule and Classification

Let C_1 and C_2 be two classes, with their respective prior probabilities being $P(C_1)$ and $P(C_2)$. Given an object x, we can compute the posterior probabilities using Bayes' rule as follows:

$$P(C_1|x) = \frac{P(x|C_1)P(C_1)}{P(x|C_1)P(C_1) + P(x|C_2)P(C_2)}$$

and

$$P(C_2|x) = \frac{P(x|C_2)P(C_2)}{P(x|C_1)P(C_1) + P(x|C_2)P(C_2)}$$

We assign x to class C_1 if $P(C_1|x) > P(C_2|x)$, else we assign x to class C_2. What we are essentially doing is to assign the test pattern x to the class with the larger posterior probability. We illustrate this with the help of a simple example.

EXAMPLE 6:   Let the two classes C_1 and C_2 be given by

- C_1: Class of Chairs $P(C_1) = 0.8$
- C_2: Class of Tables $P(C_2) = 0.2$

Let w_i be the weight in kilos of the ith object (Chair or Table); let the weight of any Chair be 2.8 kg and the weight of any Table be 6.8 kg. So, Chairs are light-weight and Tables are heavy-weight objects. Let a test pattern x that is a light-weight object be given; we can compute the posterior probabilities as follows:

$P(C_1|x) = \frac{P(x|C_1)P(C_1)}{P(x|C_1)P(C_1)+P(x|C_2)P(C_2)} = \frac{1(0.8)}{1(0.8)+0(0.2)} = 1$ ($P(x|C_1) = 1$ as Chairs are light weight)

$P(C_2|x) = \frac{P(x|C_2)P(C_2)}{P(x|C_1)P(C_1)+P(x|C_2)P(C_2)} = \frac{0(0.8)}{1(0.8)+0(0.2)} = 0$ ($P(x|C_2) = 0$ as Tables are heavier weight)

So, we classify the light-weight object x as a member of the class C_1 (Chairs) as $P(C_1|x) > P(C_2|x)$.

4.3   RANDOM VARIABLES, PROBABILITY MASS FUNCTION, PROBABILITY DENSITY FUNCTION AND CUMULATIVE DISTRIBUTION FUNCTION, EXPECTATION AND VARIANCE

We discuss some basic concepts dealing with random variables in this section.

4.3.1   Random Variables

A random variable (RV) is a mapping from the sample space to a co-domain. It is possible for the co-domain to be the set of real numbers or complex numbers. However, we restrict our discussion to the co-domain of real numbers, $\mathbb{R}$. So,

$$X : \mathcal{S} \to \mathbb{R},$$

where X is a real random variable.

EXAMPLE 7: W is a random variable that maps objects to their weights. We can have several RVs defined on the same $\mathcal{S}$. For example, H is another RV that maps objects to their heights.

We can have functions of one or more RVs. For example, we can have a function $g(W, H) = \alpha \times W + \beta \times H$, where α and β are real numbers. Such a function can also be treated as an RV. Let us consider another example.

EXAMPLE 8:   Consider tossing a coin. The sample space $\mathcal{S}_1 = \{$H,T$\}$. Let $P(H) = p$ and $P(T) = 1 - p$. Let an RV X_1 defined on $\mathcal{S}_1$ be

$$X_1(H) = 0 \text{ and } X_1(T) = 1$$

Consider tossing the coin once more; the corresponding sample space $\mathcal{S}_2$ is$\{$H,T$\}$. Consider an RV associated with the second toss as X_2 such that

$$X_2(H) = 0 \text{ and } X_2(T) = 1$$

The sample space obtained by considering both the tosses is $\mathcal{S} = \{$HH, HT, TH, TT$\}$. Let us consider RV X that is defined as $X = X_1 + X_2$. So $X(HH) = 0 + 0 = 0$, $X(HT) = 0 + 1 = 1$, $X(TH) = 1 + 0 = 1$ and $X(TT) = 1 + 1 = 2$. So, RV X is assigned

- one outcome in $\mathcal{S}$, that is, HH to 0.
- two outcomes in $\mathcal{S}$, HT and TH to 1.
- one outcome in $\mathcal{S}$, TT to 2.

4.3.2   Probability Mass Function (PMF)

So, we have

$$P(X = 0) = P(HH) = p^2$$
$$P(X = 1) = P(HT) + P(TH) = p(1 - p) + (1 - p)p = 2p(1 - p)$$
$$P(X = 2) = P(TT) = (1 - p)^2$$

Note that

$$P(X = 0) + P(X = 1) + P(X = 2) = p^2 + 2p(1 - p) + (1 - p)^2 = (p + (1 - p))^2 = 1^2 = 1$$

So, for the value in the range or for the value of RV X, the probability is non-negative and adds up to 1. Such a function $P_X(x)$ is the probability mass function (PMF) of the RV X. Note that

$$P_X(x) = \begin{cases} P(X = x) & if \ x \in Range(X) \\ 0 & Otherwise \end{cases}$$

Here, $Range(X)$ is the range of the mapping X (recall that RV is a mapping from $\mathcal{S}$ to $\mathbb{R}$.) We show a plot of the PMF of X in Fig. 4.3 assuming that $p = 0.5$.

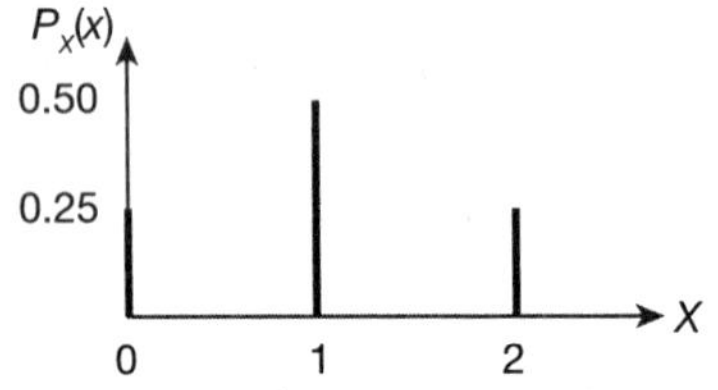

FIG. 4.3 An example to illustrate probability mass function

We will consider one more example to illustrate the PMF of multiple RVs.

EXAMPLE 9:   Let X be an RV from {H,T} to {0,1}; $X(H) = 0$ and $X(T) = 1$. Let Y be an RV from the face of a die to {1,2,3,4,5,6}. So, the joint probability $P_{X,Y}(x, y)$ values are given in Table 4.2.

TABLE 4.2 Joint probability $P_{X,Y}(x, y)$ values for two RVs X and Y

X/Y	1	2	3	4	5	6
0(H)	$\frac{1}{12}$	$\frac{1}{12}$	$\frac{1}{12}$	$\frac{1}{12}$	$\frac{1}{12}$	$\frac{1}{12}$
1(T)	$\frac{1}{12}$	$\frac{1}{12}$	$\frac{1}{12}$	$\frac{1}{12}$	$\frac{1}{12}$	$\frac{1}{12}$

Let $Z = X + Y$; then $P_Z(z)$ is given by

$$P_Z(1) = \frac{1}{12}, P_Z(2) = P_Z(3) = P_Z(4) = P_Z(5) = P_Z(6) = \frac{2}{12} = \frac{1}{6}, P_Z(7) = \frac{1}{12}$$

Note that $Z = 1$ is possible only when $X = 0$ and $Y = 1$. Similarly, $Z = 7$ is possible only when $X = 1$ and $Y = 6$. For other values of Z, there will be two possibilities. For example, $Z = 2$ can be either because of $X = 0$ and $Y = 2$ or $X = 1$ and $Y = 1$.

Given $P_{X,Y}(x,y)$, we can obtain the marginal $P_X(x)$ by summing up over all the y values for a given x. So,

$$P_X(0) = \frac{6}{12}, P_X(1) = \frac{6}{12}$$

4.3.3 Binomial Random Variable

This is a popular discrete RV. An RV is a **discrete random variable** if its range has a countable number of elements. If an RV has a finite-sized range, then it is discrete. Some examples of discrete RVs are:

- Number of training points falling in a given region
- Number of patients who tested COVID-19 positive
- Number of vehicles crossing a particular junction
- Number of consonants occurring in a speech segment

Let us consider tossing a coin n times. Let $P(T) = p$ and $P(H) = 1 - p$. Let us consider an RV X that denotes the number of tails out of n tosses of a coin.

EXAMPLE 10: Let $n = 3$. Let $\{T,H\}$ be mapped to $\{0,1\}$. Then the PMF for $n = 3$ is given by

$$P_X(0) = P(TTT) = p^3; \ P_X(1) = P(HTT) + P(THT) + P(HTT) = 3p^2(1 - p)$$

and

$$P_X(2) = P(HHT) + P(HTH) + P(THH) = 3p(1 - p)^2; \ P_X(3) = P(HHH) = (1 - p)^3$$

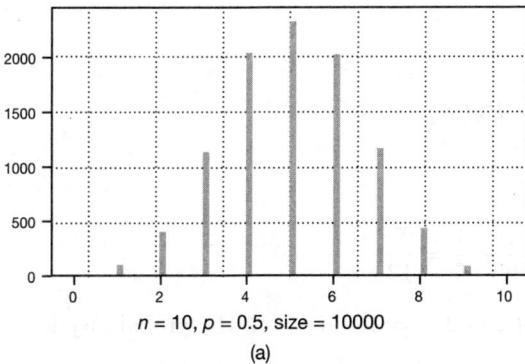

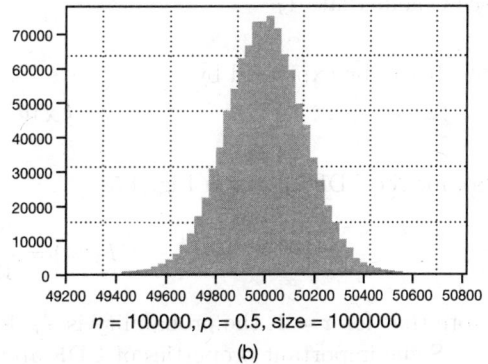

$n = 10, p = 0.5$, size = 10000

(a)

$n = 100000, p = 0.5$, size = 1000000

(b)

FIG. 4.4 Probability mass function of the binomial RV (for colour figure, please see Colour Plate 1)

Note that the coefficients are of the form $\binom{3}{0}$ ($= 1$ for p^3), $\binom{3}{1}$ ($= 3$ for $p^2(1-p)$), $\binom{3}{2}$ ($= 3$ for $p(1-p)^2$) and $\binom{3}{3}$ ($= 1$ for $(1-p)^3$). These are called **binomial coefficients**.

In general, the coefficient of $P_X(k)$ when the coin is tossed n times is $\binom{n}{k}$. Further, $P_X(k) = \binom{n}{k}p^k(1-p)^{n-k}$. Note that

- $P_X(k) \geq 0$ for $0 \leq k \leq n$, and
- $\sum_{k=0}^{n} P_X(k) = 1$.

We plot the PMF of the binomial RV in Fig. 4.4. Here, the X-axis describes the event of the number of tails obtained and the Y-axis describes the number of times the event has occurred. These frequencies are shown when the experiment is repeated 10,000 times in Fig. 4.4 (a) and a million times in Fig. 4.4 (b). The binomial makes a sharp peak as the number of tails tends to ∞.

4.3.4 Cumulative Distribution Function (CDF)

The cumulative distribution function (CDF) of an RV X is the function

$$F : \mathbb{R} \to [0, 1]$$

defined by $F_X(x) = P(X \leq x)$. We show in Fig. 4.5, a plot of a CDF.

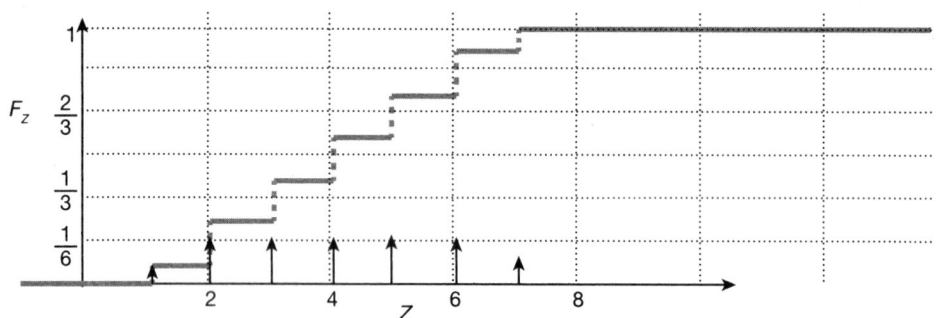

FIG. 4.5 An example of the cumulative distribution function of an RV (for colour figure, please see Colour Plate 1)

It can be expressed by

$$F_X(a) = \sum_{all \ x_i \leq a} P_X(x_i)$$

So, for the CDF shown in Fig. 4.5

$$F_Z(4) = \frac{1}{12} + \frac{1}{6} + \frac{1}{6} + \frac{1}{6} = \frac{7}{12}$$

Note that for $z = 1$, the probability is $\frac{1}{12}$; for each of $z = 2$, $z = 3$ and $z = 4$, the probability is $\frac{1}{6}$.

Some important properties of CDF are:

- $0 \leq F_Z(z) \leq 1$
- F_Z is non-decreasing
- $lim_{z \to \infty} F_Z(z) = 1$ and $lim_{z \to -\infty} F_Z(z) = 0$

4.4   CONTINUOUS RANDOM VARIABLES

If X is a continuous RV and Y is a discrete RV, then we will have probability density function (PDF) $f_X(x)$ for X against the PMF $P_Y(y)$ for the RV Y. Two necessary and sufficient properties of any PDF are:

- $f_X(x) \geq 0$ for all x; we drop the subscript X for brevity and use $f(x)$ instead.
- $\int_{-\infty}^{\infty} f(x)dx = 1$ for any continuous RV X.

EXAMPLE 11:   Let us consider a simple case of uniformly distributed RV X over the interval (a, b). Its PDF $f_X(x)$ is specified as

$$f_X(x) = \begin{cases} \frac{1}{b-a} & if\ a < x < b \\ 0 & otherwise \end{cases}$$

We show the PDF of X in Fig. 4.6. Observe that the the value of the PDF is $\frac{1}{b-a}$ in the range (a, b). Note that $f(x)$ is a PDF as

- $f(x) \geq 0$ for all x. Note that the density function has values of 0 and above. It is non-negative.
- $\int_a^b f(x)dx = \int_a^b \frac{1}{b-a}dx = 1$.

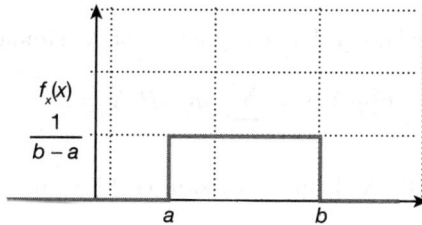

FIG. 4.6 Probability density function of a uniform RV X (for colour figure, please see Colour Plate 1)

Note that we can have CDF for both discrete and continuous RVs. Some of its essential properties are:

- $F_X(x) = P(X \leq x)$ by definition.
- If X is a discrete RV, then

$$F_X(x) = \sum_{x_i \leq x} P(X = x_i)$$

- If X is a continuous RV, then

$$F_X(x) = \int_{-\infty}^{x} f(x)dx$$

$$\frac{d}{dx}F_X(x) = f_X(x)$$

The CDF of the uniform RV X is shown in Fig. 4.7.

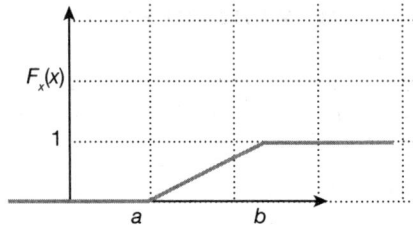

Fig. 4.7 Cumulative distribution function of the uniform RV X (for colour figure, please see Colour Plate 1)

4.4.1   Expectation of a Random Variable

Expectation or expected value of an RV characterizes the (weighted) average value. For a discrete RV X with PMF $P_X(x)$, $E[X] = \sum_{x \in R_X} x P(X = x)$.

EXAMPLE 12:   Consider tossing a coin with sample space $= \{H, T\}$. Let X be an RV such that $X(H) = 1$ and $X(T) = 0$. Let the probability of getting an H be p and the probability of getting a T be $(1 - p)$. Then

$$E[X] = 1 \times p + 0 \times (1 - p) = p$$

We can compute $E[g(X)]$, where $g(X)$ is a function of X, similarly.

$$E[g(X)] = \sum_{x \in R_X} g(x) P(X = x)$$

EXAMPLE 13:   Consider the RV X defined in Example 12. Consider $g(X) = X^2$. So,

$$E[g(X)] = E[X^2] = 1^2 \times p + 0^2 \times (1 - p) = p$$

It is possible to show that if X and Y are two RVs, then $E[X + Y] = E[X] + E[Y]$. We can show it as follows:

$$E[X + Y] = \sum_{x \in R_X} \sum_{y \in R_Y} P_{X,Y}(X = x, Y = y)$$

$$= \sum_x \sum_y x P_{X,Y}(X = x, Y = y) + \sum_x \sum_y y P_{X,Y}(X = x, Y = y)$$

$$= \sum_x x \sum_y P_{X,Y}(X = x, Y = y) + \sum_y y \sum_x P_{X,Y}(X = x, Y = y)$$

$$= \sum_x x P_X(x) + \sum_y y P_Y(y) = E[X] + E[Y]$$

Consider two discrete random variables X and Y. We say that X and Y are independent if $P(X = x, Y = y) = P(X = x)P(Y = y)$, for all x, y. If X and Y are continuous RVs and independent, then

$$P_{X,Y}(x, y) = P_X(x)P_Y(y)$$

If X is a continuous RV, then $E[X] = \int_{x \in R_X} x p_X(x) dx$. Two RVs are uncorrelated if $E[XY] = E[X] \times E[Y]$. If X and Y are independent, then they are uncorrelated, that is,

$E[XY] = E[X]E[Y]$. It can be shown as follows: $E[XY] = \int_x \int_y xy P_{X,Y}(x,y) = \int_x \int_y xy P_X(x) P_Y(y)$ (as they are independent) $= \int_x x P_X(x) \int_y y P_Y(y) = E[X]E[Y]$. However, two uncorrelated RVs X and Y need not be independent.

4.4.2 Variance of a Random Variable

Variance of an RV X characterizes the spread around its mean value $E[X] = \mu$.

It is denoted by σ^2. It is

$$Var[X] = E[(X_\mu)^2 = E[X^2] - \mu^2$$

It is also called the *second central moment* as it is centred around the mean; second because of the square term while the expected values are called moments. σ is called the **standard deviation.**

In general, $Var[X+Y] \neq Var[X] + Var[Y]$. However, variance of the sum is equal to the sum of the variances if X and Y are uncorrelated. It is possible to show it as follows:

Let $X + Y = Z$. So,

$$Var[X+Y] = Var[Z] = E[Z^2] - (E[Z])^2$$

$$= E[(X+Y)^2] - (E[X] + E[Y])^2) \ (as \ Z = X + Y)$$

$$= E[X^2 + Y^2 + 2XY] - E[X]^2 - E[Y]^2 - 2E[X]E[Y]$$

$$= (E[X^2] - (E[X])^2) + (E[Y^2 - (E[Y])^2) + 2E[XY] - 2E[X]E[Y] \ (rearranging)$$

$$= Var[X] + Var[Y] + 2(E[XY] - E[X]E[Y])$$

$$= Var[X] + Var[Y] + 2 \times 0 \ (X \ and \ Y \ are \ uncorrelated)$$

$$= Var[X] + Var[Y]$$

EXAMPLE 14: We will consider a simple discrete RV X that is Bernoulli distributed; it has two outcomes. Consider a single toss of a coin with $P(H) = p$ and $P(T) = 1 - p$. Let $X(H) = 1$ and $X(T) = 0$. So, $E[X] = 1 \times p + 0 \times (1 - p) = p$.

Consider $E[X^2] = 1^2 \times p + 0^2 \times (1 - p) = p$. We know that $Var[X] = E[X^2] - (E[X])^2 = p - p^2 = p(1 - p)$.

4.4.3 Normal Distribution

It is one of the most popularly used continuous RVs in ML. Some of its important properties are as follows:

• We say that an RV X is *normally distributed* if the PDF of X is given by

$$f_X(x) = \frac{1}{\sqrt{2\pi\sigma^2}} e^{-\frac{(x-\mu)^2}{2\sigma^2}} \ for \ -\infty < x < \infty$$

This is also called the **Gaussian RV** and it can be uniquely specified by the parameters μ, its mean, and σ^2, its variance. We show a plot of the normal PDF in Fig. 4.8.

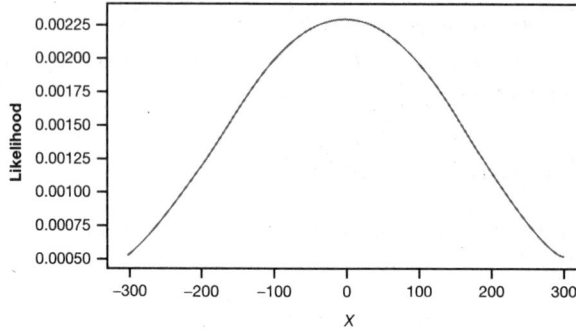

FIG. 4.8 Probability density function of the Gaussian RV X

- It is symmetric around its mean μ. For example, its likelihood is equal for $x = \mu + a$ and $x = \mu - a$ for some real number a.
- The value of its variance σ^2 determines the spread of the curve. If $\sigma = 0$, the curve becomes a Dirac delta function centred around μ. If $\sigma >> 0$ (very large), then it will spread too thin. Most of the samples fall in the region $\mu - 2\sigma$ to $\mu + 2\sigma$, as shown in Fig. 4.9. So, 68.3% of samples fall in the range $\mu - \sigma$ to $\mu + \sigma$ whereas 95.4% fall in the range $\mu - 2\sigma$ to $\mu + 2\sigma$ and 99.7% fall in the range $\mu - 3\sigma$ to $\mu + 3\sigma$.

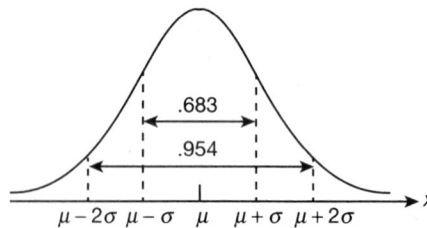

FIG. 4.9 The role of σ in the spread

- The popularity of this univariate normal RV in ML is attributed to its analytical behaviour. If we take the logarithm of the PDF, we get a quadratic

$$ln(f_X(x)) = c - \frac{(x - \mu)^2}{2\sigma^2},$$

where c is given by $-\frac{1}{2}ln(2\pi\sigma)$.
- The multivariate normal RV is specified by its PDF

$$\frac{1}{(2\pi)^{\frac{d}{2}}|\Sigma|^{\frac{1}{2}}} e^{-\frac{(x - \mu)^t \Sigma^{-1}(x - \mu)}{2}},$$

where x and μ are d-dimensional vectors, $|\Sigma|$ is the determinant of the covariance matrix Σ, Σ is a $d \times d$ symmetric matrix. The quantity $(x - \mu)^t \Sigma^{-1}(x - \mu)$ is the squared Mahalanobis distance between x and μ.
- The distribution of points in a two-dimensional case is shown in Fig. 4.10.

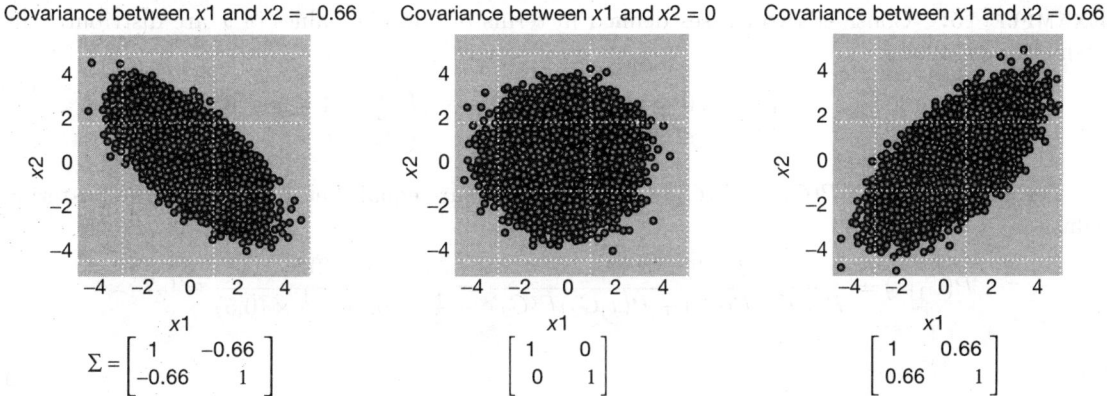

FIG. 4.10 Normally distributed two-dimensional points (for colour figure, please see Colour Plate 1)

Note that when Σ is a diagonal matrix, we have a circular shape for the distribution of points (Fig. 4.10 (b)). When the off-diagonal entries are non-zero, we have elliptical regions (Fig. 4.10 (a) and (c)) with different orientations based the polarity of these entries.

4.5   THE BAYES CLASSIFIER AND ITS OPTIMALITY

We have seen in Section 4.2.4 how posterior probabilities are obtained from prior probabilities. We have seen that

$$P(C_1|x) = \frac{P(x|C_1)P(C_1)}{P(x)} \Rightarrow \text{Posterior} = \frac{\text{Likelihood} \times \text{Prior}}{\text{Evidence}}$$

Note that $P(x) = P(x|C_1)P(C_1) + P(x|C_2)P(C_2)$; it is a normalizer to ensure that $P(C_1|x) + P(C_2|x) = 1$.

The Bayes classifier assigns a test pattern x to C_1 if $P(C_1|x) > P(C_2|x)$, else to C_2. So, for a given x, the **probability of error** is

$$P(error|x) = min(P(C_1|x), P(C_2|x))$$

So, the **average or expected error** across all possible values of x is

$$\int_x P(error|x)f(x)dx = \int_x min(P(C_1|x), P(C_2|x))f(x)dx$$

Here, $f(x)$ is the PDF of x and it is fixed; note that for every x, we take a decision so that $P(error|x)$ is minimum. So, the Bayes classifier is optimal in the sense that it minimizes the average probability of error or error rate. So, it is the *minimum error rate classifier*.

Note that

$$P(C_1|x) > P(C_2|x) \Rightarrow P(x|C_1)P(C_1) > P(x|C_2)P(C_2)$$

We consider some examples to illustrate the use of the Bayes classifier.

EXAMPLE 15:   Consider two classes defined in terms of how the values of x are distributed in each class as follows:

$$P(x|C_1) = \begin{cases} \frac{1}{2} & 0 \le x \le 2 \\ 0 & else \end{cases} \qquad P(x|C_2) = \begin{cases} \frac{1}{5} & 1 \le x \le 6 \\ 0 & else \end{cases}$$

Let us assume that $P(C_1) = P(C_2) = 0.5$; the priors are equal. Let $x = 2$. Then the posterior values

$$P(C_1|x) = \frac{P(x|C_1)P(C_1)}{P(x|C_1)P(C_1) + P(x|C_2)P(C_2)} = \frac{\frac{1}{2} \times (0.5)}{\frac{1}{2} \times (0.5) + \frac{1}{5} \times (0.5)} \approx 0.7$$

and

$$P(C_2|x) = \frac{P(x|C_2)P(C_2)}{P(x|C_1)P(C_1) + P(x|C_2)P(C_2)} = \frac{\frac{1}{5} \times (0.5)}{\frac{1}{2} \times (0.5) + \frac{1}{5} \times (0.5)} \approx 0.3$$

So, the pattern x with a value of 2 is assigned to class C_1 by comparing the posteriors as $P(C_1|x) > P(C_2|x)$.

EXAMPLE 16:   Let two classes be normally distributed with the same variance $\sigma^2 = 1$. Let the mean μ_1 of Class 1 be 2 and the mean of Class 2, μ_2, be 4. Let the prior probabilities be equal for the two classes, that is, $P(C_1) = P(C_2) = 0.5$. Let the test pattern be $x = 2$. Note that we can compare either the posterior probabilities or the numerators of the posteriors as the denominators are the same for both the posteriors. So, we consider the numerators in the posteriors for C_1 and C_2 (they are $f(x|C_1)P(C_1)$ and $f(x|C_2)P(C_2)$). These numerator quantities are as follows:

$$f(x = 2|C_1) \times P(C_1) = \frac{1}{\sqrt{2\pi 1^2}} e^{-\frac{1}{2}\frac{(2-2)^2}{1^2}} \times 0.5 = \frac{0.5}{\sqrt{2\pi}}$$

$$f(x = 2|C_2) \times P(C_2) = \frac{1}{\sqrt{2\pi 1^2}} e^{-\frac{1}{2}\frac{(2-4)^2}{1^2}} \times 0.5 = \frac{0.5}{\sqrt{2\pi e^2}}$$

So, $P(C_1|x) \propto \frac{0.5}{\sqrt{2\pi}}$ and $P(C_2|x) \propto \frac{0.5}{\sqrt{2\pi e^2}} \Rightarrow P(C_1|x) > P(C_2|x)$

So, assign the test pattern $x = 2$ to Class C_1.

It is possible to extend these ideas to d-dimensional vectors $(d > 1)$. Recall that multi-variate normal, for example, is characterized by **mean vector μ and the covariance matrix** Σ. If the vectors are d-dimensional, then μ is a d-dimensional vector and Σ is a $d \times d$ symmetric matrix, that is, $\Sigma_{i,j} = \Sigma_{j,i}$. All the diagonal entries are variances; $\Sigma_{i,i}$ is the variance of the ith feature; $\Sigma_{i,j}$ is the covariance between the ith and j^{th} features. So, the decision making in the d-dimensional case is as follows:

- If $f(x|C_1)P(C_1) > f(x|C_2)P(C_2)$, assign x to C_1, else assign x to C_2.
- If $P(C_1) = P(C_2)$, we need to compare only the likelihood values $f(x|C_1)$ and $f(x|C_2)$.
- If the covariance matrices are equal, that is, $\Sigma_1 = \Sigma_2 = \sigma^2 I$, then the covariance matrices are diagonal and all the diagonal entries are equal to σ^2.
- Under the given conditions, $f(x|C_1) > f(x|C_2) \Rightarrow e^{-\frac{1}{2}(x-\mu_1)^t \frac{I}{\sigma^2}(x-\mu_1)} > e^{-\frac{1}{2}(x-\mu_2)^t \frac{I}{\sigma^2}(x-\mu_2)} \Rightarrow (x - \mu_1)^t(x - \mu_1) > (x - \mu_2)^t(x - \mu_2)$.
  This means assign x to C_1 if the squared Euclidean distance between x and μ_1 is less than the squared Euclidean distance between x and μ_2; equivalently assign x to that class whose mean is closer to x based on Euclidean distance.

This is depicted in Fig. 4.11. Note that the decision boundary (the broken line) that separates the two classes is the perpendicular bisector of the line joining the two means. Any test pattern x falling to the left of the decision boundary is classified as belonging to C_1; points on the right-hand side of the decision boundary are classified as belonging to Class C_2.

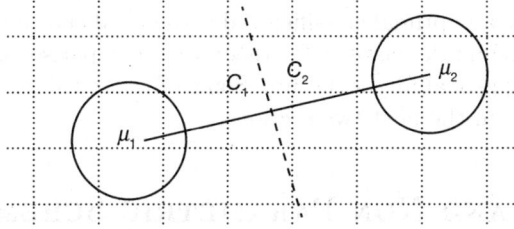

FIG. 4.11 Minimal distance classifier

This simple classifier is called the **minimal distance classifier (MDC)** and it is optimal when the priors are equal and the classes are normally distributed with the same covariance matrix; the covariance matrix is diagonal with the same entries in the diagonal locations.

If $\Sigma_1 = \Sigma_2$, it is possible to show that assigning x to C_1 is optimal if the squared Mahalanobis distance between x and μ_1 is smaller than that between x and μ_2.

4.5.1   Multi-Class Classification

We have discussed the use of the Bayes classifier in the two-class case. It can be easily used to deal with multi-class cases, that is, when the number of classes is more than 2. It may be described as follows:

- Let the classes be $C_1, C_2, \cdots, C_q$, where $q \geq 2$.
- Let the prior probabilities be $P(C_1), P(C_2), \cdots, P(C_q)$.
- Let x be the test pattern to be classified as belonging to one of these q classes.
- Compute the posterior probabilities using Bayes' rule

$$P(C_i|x) = \frac{P(x|C_i)P(C_i)}{\sum_{j=1}^{q} P(x|C_j)P(C_j)}, \text{ for } i = 1, 2, \cdots, q$$

- Assign the test pattern x to class C_l if

$$P(C_l|x) \geq P(C_i|x), \text{ for } i = 1, 2, \cdots, q$$

- In the case of a tie (two or more of the largest-valued posteriors are equal), assign arbitrarily to any one of the corresponding classes. In practice, breaking the tie arbitrarily is the prescription suggested for any ML model.
- In this case, the probability of error is the sum of the posterior probabilities of the remaining $q-1$. We know that the posteriors across all the q classes add up to 1, that is, $\sum_{i=1}^{q} P(C_i|x) = 1$. So, if x is assigned to class C_l, then the probability of error is $1 - P(C_l|x)$.
- In this case also, we have average probability of error as

$$\int_x P(error|x)P(x)dx = \int_x (1 - P(C_l|x))P(x)dx$$

This is the minimum possible because for every x, we are choosing the class that has the largest posterior. So, $P(C_l|x)$ is the largest and $1 - P(C_l|x)$ is minimized for x. So, even in the multi-class case, the Bayes classifier is optimal by being the minimum error rate classifier. It can deal with a mix of both categorical and numerical attributes provided the required probabilities are known.

So, the Bayes classifier is an optimal classifier and is the ideal choice for classification. However, it is used more as a benchmark classifier for theoretical comparisons. In practice, it is difficult to obtain the underlying probability structure. Some of related simplifications that are popular in practice are discussed in the next two sections.

4.6 PARAMETRIC AND NON-PARAMETRIC SCHEMES FOR DENSITY ESTIMATION

The Bayes classifier is an optimal classifier. It is versatile in terms of being used in applications involving mixed variables. However, a major bottleneck in its effective use is the assumption that the underlying probability structure is available. We need to have the prior probabilities and the PDF or PMF for each class. We will consider the estimation of prior probabilities.

EXAMPLE 17: Consider a tweet, and out of 100 people in a community, let 10 from set $\{1, 3, 12, 21, 33, 54, 66, 75, 84, 93\}$ have retweeted while the remaining 90 did not.

It is possible to view the retweeting pattern x as a binary string of length 100, where the ith bit is 1 if the ith person has retweeted for $i = 1, 2, \cdots, 100$, else it is 0.

Let p be the probability of retweet; then the corresponding probability may be captured by $p^{x^i}(1 - p)^{1-x^i}$. Note that this quantity selects p if $x^i = 1$ and $(1-p)$ if $x^i = 0$, where x^i is the ith bit of x.

We assume that people retweet independently; then the joint probability is the product of the individual probabilities. So, the joint probability is $p \cdots (1 - p) \cdots p \cdots (1 - p) = p^{10}(1 - p)^{90}$. This is the likelihood of 10 out of 100 people retweeting and the remaining 90 not retweeting. The logarithm of the likelihood is $l(p) = 10 log p + 90 log(1 - p)$. Note that the maximum value of the likelihood is the same as the maximum value of the logarithm of the likelihood as logarithm is a monotonic function.

The derivative of $l(p)$ with respect to p gives us $\frac{10}{p} - \frac{90}{1-p} = 0 \Rightarrow 100p = 10$. So, $p = 0.1 = \frac{10}{100}$; it is actually an estimate of p called the **maximum likelihood estimate (MLE)**. If k out of n people have retweeted, then MLE of p is $p = \frac{k}{n}$. It is called the maximum likelihood estimate because the estimate maximizes the joint probability or likelihood.

4.6.1 Parametric Schemes

Here, we typically assume that the form of the underlying PDF or PMF is known and estimate the parameters involved. In Example 15, we looked at n independent trials of a Bernoulli RV which amounted to the Binomial distribution with $E[k] = np \Rightarrow \hat{p} = \frac{E[k]}{n} = \frac{k}{n}$ (refer to Exercise 10 at the end of this chapter for more details on $E[k] = np$.)

Maximum Likelihood Estimation (MLE)

The maximum likelihood estimate appears to be a simple and intuitively appealing scheme for dealing with estimation of parameters. We will consider the case of a continuous RV.

EXAMPLE 18:   Let us consider the univariate normal PDF shown in Fig. 4.12.

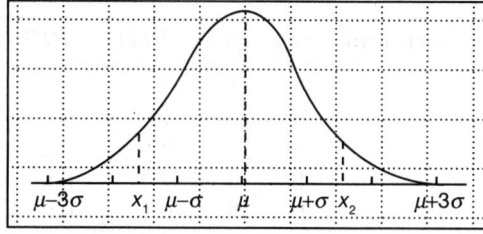

FIG. 4.12 Estimation of parameters of the normal distribution

Let there be one sample x_1 available for estimating the MLE. The log likelihood is

$$l(\mu) = c_1 - \frac{1}{2}\frac{(x_1 - \mu)^2}{\sigma^2},$$

where c_1 is a constant with respect to μ

$$\frac{dl(\mu)}{d\mu} = 0 \Rightarrow \hat{\mu} = x_1$$

If there are two samples x_1 and x_2, then the log likelihood is given by

$$l(\mu) = c_2 - \frac{1}{2}\frac{(x_1 - \mu)^2}{\sigma^2} - \frac{1}{2}\frac{(x_2 - \mu)^2}{\sigma^2}$$

Now

$$\frac{dl(\mu)}{d\mu} = 0 \Rightarrow \hat{\mu} = \frac{(x_1 + x_2)}{2}$$

If we are given n training patterns, $x_1, x_2, \ldots, x_n$, then log likelihood is

$$l(\mu) = c_n - \sum_{i=1}^{n}\frac{1}{2}\frac{(x_i - \mu)^2}{\sigma^2}$$

(c_n is a constant with respect to μ.) So,

$$\frac{dl(\mu)}{d\mu} = 0 \Rightarrow \hat{\mu} = \sum_{i=1}^{n}\frac{x_i}{n}$$

However, there can be problems when the number of data points, n, is small. We will illustrate this with an example.

EXAMPLE 19:   Consider tossing a fair coin 10 times. Let the number of heads, n_H, and number of tails, n_T, out of 10 tosses be as given in Table 4.3. We can conduct this experiment, of tossing a coin 10 times, repeatedly and record the values of n_H and n_T in the table.

TABLE 4.3 Estimates of $P(H)$ and $P(T)$ based on 10 tosses of a fair coin using the maximum likelihood scheme

Experiment	n_H	n_T	$P(H)$	$P(T)$
1	0	10	0	1
2	10	0	1	0
3	8	2	0.8	0.2
4	4	6	0.4	0.6
5	1	9	0.1	0.9
6	7	3	0.7	0.3

We have conducted the experiment 6 times and recorded the values in 6 rows of the table. Note that the estimates of $P(H)$ and $P(T)$ using MLE and shown in the table vary significantly. In the first case, the first row, the estimate of $P(H)$ is given by $\frac{n_H}{n} = 0$ and in the second row, the estimate of $P(T)$ is given by $\frac{n_T}{n} = 0$. Such estimates resulting in 0 (zero) values are not useful in practice. Note that the estimated values of $P(H)$ and $P(T)$ vary significantly among the different experiments.

One way to deal with the 0 valued estimates is by integrating prior knowledge of the parameter. This approach is called the **Bayesian estimation (BE)**.

For example, we know that for a fair coin, $P(H) = P(T) = \frac{1}{2}$ a priori. If we want to integrate it into the estimates of $P(H)$ and $P(T)$, then they must have the following form

$$P(H) = \frac{n_H + 1}{n + 2} \text{ and } P(T) = \frac{n_T + 1}{n + 2}$$

These estimates tell us that if no experimentation is carried out, that is, if $n = 0$, then $n_H = n_T = 0$. In such a case, we get the values of $P(H) = P(T) = \frac{1}{2}$ which is intuitively appealing.

If we want to compare MLE with BE, we get the following:

- MLE: $P(H) = \frac{n_H}{n}$ and $P(T) = \frac{n_T}{n}$.
- BE: $P(H) = \frac{n_H+1}{n+2}$ and $P(T) = \frac{n_T+1}{n+2}$.

EXAMPLE 20: The values obtained using the Bayesian scheme are shown in Table 4.4. However, as $n \to \infty$, we know that $n_H \to \infty$ and $n_T \to \infty$. So, in the case of BE, $P(H) = \frac{n_H}{n}$ and $P(T) = \frac{n_t}{n}$. We get the same estimates as the MLE scheme.

So, when we have large training data, that is, when $n \to \infty$, there is not much to choose between MLE and BE; they give the same estimates.

We can formalize the above arguments with respect to BE as follows:

- Note that the likelihood is of the form

$$p^{\sum_{i=1}^{n} x_i}(1-p)^{\sum_{i=1}^{n}(1-p)^{1-x_i}} = p^{\sum_{i=1}^{n} x_i}(1-p)^{n-\sum_{i=1}^{n} p^{1-x_i}}$$

TABLE 4.4 Estimates of $P(H)$ and $P(T)$ based on 10 tosses of a fair coin using the maximum likelihood scheme

Experiment	n_H	n_T	$P(H)$	$P(T)$
1	0	10	$\frac{1}{12}$	$\frac{11}{12}$
2	10	0	$\frac{11}{12}$	$\frac{1}{12}$
3	8	2	0.75	0.25
4	4	6	$\frac{5}{12}$	$\frac{7}{12}$
5	1	9	$\frac{2}{12}$	$\frac{10}{12}$
6	7	3	$\frac{8}{12}$	$\frac{4}{12}$

So, a convenient prior that simplifies the computations is the **conjugate prior** for the Bernoulli data and is given by the beta distribution specified by:

$$Beta(a, b) = \frac{\Gamma(a+b)}{\Gamma(a)\Gamma(b)} p^{a-1}(1-p)^{b-1},$$

where $p \in [0, 1]$ and $a, b \geq 1$.
- The advantage of the conjugate prior is that the resulting posterior will also be beta distributed.
- Note that the posterior is proportional to the product of likelihood and the prior. We have

$$Posterior \propto p^{\sum_{i=1}^{n} x_i}(1-p)^{n-\sum_{i=1}^{n} p^{1-x_i}} \times p^{a-1}(1-p)^{b-1}$$

So, the posterior is $Beta(\sum_{i=1}^{n} x_i + a, n + b - \sum_{i=1}^{n} x_i)$.
- The estimate of p that maximizes the posterior is $\hat{p} = \dfrac{\sum_{i=1}^{n} x_i + a - 1}{n + a + b - 2}$.
- If $a = b = 1$, then it is the same as the MLE, that is, $\hat{p} = \frac{1}{n}\sum_{i=1}^{n} x_i$; when $a = b = 1$, the beta distribution is flat, as shown in Fig. 4.13.

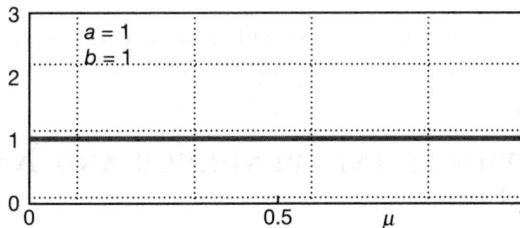

FIG. 4.13 Beta distribution when $a = b = 1$

Note that as $n \to \infty$, $\hat{p} = \dfrac{\sum_{i=1}^{n} x_i + a - 1}{n + a + b - 2} \to \dfrac{\sum_{i=1}^{n} x_i}{n}$ as the numerator tends to $\sum_{i=1}^{n} x_i$ and the denominator tends to n. In such a case, again the BE and MLE are equal.

A quick comparison of the MLE and BE schemes is given below:

- Both MLE and BE are *parametric estimation schemes*. Both of them assume that the functional form of the underlying density is known and the parameters need to be estimated. For example, if we assume that the density is binomial, we need to estimate the value of p that corresponds to the probability of success; the probability of failure is $1 - p$.
- Both of them assume that the training data points are drawn *independently* from the unknown density.
- MLE assumes that the parameters are unknown but are deterministic quantities. For example, the estimate of p in n Bernoulli trials (binomial) is $\hat{p} = \frac{k}{n}$, where k is the number of successes out of n trials. However, BE assumes that the unknown parameters are random; for example, in the case of Bernoulli data, we assumed that the unknown parameter p is an RV and it has beta density.
- In the case of normal density, MLE assumes that the unknown mean is deterministic and it estimates mean, $\hat{\mu}$, using the training data only. The estimate is $\hat{\mu} = \frac{1}{n} \sum_{i=1}^{n} x_i$, as discussed in an earlier section.
- In the case of BE, for estimating the mean of the normal:
  - The conjugate prior density is normal with mean μ_{in} and variance σ_{in}^2.
  - Because of the use of conjugate prior, the posterior is also normal. If the posterior is normal with mean μ_f and variance σ_f^2, then it is possible to show that

  $$\mu_f = \frac{n\sigma_{in}^2}{n\sigma_{in}^2 + \sigma^2} \times \mu_n + \frac{\sigma^2}{n\sigma_{in}^2 + \sigma^2} \times \mu_{in},$$

  where the data is normally distributed with mean μ and variance σ^2 and μ_n is the sample mean or the MLE.
  - Note that the estimate of the mean by the BE scheme coincides again with the estimate of the MLE scheme when $n \to \infty$. Then we have $\mu_f = \mu_n$ as $\frac{\sigma^2}{n\sigma_{in}^2 + \sigma^2} \to 0$ as $n \to \infty$ and $\frac{n\sigma_{in}^2}{n\sigma_{in}^2 + \sigma^2} \to \frac{n\sigma_{in}^2}{n\sigma_{in}^2} = 1$ as $n \to \infty$.
  - Note that BE gives the estimate of the mean as a weighted combination of the MLE value μ_n and the mean of the prior given by μ_{in}. These weights are $\frac{n\sigma_{in}^2}{n\sigma_{in}^2 + \sigma^2}$ and $\frac{\sigma^2}{n\sigma_{in}^2 + \sigma^2}$ and they add up to 1; such a weighted combination is called a **convex combination**.
  - So, BE is a generalized version, of which MLE may be viewed as a special case.

4.7 CLASS CONDITIONAL INDEPENDENCE AND NAÏVE BAYES CLASSIFIER

In this section, we examine the difficulties associated with the practical usage of the Bayes classifier and then provide a simplified scheme for estimating the probability structure.

4.7.1 Estimation of the Probability Structure

There are two different schemes for the estimation of parameters: non-parametric schemes and parametric schemes.

Non-Parametric Schemes

Here, the training data is used to directly estimate the PDF. We will consider the non-parametric schemes first to estimate the probability density of a class.

Let n independently drawn training patterns be given from a class. Let each of them be an l-dimensional vector. Let the probability that any one of them falls in a small region, R, in the l-dimensional space be P_R. Let some k out of the n patterns fall in R. The RV here is binomially distributed as out of n independent trials, some k fall in R and the remaining $n - k$ fall outside R. The expected value of k, $E[k] = n \times P_R$ (refer to Exercise 10 at the end of this chapter). Using its expected/average value as an estimate of k, $\hat{k}$, we get $\hat{k} = n \times P_R$. So, $P_R = \frac{\hat{k}}{n}$.

Assume that $k = \hat{k}$; this assumption is valid when the training data size n is large. Then

$$k = n \times P_R \Rightarrow P_R = \frac{k}{n}$$

If the unknown PDF is $p_X(x)$, then P_R is obtained by taking the integral of the PDF in the region R. So, $P_R = \int_R p(X)dx$. But we have $P_R = \frac{k}{n}$. So, from these two we get

$$k = n \times \int_R p_X(x)$$

If we assume that $p_X(x)$ is some constant p, in the region considered, because the region R is very small, then

$$k = n \times p \times V_R,$$

where we get

$$\int_R p_X(x)dx = p \int_R dx = pV_R$$

based on p being constant and V_R being the volume of the region. So, the estimate of the PDF, p, in a small region is given by

$$p = \frac{1}{n}\frac{k}{V_R}$$

This estimate makes sense when the value n is large and the region R is very small. So, the non-parametric schemes demand very large value for n with the density being constant in R. Hence, they are not widely used in practice.

Parametric Schemes

Here, we assume that the functional form of the PDF is given and we need to estimate the underlying parameters. There are two popular schemes:

- **Maximum Likelihood Estimation (MLE):** Here, likelihood corresponding to the n independently drawn training patterns is maximized to find the estimates of the underlying parameters. The estimate obtained is such that the probability of generating the given patterns from the resulting distribution is maximum.
- **Bayesian Estimation (BE):** In this case, in addition to the assumptions made by MLE, the parameters are assumed to be RVs with known prior distribution based on domain knowledge. The priors are converted into posterior probabilities using the likelihood of the training data. A popular scheme is to use the maximum a posteriori (MAP) estimate.

We start with an example to make the ideas clear.

EXAMPLE 21: Consider the data shown in Table 4.5. There are two classes, C_0 and C_1, and three training patterns from each of the classes. Each pattern is characterized by three binary features. There is a test pattern in the form of pattern 7. We will use MLE and BE to estimate the probabilities using the 6 training patterns and use the Bayes classifier to classify pattern 7.

TABLE 4.5 Data set to illustrate the difficulties associated with the Bayes classifier

Pattern	Feature1	Feature2	Feature3	Class
1	0	0	0	C_0
2	1	0	1	C_1
3	1	0	0	C_0
4	1	1	1	C_1
5	0	1	1	C_1
6	0	1	1	C_0
7	1	0	1	?

Let us consider the Bayes classifier. Note that the prior probabilities are $P(C_0) = P(C_1) = \frac{1}{2}$ using either MLE or BE.

Consider pattern 7. The corresponding posterior probabilities are:

- Using Bayes' rule, we have $P(C_0|Feature1 = 1, Feature2 = 0, Feature3 = 1)$
  $\propto P(C_0) \times P(Feature1 = 1, Feature2 = 0, Feature3 = 1|C_0)$
  $\propto P(C_0) \times P(Feature1 = 1|C_0) \times P(Feature2 = 0|Feature1 = 1, C_0) \times P(Feature3 = 1|Feature1 = 1, Feature2 = 0, C_0)$.

- Similarly, using Bayes' rule, we get $P(C_1|Feature1 = 1, Feature2 = 0, Feature3 = 1) \propto P(C_1) \times P(Feature1 = 1, Feature2 = 0, Feature3 = 1|C_1)$
  $\propto P(C_1) \times P(Feature1 = 1|C_1) \times P(Feature2 = 0|Feature1 = 1, C_1) \times P(Feature3 = 1|Feature1 = 1, Feature2 = 0, C_1)$

1. The MLE estimates are as follows:

   - For class C_0:
     - $P(Feature1 = 1|C_0) = \frac{1}{3}$
     - $P(Feature2 = 0|Feature1 = 1, C_0) = \frac{1}{1} = 1$
     - $P(Feature3 = 1|Feature1 = 1, Feature2 = 0, C_0) = \frac{0}{1} = 0$

   - For class C_1:
     - $P(Feature1 = 1|C_1) = \frac{2}{3}$
     - $P(Feature2 = 0|Feature1 = 1, C_1) = \frac{1}{2}$
     - $P(Feature3 = 1|Feature1 = 1, Feature2 = 0, C_1) = \frac{1}{1} = 1$

   So, using the MLE estimates, $P(C_0|Feature1 = 1, Feature2 = 0, Feature3 = 1) \propto \frac{1}{2} \times \frac{1}{3} \times 1 \times 0 = 0$.

   Using the MLE estimates for class C_1, we have $P(C_1|Feature1 = 1, Feature2 = 0, Feature3 = 1) \propto \frac{1}{2} \times \frac{2}{3} \times \frac{1}{2} \times 1 = \frac{1}{6}$.

   So, $P(C_1|Feature1 = 1, Feature2 = 0, Feature3 = 1)(= \frac{1}{6} > P(C_0|Feature1 = 1, Feature2 = 0, Feature3 = 1)(= 0)$.

   So, pattern 7 is assigned to C_1 by employing the estimates obtained using the MLE scheme.

2. The BE estimates are as follows:

- For Class C_0:
    - $P(Feature1 = 1|C_0) = \frac{1+1}{3+2} = \frac{2}{5}$
    - $P(Feature2 = 0|Feature1 = 1, C_0) = \frac{1+1}{1+2} = \frac{2}{3}$
    - $P(Feature3 = 1|Feature1 = 1, Feature2 = 0, C_0) = \frac{0+1}{1+2} = \frac{1}{3}$
- For Class C_1:
    - $P(Feature1 = 1|C_1) = \frac{2+1}{3+2} = \frac{3}{5}$
    - $P(Feature2 = 0|Feature1 = 1, C_1) = \frac{1+1}{2+2} = \frac{1}{2}$
    - $P(Feature3 = 1|Feature1 = 1, Feature2 = 0, C_1) = \frac{1+1}{1+2} = \frac{2}{3}$

So, $P(C_0|Feature1 = 1, Feature2 = 0, Feature3 = 1) \propto \frac{1}{2} \times \frac{2}{5} \times \frac{2}{3} \times \frac{1}{3} = \frac{2}{45}$.
Similarly, $P(C_1|Feature1 = 1, Feature2 = 0, Feature3 = 1) \propto \frac{1}{2} \times \frac{3}{5} \times \frac{1}{2} \times \frac{2}{3} = \frac{1}{10}$.
So, $P(C_1|Feature1 = 1, Feature2 = 0, Feature3 = 1) > P(C_0|Feature1 = 1, Feature2 = 0, Feature3 = 1)$.
So pattern 7 is assigned to C_1 by employing the estimates obtained using the BE scheme.

Some observations based on Example 21:

- Both the MLE and BE schemes have assigned pattern 7 to the same class, that is, C_1. This is not the case in general. We will examine it in Exercise 12 at the end of this chapter.
- It is possible that the MLE scheme estimates zero posterior probabilities for two or more classes, leading to a difficulty in taking a meaningful decision. This problem will not be encountered when we use BE. This property also will be examined in Exercise 12 at the end of this chapter.
- On large data sets, that is, when $n \to \infty$, the BE scheme gives the same estimate as the MLE scheme. So, the recommendation is to use the BE scheme for estimation when the training data is small by integrating the domain knowledge in the form of a suitable prior.
- On larger sized training data, it is good to use the MLE estimate as it depends solely on the data.

4.7.2   Naïve Bayes Classifier (NBC)

We have discussed the difficulties associated with the use of the Bayes classifier in practice. One simplification that is popular is based on *class-conditional independence*. The resulting classifier is called the naïve Bayes classifier as it is a Bayes classifier with some simplification. It may be explained using an example.

EXAMPLE 22:   Consider the data used in Example 21 and Table 4.5. Let us again consider the posterior probabilities for pattern 7. They are:

$$P(C_0|pattern\ 7) = \frac{P(Feature1=1, Feature2=0, Feature3=1|C_0) \times P(C_0)}{P(Feature1=1, Feature2=0, Feature3=1)} \text{ and}$$

$$P(C_1|pattern\ 7) = \frac{P(Feature1=1, Feature2=0, Feature3=1|C_1) \times P(C_1)}{P(Feature1=1, Feature2=0, Feature3=1)}$$

Note that both the posteriors have the same denominator. So, instead of comparing the posteriors, we can compare their numerators for the sake of simplicity as was done in the previous subsection.

We have $P(C_0) = P(C_1) = 0.5$. We need to compute the likelihood values for the two classes. We can simplify the computation using class-conditional independence as follows:

- For Class C_0, $P(Feature1 = 1, Feature2 = 0, Feature3 = 1|C_0)$
  $= P(Feature1 = 1|C_0) \times P(Feature2 = 0|C_0) \times P(Feature3 = 1|C_0)$
- Similarly, for C_1, $P(Feature1 = 1, Feature2 = 0, Feature3 = 1|C_1)$
  $= P(Feature1 = 1|C_1) \times P(Feature2 = 0|C_1) \times P(Feature3 = 1|C_1)$

What we have done is to write the probability of a conjunction conditioned on Class C_0 or C_1 as the product of probabilities of individual conjuncts that are conditioned on the class. This is **class-conditional independence.**

The required probabilities may be calculated using the MLE scheme for Class C_0 as:

- $P(Feature1 = 1|C_0) = \frac{1}{3}$.
- $P(Feature2 = 0|C_0) = \frac{2}{3}$.
- $P(Feature3 = 1|C_0) = \frac{1}{3}$.
  So, $P(C_0|pattern\ 7) \propto \frac{1}{2} \times \frac{1}{3} \times \frac{2}{3} \times \frac{1}{3} = \frac{1}{27}$

Similarly, for C_1 we have

- $P(Feature1 = 1|C_1) = \frac{2}{3}$.
- $P(Feature2 = 0|C_1) = \frac{1}{3}$.
- $P(Feature3 = 1|C_1) = \frac{3}{3} = 1$.
  So, $P(C_1|pattern\ 7) \propto \frac{1}{2} \times \frac{2}{3} \times \frac{1}{3} \times 1 = \frac{1}{9}$

So, we assign pattern 7 to C_1 as $P(C_1|pattern\ 7) > P(C_0|pattern\ 7)$

Some general observations from Example 22:

- In this example, NBC has made the same decision as the Bayes classifier. However, they can give different results in general.
- NBC employs class-conditional independence. It is given in general as

$$P(f_1 = v_1, f_2 = v_2, \cdots, f_l = v_l|C) = P(f_1 = v_1|C) \times P(f_2 = v_2|C) \times \cdots \times P(f_l = v_l|C),$$

  where C is the class and $f_i = v_i$ means feature i (f_i) will have value v_i.
- The decisions of NBC coincide with that of the Bayes classifier if class-conditional independence holds.
- NBC has simplified the computation and even the MLE scheme estimates did not create a problem here as the estimates of these simpler probabilities are non-zero.
- We have used the MLE scheme to estimate the probabilities. It is possible to use the BE scheme to ensure that we do not have zero-valued estimates. This is left to the reader as an exercise.

SUMMARY

In this chapter, we examined ML models based on Bayes' rule. Some important issues discussed are as follows:

- Bayes' rule plays an important role in understanding the Bayes classifier.
- The Bayes classifier is an optimal classifier. It minimizes the probability of error.
- Akin to the classifiers based on DTs, the Bayes classifier can be used when the data set has mixed type features, that is, categorical and numerical. However, we require the underlying probability structure to use the Bayes classifier effectively.
- It is possible to estimate the probability structure using training data with the help of the maximum likelihood approach or the Bayesian scheme.

- Estimation of the probability structure can be simplified by assuming class-conditional independence.
- NBC makes use of class-conditional independence and is a popular choice when the data set has mixed type features.
- NBC is a linear classifier and hence is a good candidate for providing explanation to the users and domain experts.

EXERCISES

1. Consider the discussion immediately after Example 1. Show that KNN is equivalent to prior probability-based classification, when $k = n$ and the probabilities are estimated based on ratio of frequencies.

2. Consider tossing a coin twice. What is the sample space? Find the probability that the first toss results in a tail. What is the probability that the second toss results in a head?

3. Show that $P(A \cap B) = P(A|B)P(B) = P(B|A)P(A)$.

4. If events A and B are independent, then show that A^c and B are independent.

5. Suppose a fair coin is tossed three times. Let A be the event that we get two heads in these three tosses. Let B be the event that the first toss shows up heads. Obtain

   a. $P(A)$ and $P(B)$
   b. $P(A|B)$ and $P(B|A)$

6. Consider the joint probability function $P_{X,Y}(x, y)$ given in Table 4.2. Obtain the marginal probability values of $P_Y(y)$ for all possible values of y.

7. Consider the discussion on binomial RV at the end of Section 4.3.3. The probability of k tails out of n tosses is given by $P_X(k)$. Show that

   a. $P_X(k) \geq 0$ for $0 \leq k \leq n$, and
   b. $\sum_{k=0}^{n} P_X(k) = 1$.

8. Consider the discussion before and in Example 13 of computing the expected value of a function of an RV. Let X be an RV with its expectation $E[X]$. If $h(X) = a \times X + b$, where a and b are some constants, find $E[h(X)]$.

9. Consider an RV X that is uniformly distributed in the range (0,1). Plot its PDF, $f_X(x)$, and CDF, $F_X(x)$.

10. Given that the mean of a Bernoulli-distributed RV is p and variance is $p(1-p)$ (refer Example 14), show that the mean of the binomial RV is np and its variance is $np(1-p)$ using the fact that binomial RV corresponds to n independent trails of the Bernoulli RV. If a coin is tossed n times (independently), getting k heads is binomially distributed. So, mean of binomial RV is $E[k]$ and variance is $E[(k - E[k])^2]$.

11. Show that if two classes, C_1 and C_2, are normally distributed with equal priors and covariance matrices being equal to Σ, then it is optimal to assign x to C_1 if

$$(x - \mu_1)^t \Sigma^{-1}(x - \mu_1) < (x - \mu_2)\Sigma^{-1}(x - \mu_2).$$

12. Consider the discussion in Example 22. Classify pattern 7 using NBC and estimate the probabilities using the BE scheme. Assume that prior probabilities are equal to $\frac{1}{2}$ for both the classes.

13. Consider the training data used in Example 21 and shown in Table 4.5. Classify the test pattern for which $Feature1 = 1$, $Feature2 = 1$ and $Feature3 = 0$ using the Bayes classifier.

    a. Use the MLE scheme for estimating the probabilities. Is there any problem?

    b. Use the BE scheme for estimating the probabilities.

14. Solve Q13 using NBC instead of the Bayes classifier. Use the MLE and BE schemes to estimate the probabilities.

PRACTICAL EXERCISE

1. Download the Olivetti Face data set. There are 40 classes (corresponding to 40 people), each class having 10 faces of the individual; so there are a total of 400 images. Here, each face is viewed as an image of size 64 × 64 (= 4096) pixels; each pixel has values 0 to 255 which are ultimately converted into floating numbers in the range [0,1]. Visit `https://scikit-learn.org/0.19/datasets/olivetti\_faces.html` for more details.

   Split the data sets into train and test parts. Perform this splitting randomly 10 times and report the average accuracy. You may vary the test and train data set sizes. Use NBC to classify the test data set. Obtain the accuracy on the test data.

Bibliography

- Murty, MN and Susheela Devi, V. 2015 *Introduction to Pattern Recognition and Machine Learning*, World Scientific Publishing Co. Pte. Ltd.: Singapore.

- Duda, RO, Hart, PE and Stork, DG. 2007. *Pattern Classification*, New York: John Wiley & Sons.

- Han, J, Kamber, M and Pei, J. 2012. *Data Mining Concepts and Techniques*, Morgan Kaufmann Publishers, Waltham.

- Witten, IH, Frank, E and Hall, MA. 2011. *Data Mining: Practical Machine Learning Tools and Techniques, Third Edition*, Morgan Kaufmann Publishers, Burlington.

- Tan, P-N, Steinbach, M, Karpatne, A and Kumar, V. 2018. *Introduction to Data Mining, Second Edition*, Pearson.

CHAPTER 5

Machine Learning Based on Frequent Itemsets

5.1 INTRODUCTION TO THE FREQUENT ITEMSET APPROACH

The presence of recurrent arrangements in a data set characterizes frequent patterns. These patterns can be itemsets, sub-sequences or sub-structures. For instance, a **frequent itemset** refers to a group of items like potato and onion, which are frequently bought together in a transaction data set. Similarly, a frequently recurring sequence such as purchasing a laptop, followed by headphones and then a cooling pad is known as a **frequent sequential pattern**.

In situations where web pages and their navigation paths form a graph, **frequent sub-structures**, which are the pages that are frequently accessed by the viewer, can be considered as projections from the initial graph.

The detection of such frequent patterns is crucial in uncovering associations, correlations and other interesting relationships in the data. Furthermore, it is useful in performing data classification, clustering and other data mining tasks; it has been shown that frequent itemsets are discriminative. As a result, frequent pattern mining has become a crucial data mining task and a focused area of research in machine learning.

This chapter aims to introduce frequent patterns, associations and correlation concepts and explore their efficient mining techniques.

5.2 FREQUENT ITEMSETS

Frequent itemset mining involves uncovering connections and correlations between items in large transactional or relational data sets. With vast amounts of data being continuously gathered and stored, many industries are interested in discovering such patterns in their databases.

Market basket analysis is a typical example of frequent itemset mining. This process examines customer buying patterns by identifying connections between the different items that customers place in their shopping baskets.

The task is to derive important associations among items such that the presence or absence of some items in a transaction will imply the presence or absence of some other items in the same transaction. For example, 'Alsatian_dog $\implies$ brand_a_biscuits' is an association rule which shows the existence of an association between Alsatian_dog and brand_a_biscuits. Another example, 'asthma_patient $\nRightarrow$ watermelon' means asthma_patient and watermelon are negatively associated. The food consumed by an asthma patient should not include watermelon.

Here the aim is to find associations between collection of items based on their frequency of co-occurrence. Mining for association rules consists of determining the correlation between items in transactions belonging to a transaction data set.

Let I be a set of items and D a set of transactions (patterns), where each transaction has a unique identifier (*tid*) and contains a set of items. A set of items is also called an **itemset**. An itemset with k items is called a k-itemset. The **support** of an itemset X, Support(X), is the ratio of the number of transactions in which it occurs as a subset to the total number of transactions in the database.

An itemset is **frequent** if its support is more than a user-specified minimum support (σ) value. An **association rule** is an expression of the form $\mathfrak{A} \implies \mathfrak{C}$, where $\mathfrak{A}$ and $\mathfrak{C}$ are sets of items. The left-hand side of the rule ($\mathfrak{A}$) is called the **antecedent** and the right-hand side of the rule ($\mathfrak{C}$) is called the **consequent**. In literature, a constraint $\mathfrak{A} \cap \mathfrak{C} = \emptyset$ is tacitly assumed. Every association rule must satisfy two user-specified constraints, one is support (σ) and the other is confidence(c). The support of the rule $\mathfrak{A} \implies \mathfrak{C}$ is defined as the fraction of tuples that contain both $\mathfrak{A}$ and $\mathfrak{C}$. In other words, it is Support($\mathfrak{A} \cup \mathfrak{C}$). The confidence of the rule is defined as Support($\mathfrak{A} \cup \mathfrak{C}$) / Support($\mathfrak{A}$). The goal is to find all the rules that satisfy minimum support and minimum confidence as specified by the user.

Note that the implication ($\implies$) is not a logical implication. This is because for every presence of $\mathfrak{A}$, $\mathfrak{C}$ may not be present in the same pattern; but they are present together by a fraction 'Support($\mathfrak{A} \cup \mathfrak{C}$)'.

Association rule mining consists of two parts:

1. **Discovery of all frequent (large) itemsets**. This refers to the set of items that have support greater than or equal to a user-defined minimum support (σ).
2. Use of the frequent itemsets to **generate association rules** using the user-defined confidence factor (c). That is, constructing rules of the form $\mathfrak{A} \implies \mathfrak{C}$ from the frequent itemsets, such that the ratio of the number of tuples having itemsets $\mathfrak{A}$ and $\mathfrak{C}$ together to the number of tuples having itemsets $\mathfrak{A}$ in the transaction database is greater than or equal to c.

Note that the second step is less costly in terms of computation than the first; the overall performance of association rule mining is determined by the first step.

EXAMPLE 1: Consider the un-ordered transaction database shown in Table 5.1. Let the user-defined minimum support (σ) be 0.3 and the user-defined confidence factor (c) be 0.6.

In this transaction database, an example for 3-itemset is {Potato, Onion, Tomato}. The Support(Onion, Tomato, Ginger) = $\mathfrak{s} = \frac{2}{5} = 0.4$. Since $\mathfrak{s} > \sigma$ (which is 0.3), {Onion, Tomato, Ginger} is a frequent 3-itemset.

Further, we can have an association rule, $\{Onion, Tomato\} \implies \{Ginger\}$, because Support (Onion, Tomato, Ginger)/Support(Onion, Tomato) = $\mathfrak{c} = \frac{2}{3} = 0.67$, which is greater than the user-defined confidence, c (which is 0.6) With the same 3-itemsets, {Onion, Tomato, Ginger}, under the same σ and c, we can have the association rules shown in Table 5.2.

TABLE 5.1 An example of a transaction database

Transaction ID	Items purchased
t_1	Potato, Onion
t_2	Potato, Tomato, Ginger, Cucumber
t_3	Onion, Tomato, Ginger, Chilli
t_4	Potato, Onion, Tomato, Ginger
t_5	Potato, Onion, Tomato, Chilli

TABLE 5.2 Other association rules

$\{Onion, Tomato\} \Longrightarrow \{Ginger\}$	with $s = 0.4$, $c = 0.67$
$\{Onion, Ginger\} \Longrightarrow \{Tomato\}$	with $s = 0.4$, $c = 1$
$\{Tomato, Ginger\} \Longrightarrow \{Onion\}$	with $s = 0.4$, $c = 0.67$
$\{Ginger\} \Longrightarrow \{Onion, Tomato\}$	with $s = 0.4$, $c = 0.67$
$\{Tomato\} \Longrightarrow \{Onion, Ginger\}$	with $s = 0.4$, $c = 0.5$
$\{Onion\} \Longrightarrow \{Tomato, Ginger\}$	with $s = 0.4$, $c = 0.5$

Observations:

- All of the above rules are binary partitions of the same itemset: {Onion, Tomato, Ginger}.
- Rules originating from the same itemset have identical support but can have different confidence ($s = 2/5$, $c = 2/3, 2/2, 2/3, 2/3, 2/4, 2/4$)
- Thus, we can decouple the support and confidence requirements.

5.3 FREQUENT ITEMSET GENERATION

One way to find the frequent itemsets is to find all the candidate itemsets first and then check whether they are frequent or not. Figure 5.1 shows the candidate itemsets for four items, A, B, C and D.

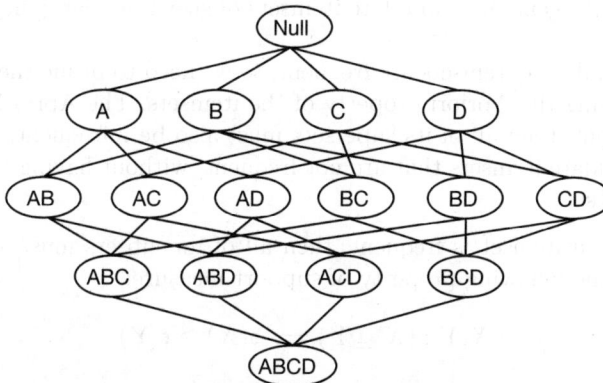

FIG. 5.1 Candidate itemsets for four items

Each itemset in the lattice is a candidate frequent itemset. One needs to scan the entire data set to count the support of each itemset. Note that with d distinct items, the possible candidate itemsets generated is 2^d. Further, with n transactions (patterns), the complexity of the brute-force algorithm to find all the frequent itemsets is $\mathcal{O}(n2^dT)$, where T is the average length of the transaction (pattern). This is computationally very expensive.

5.3.1 Frequent Itemset Generation Strategies

- **Reduce the number of candidates generated:** As discussed earlier, the complete search needs to generate 2^d candidates. One can use a pruning technique to reduce unnecessary candidate generation. The Apriori algorithm, in Section 5.4 discusses this concept.
- **Reduce the number of comparisons:** It is not necessary to match every candidate itemset with every transaction. Only the itemsets that are relevant to the transaction need to be compared. This can be done using efficient data structures for storing the candidates. The frequent pattern (FP) tree-based algorithm in Section 5.5 discusses this concept.

5.4 Apriori Algorithm

The Apriori algorithm is a classic algorithm for frequent itemset generation in association rule mining. It works by iteratively generating candidate itemsets of increasing size and counting their support in the transaction data set to identify frequent itemsets.

The algorithm can be broken down into two main steps:

1. Generating candidate itemsets
2. Pruning infrequent itemsets

Let us look at each of these steps in more detail.

Step 1: Generating Candidate Itemsets

The first step of the Apriori algorithm is to generate candidate itemsets of size k, given frequent itemsets of size $k-1$. This is done by performing a join and a prune operation.

The join operation involves taking two frequent itemsets of size $k-1$ and joining them together to form a candidate itemset of size k. For example, suppose we have two frequent itemsets of size 3: {ABC} and {BCD}. To generate candidate itemsets of size 4, we can join these two itemsets to obtain: {ABCD}.

However, not all candidate itemsets are frequent; so we need to prune the infrequent ones. The pruning can be done using the Apriori property of the itemsets. The Apriori property states that if an itemset is infrequent, then all of its supersets must also be infrequent. This property can be used to prune the candidate itemsets that are not frequent, without having to count their support in the transaction data set.

Apriori principle: If an itemset is frequent, then all of its subsets must also be frequent. This principle holds due to the following property of support measure:

$$\forall X, Y : (X \subseteq Y) \implies \mathfrak{s}(X) \geq \mathfrak{s}(Y)$$

Note that the support of a set never exceeds its subset. This property is also known as the **anti-monotone property of support**.

EXAMPLE 2:   Suppose we have transactions based on four items, A, B, C and D. After the scan, we come to know that the 2-itemset AB is infrequent; then we can prune all the candidates ABC, ABD and ABCD without having to count its support in the transaction database, because it must also be infrequent according to the Apriori property. This is shown in Fig. 5.2.

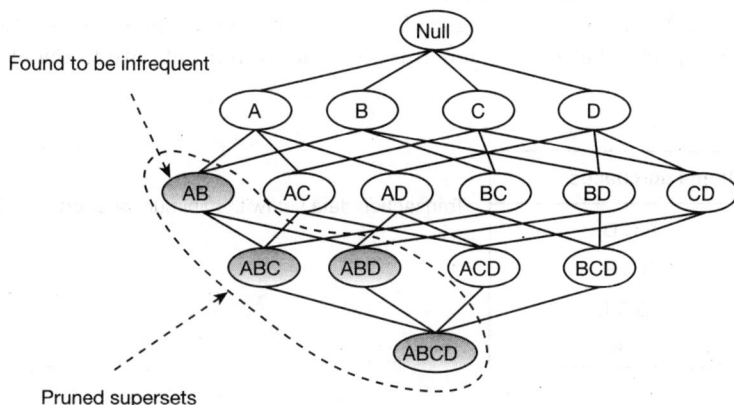

FIG. 5.2 Pruning of candidates for no further consideration

Step 2: Pruning Infrequent Itemsets

The second step of the Apriori algorithm is to prune infrequent itemsets. This is done by counting the support of each candidate itemset in the transaction database, and keeping only those itemsets that meet the minimum user-defined support (σ).

The Apriori algorithm works by iteratively generating candidate itemsets of increasing size, and pruning infrequent itemsets at each iteration until no more frequent itemsets can be found. Figure 5.3 shows the flowchart of the Apriori algorithm.

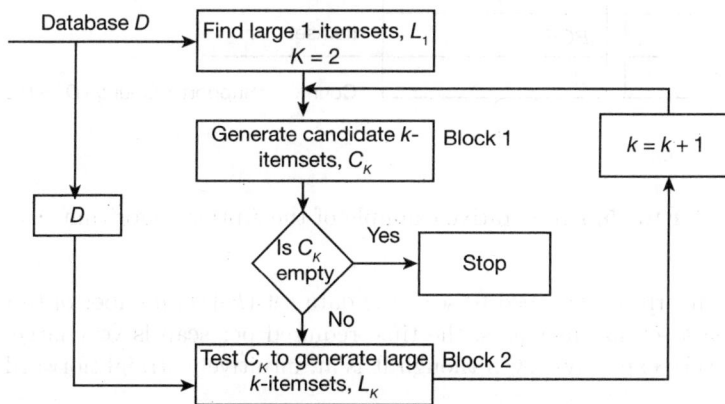

FIG. 5.3 Apriori algorithm

Note that Block 1 in Fig. 5.3 is called the **generation phase**; Block 2 is called the **test phase**. In the generation phase, k-candidate itemsets are generated from the previous stage ($k - 1$)large/frequent itemsets. In the test phase, the data set is scanned to check whether the candidate itemsets are large (frequent) or not based on the user-defined minimum support (σ).

EXAMPLE 3: Figure 5.4 shows an example for the Apriori algorithm. Figure 5.4 (a) shows the transaction data set ($\mathcal{D}$) with $\sigma = 0.5$. Figure 5.4 (b) shows the candidate generation step followed by the test phase to check whether the generated candidate itemsets are frequent or not.

TID	Items purchased
100	A C D
200	B C E
300	A B C E
400	B E

Transaction data set (with minimum support (σ) = 2/$|D|$ = 0.5)

(a)

Candidate 1-itemsets				
{A}	{B}	{C}	{D}	{E}
2	3	3	1	3

Large 1-itemsets				
{A}	{B}	{C}	{E}	Itemset
2	3	3	3	Count

Support = Count/$|D|$ > 0.5

Candidate 2-itemsets					
{AB}	{AC}	{AE}	{BC}	{BE}	{CE}
1	2	1	2	3	2

Large 2-itemsets				
{AC}	{BC}	{BE}	{CE}	Itemset
2	2	3	2	Count

Support = Count/$|D|$ > 0.5

Candidate 3-itemsets
{BCE}
2

Large 3-itemsets	
{BCE}	Itemset
2	Count

Support = Count/$|D|$ > 0.5

(b)

FIG. 5.4 Illustrative example of the Apriori algorithm

In the Apriori algorithm, we need to scan the data set ($k + 1$) number of times for k-frequent itemsets. As the data set size increases, the time required per scan is very large. This makes the Apriori algorithm very expensive, even though it is an intuitively straightforward algorithm.

5.5 FREQUENT PATTERN TREE AND VARIANTS

It has been observed that in the Apriori algorithm, the following are the two main issues:

- Generation of k-candidate itemsets from large $(k-1)$ itemsets. For example, with 10^4 large 1-itemsets, the number of 2-itemset candidates is about 10^7 which is very large.
- The number of data set or database scans. To mine frequent k-itemsets, we need to scan the database $(k+1)$ times. With huge databases, each scan will take a considerable amount of time, thus making the Apriori algorithm computationally expensive.

These issues are handled effectively by making use of the tree data structure, frequent pattern (FP) tree. In this section, we discuss the formation of the FP tree and how this structure can be used for generation of all large itemsets.

5.5.1 FP Tree-Based Frequent Itemset Generation

The frequent pattern tree (FP tree) is generated from the transactions in the database. It is a compressed tree configuration that aids in discovering associations between items within the transaction database. Essentially, the existence of certain items in a transaction indicates the likelihood of other items co-occurring within that same transaction.

The methodology for construction of an FP tree and generation of itemsets for the FP tree is discussed in the following steps.

Step 1: Scan the database to find the frequency (support count) of each item. Discard any item whose frequency is less than the minimum user-defined support (σ). The resulting set is the frequent 1-itemset, L_1.

Step 2: Sort L_1 in descending order of frequency.

Step 3: Construct the FP tree by adding the transactions one by one, as follows:

  a) Start with an empty root node.

  b) In each transaction, filter out the infrequent items and sort the remaining items using L_1 in descending order of frequency. Let it be t.

  c) For each item in t, check if it is already a child of the current node. If yes, increment the frequency of the existing node; if not, add a new node for the item.

  d) After processing all the items in t, repeat the process from Step 3b) for the next transaction.

Step 4: For each frequent item in L_1, starting from the least frequent item, construct its conditional pattern base and conditional FP tree as follows:

  a) Starting from the bottom of the FP tree, trace the path from the item node to the root node, and for each node on the path (excluding the item node itself), collect the set of items in the path as well as the frequency of the node.

  b) Consider the item node and its corresponding transactions from the FP tree to obtain the conditional pattern base.

  c) Construct the conditional FP tree by applying the same steps as in the main FP tree construction algorithm, but using the conditional pattern base as the input database.

Step 5: Recursively mine the conditional FP tree for each frequent item to generate all frequent itemsets containing the current itemset and the frequent item.

Step 6: Remove the frequent item from the current itemset and repeat the process with the next frequent item.

Step 7: Combine all the frequent itemsets obtained from the recursive mining of the conditional FP trees to obtain the final set of frequent itemsets.

EXAMPLE 4: Consider the transaction database ($\mathcal{D}$) shown in Table 5.3 which is the same as that of the transaction data set ($\mathcal{D}$) considered in Fig. 5.4 (a) with $\sigma = 0.5$ (the user-defined support count is 2). The frequent 1-itemset for this example is given in Table 5.4.

TABLE 5.3 An example of a transaction database

Transaction ID	Items purchased
t_1	A, C, D
t_2	B, C, E
t_3	A, B, C, E
t_4	B, E

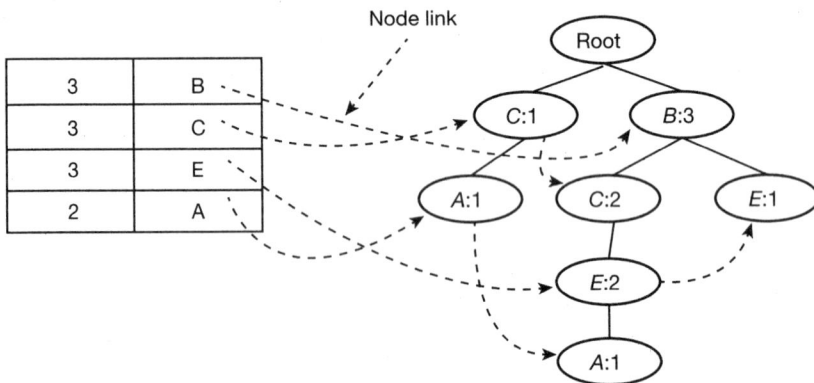

FIG. 5.5 FP tree with node link from L_1

TABLE 5.4 L_1 in descending order of frequency with lexicographic order for breaking ties

Item	B	C	E	A
Frequency	3	3	3	2

The FP tree constructed with the node links from L_1 is shown in Fig. 5.5. The conditional pattern base, conditional FP tree and frequent pattern generated are shown in Table 5.5.

TABLE 5.5 Conditional pattern base, conditional FP tree and frequent patterns

Item	Conditional pattern base	Conditional FP tree	Frequent patterns generated
A	{{C:1}, { E:1, C:1, B:1}}	{{C:2}}	{A,C:2}
E	{{C:2, B:2}, {B:1}	{C:2, B:3}	{C,E:2}, {B,E:3}, {B,C,E:2}
C	{B:2}	{B:2}	{B,C:2}
B	-	-	-

In Table 5.5, in the first row, since E and B have frequency 1 which is less than σ (or user-defined support count, 2), they are not considered; {C:1} and {C:1} are merged to {C:2} in the conditional FP tree. In the second row, {B:2} and {B:1} are merged to {B:3} in the conditional FP tree. The final frequent patterns (with number of itemsets > 1) are listed in the third column of Table 5.5.

Note that the number of database or data set scans using the FP tree-based frequent itemset generation algorithm is 2. That is, one database scan is for generating large 1-itemsets and one more to construct the FP tree.

Later, the FP tree (abstraction of the database) is scanned to generate all the frequent itemsets. Since the number of frequent items and, thereby itemsets, is much less than the number of items and candidate itemsets, the abstraction generated is a compressed representation of the database.

Is it possible to construct the abstraction (compressed representation of database) with fewer than two database scans? The answer is yes; the PC tree structure discussed in the following section shows how such abstraction can be constructed using one database scan and how such a structure can be used for efficient classification and clustering problems.

5.5.2   Pattern Count (PC) Tree-Based Frequent Itemset Generation

Pattern count tree, PC tree for short, is a data structure which is used to store all the patterns that occur in the transaction of a transaction database, where a count field is associated with each item in every pattern.

In the PC tree, all the patterns in the transaction database are stored. The count field is responsible for a compact realization of the database. Each node of the tree consists of the following structure: item-name, count and two pointers called child (c) and sibling (s). Figure 5.6 shows the PC tree node structure.

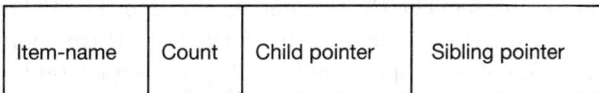

Item-name	Count	Child pointer	Sibling pointer

FIG. 5.6 PC tree node structure

In the node, the *Item-name* field specifies which item the node represents, the *Count* field specifies the number of transactions represented by a portion of the path reaching this node, *Child pointer* (c-pointer) represents the pointer to the following pattern; and *Sibling pointer* (s-pointer) points to the node which indicates the subsequent other patterns from the node under consideration.

Construction of a PC Tree:

1. Construct the root of the PC tree, T.
2. For each transaction t in the transaction database DB:

> Begin
> $prnt = T$
> For each item i in t:
> > Begin
> > If any one of the child nodes, say, *chld* of *prnt*, has *Item-name* = item number of i
> > - Increment the *Count* field value of *chld*.
> > - Set *prnt = chld*.
> > Else
> > - Create a new node, *nw*.
> > - Set the *Item-name* field of *nw* to the item number of i.
> > - Set the *Count* field of *nw* to 1.
> > - Attach *nw* as one of the child nodes of *prnt*.
> > - Set *prnt = nw*.
> > End.
> End.

The following example explains the construction of a PC tree.

EXAMPLE 5: Consider the transaction database shown in Table 5.6. We use the equivalent

TABLE 5.6 Transaction database, D

Transaction ID	Item numbers of items purchased (transactions)
t_1	19, 40, 125, 179, 510, 527, 790, 795, 799
t_2	19, 40, 179
t_3	19, 40, 125, 510, 520
t_4	527, 740
t_5	527, 740, 795
t_6	510, 527, 790, 795

binary tree representation to represent the PC tree and its variants. The PC tree for the transaction database, D, is shown in Fig. 5.7 (a). Figure 5.7 (b) shows the FP tree, for the same transaction database D shown in Table 5.6 with support value = 3. [The header table and the link between the same nodes in the tree are not shown for simplicity].

The PC and FP trees share the following properties:

- Let $A_1, A_2, \ldots, A_k$ be the item names of nodes in a path of a PC/FP tree. Let $C_1, C_2, \ldots, C_k$ be the values in the *Count* fields of the above nodes. Then, $C_1 \geq C_2 \cdots \geq C_k$.
- Let the ordered set $\mathfrak{A} = \{A_{i1}, A_{i2}, \ldots, A_{il}\}$ correspond to a path in the PC/FP tree from the root till some leaf node A_{il}. Then, any subset of $\mathfrak{A}$ of size k $(\leq l)$ is a k-itemset.
  For example, corresponding to the path $\{510, 527, 790, 795\}$ in the PC tree shown in Fig. 5.7 (a), the possible 3-itemsets are $\{510, 527, 790\}$, $\{510, 527, 795\}$, $\{510, 790, 795\}$, $\{527, 790, 795\}$.

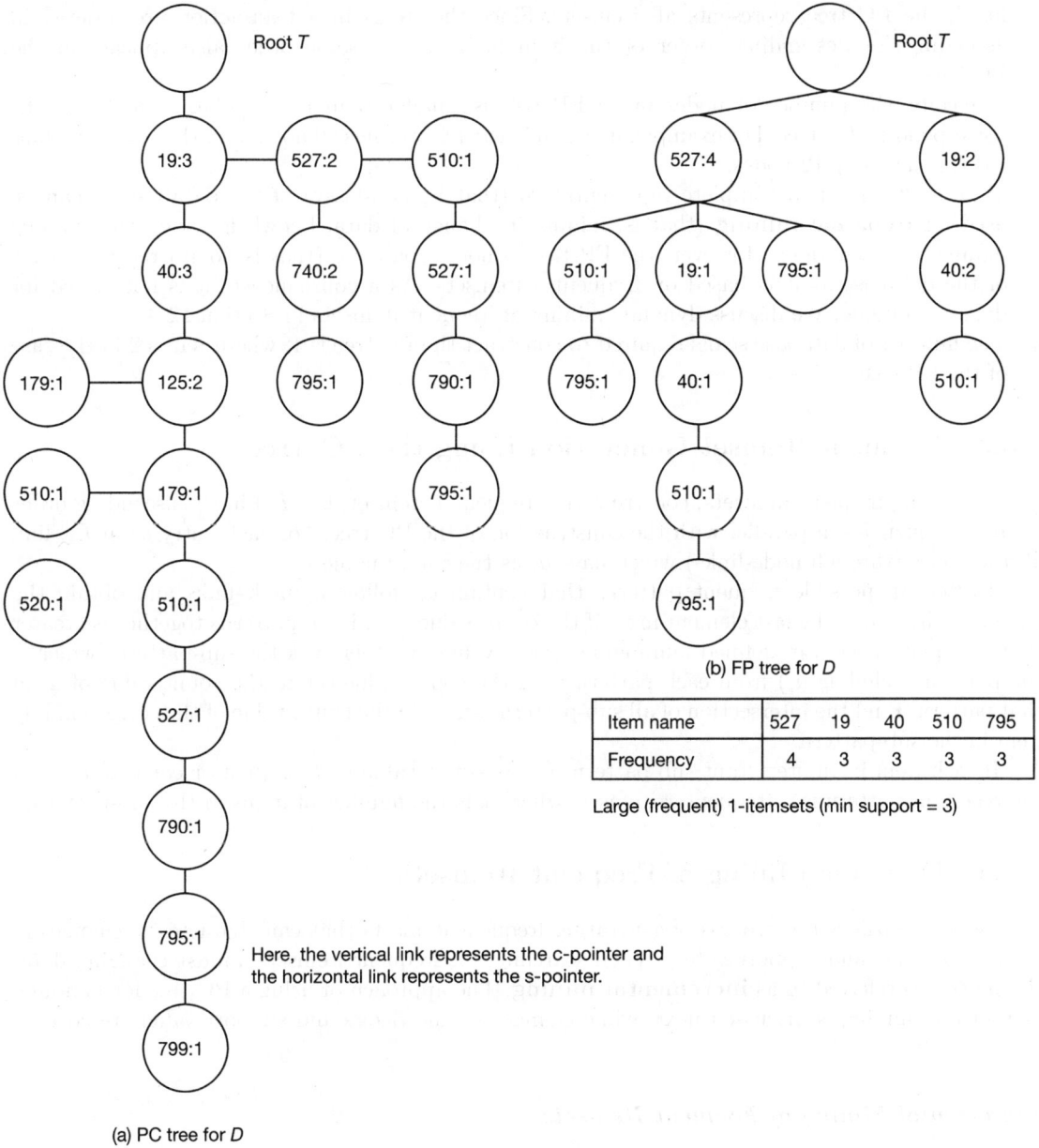

(b) FP tree for D

Item name	527	19	40	510	795
Frequency	4	3	3	3	3

Large (frequent) 1-itemsets (min support = 3)

Here, the vertical link represents the c-pointer and the horizontal link represents the s-pointer.

(a) PC tree for D

FIG. 5.7 Tree structures for storing patterns

The differences between the PC and FP trees are as follows:

- A pattern in an FP tree is based on frequent 1-itemsets while the ordering of items in the transaction is unimportant. The items in the patterns are ordered in support-descending order. More frequently occurring items are arranged closer to the root of the FP tree. On the other hand, the PC tree represents all itemsets. Since the items in a transaction are ordered in ascending (or descending) order of the item number, the same order also appears in the PC tree.

- Typically, the number of nodes in an FP tree is smaller than the number of nodes in the corresponding PC tree. For example, in Fig. 5.7, the PC tree has 20 nodes and the corresponding FP tree has only 12 nodes.
- Since a PC tree is a complete representation (that is, non-lossy) of the database, it can be used for **dynamic mining** (that is, mining in change of data, knowledge and other mining parameters scenarios). However, the FP tree is not a complete (that is, lossy) representation of the database as it is based on frequent 1-itemsets. As a consequence, it is not suited for dynamic mining. We discuss dynamic mining of frequent itemsets in Section 5.5.4.
- The number of database scans required to construct the PC tree is 1, whereas it is 2 in the case of the FP tree.

5.5.3   Frequent Itemset Generation Using the PC Tree

For generating frequent itemsets, construct the frequent 1-itemset list (L_1) in the ascending order of their frequencies in parallel with the construction of the PC tree. For each entry a_i in L_1, link all the nodes (through node-links) which have a_i as their item name.

To find all possible frequent patterns that contain a_i, follow a_i node-links and obtain the patterns where a_i is the last element in it. If the count value of a_i in all patterns together is greater than or equal to the user-defined minimum support value (σ), then pick the sub-pattern (which is the pattern excluding a_i) from each pattern with the count value set to the count value of a_i in that pattern. Find the intersection of all sub-patterns and add the count value of the corresponding item in the sub-pattern.

If each item in the resultant sub-pattern has its count value greater than or equal to σ, then the resultant pattern is a frequent x-itemset, where x is the number of items in the sub-pattern.

5.5.4   Dynamic Mining of Frequent Itemsets

This section outlines the process of generating frequent itemsets that can characterize changes in data, knowledge and support value. When searching for frequent itemsets amidst changing data, the process is referred to as **incremental mining**. The approach of using a PC tree for handling incremental mining is discussed next, while changes in knowledge and support value are covered subsequently.

Incremental Mining of Frequent Itemsets

Incremental mining is a technique used for generating frequent itemsets in a dynamic transaction database, which undergoes continuous updates such as addition or deletion of transactions. This technique is challenging because these updates may invalidate existing frequent itemsets or make some previously infrequent itemsets become frequent.

Real-life databases such as those used in supermarkets and reservation systems are dynamic in nature and require constant updates. In this context, the PC tree has been shown to be a suitable structure for incremental mining of frequent itemsets, as it handles both addition and deletion of transactions.

The three operations that can result in changes to a database are adding a new transaction, deleting an existing transaction and modifying an existing transaction by deleting some of its items.

Incremental mining algorithms must handle addition and deletion of transactions. The PC tree supports incremental mining by changing the structure of the tree as new transactions are added PC-tree without the need to scan the database again. Changes to the PC tree are minor and involve updating the count value of some existing nodes or adding new nodes.

For example, consider the addition of transaction, t_7, given in Table 5.7 to the database shown in Table 5.6.

TABLE 5.7 Transaction, t_7

Transaction ID	Item numbers of items purchased (transactions)
t_7	527, 740, 900

The inclusion of t_7 in the PC tree shown in Fig. 5.7 (a) leads to changes in the path below node $C1$. The modified path below C_1 is shown in Fig. 5.8.

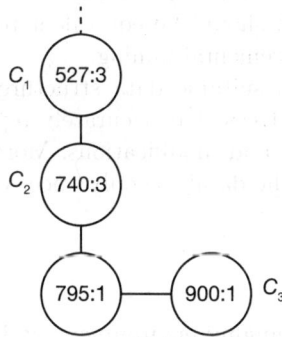

FIG. 5.8 Modified path below C_1 of Fig. 5.7 (a)

Observe that there is both a change in the count value of nodes C_1 and the C_2 and the addition of node C_3 as a child node of C_2.

Consider the deletion of transaction t_5 from the PC tree shown in Fig. 5.7. We need to nullify the effect of t_5. This is achieved by decrementing the count value of the nodes having item numbers 527 and 740 and deleting the node having item number 795, since its modified count value is 0.

Now consider the deletion of item 527 from transaction t_6. This is achieved by deleting transaction t_6 : {510, 527, 790, 795} and adding the new transaction t_6' :{510, 790, 795} to the PC tree shown in Fig. 5.7. This leads to changes in the path below node C_4 in Fig. 5.7. The modified path below C_4 is shown in Fig. 5.9.

Observe that the node with item number 527 in the path below node C_4 of Fig. 5.7 does not exist in the modified path shown in Fig. 5.9.

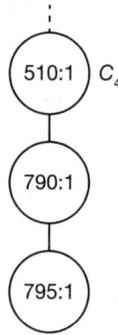

FIG. 5.9 Modified path below node C_4 of Fig. 5.7 (a)

The process of dynamic updation involves updating a database, denoted as DB, to a new database denoted as $D\acute{B}$. This is achieved by adding a set of transactions, denoted as db_1, while simultaneously deleting another set of transactions, denoted as db_2, from the original database.

Incremental mining, on the other hand, requires updating the frequent-itemset equivalent representation, denoted as R, of DB to a new representation, denoted as $\acute{R}$, using only R, db_1 and db_2. The resulting representation, $\acute{R}$, should be equivalent to $D\acute{B}$ in terms of frequent itemsets, and is considered appropriate for incremental mining.

In this regard, the PC tree is a suitable data structure for incremental mining because it enables the creation of updated PC trees that accurately represent modified databases resulting from transaction additions, deletions and modifications. Moreover, the PC tree can be efficiently updated incrementally by scanning the database only once, which makes it a highly efficient tool for incremental mining.

Change in Knowledge

In order to mine in a constantly changing environment, it is important to be able to adapt to changes in knowledge representation during the mining process. This is primarily done through an 'is_a' hierarchy, represented as a tree structure.

This means that when knowledge changes, existing nodes in the tree may need to be deleted or new nodes created, and children nodes may need to be moved within the hierarchy. An example of this is shown in Fig. 5.10.

If K_1 represents the current knowledge hierarchy and PC is the PC tree for the database D, a generalized PC tree (GPC tree) can be constructed by applying K_1 to the transactions obtained from PC, without accessing D.

If the knowledge representation changes from K_1 to K_2, a new generalized PC tree can be constructed by applying K_2 to PC, without accessing D. Therefore, the PC tree is capable of handling dynamic knowledge.

Change in Support Value

The support value used in mining processes to discover frequent itemsets is often arbitrarily fixed since users are unable to determine the appropriate support value for discovering these unknown patterns. Although the mining process aims to find unknown patterns, not all of these patterns

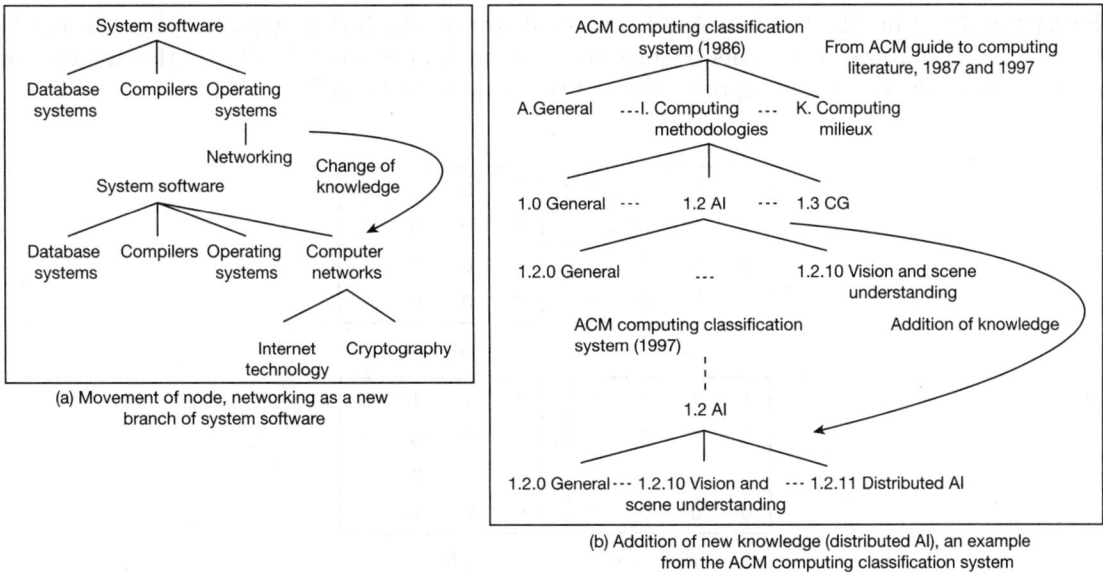

FIG. 5.10 Examples of change in knowledge

are useful and some may even be trivial. As a result, certain algorithms introduce variations in the user-defined support value to mine patterns. The PC tree, which is comprehensive in terms of the database, is capable of handling this situation as well.

5.6 CLASSIFICATION RULE MINING

There are two significant techniques in data mining, namely, classification rule mining and association rule mining.

Association rule mining aims to discover all the rules in a database that meet minimum support and confidence requirements, without a predetermined target. Association rule mining was discussed in the previous section. On the other hand, classification rule mining has a pre-defined target, that is, the class, which is the only target.

Let us assume that D is the data set, I is the set of all items in D and Y is the set of class labels. A **class association rule (CAR)** follows the form $X \Longrightarrow y$, where X belongs to I and y belongs to Y.

A rule $X \Longrightarrow y$ is said to hold in D with a confidence of $c\%$ if $c\%$ of the cases in D that contain X are labelled with class y.

The rule $X \Longrightarrow y$ has support s in D if $s\%$ of the cases in D contain X and are labelled with class y. If the support of the CAR is greater than a user-defined support (σ), the rule is considered *frequent*. If the confidence is greater than a user-defined confidence (c), the rule is deemed *accurate*.

EXAMPLE 6: Age $< 25 \wedge$ credit $=$ 'good' $\Longrightarrow$ iPhone (sup=30%, conf=80%) is a generic example of a classification association rule.

EXAMPLE 7: Consider the four 4×4 squares shown in Fig. 5.11 to represent digits 1 and 7, where '1' in each square represents the existence of a black pixel and '0' indicates the existence of a white pixel. The row-major representation of the same is shown in Fig. 5.12.

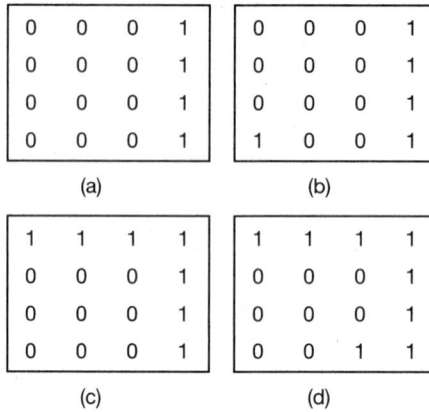

0	0	0	1
0	0	0	1
0	0	0	1
0	0	0	1

(a)

0	0	0	1
0	0	0	1
0	0	0	1
1	0	0	1

(b)

1	1	1	1
0	0	0	1
0	0	0	1
0	0	0	1

(c)

1	1	1	1
0	0	0	1
0	0	0	1
0	0	1	1

(d)

FIG. 5.11 Digits '1' and '7'

Figure label	Row-major representation
(a)	0001000100010001
(b)	0001000100011001
(c)	1111000100010001
(d)	1111000100010011

FIG. 5.12 Row-major representation of digits '1' and '7'

In this example, $\{4, 8, 12, 16\} \implies$ digit:'1' and $\{1,2,3,4,8,12,16\} \implies$ digit:'7' are possible classification association rules. Note that the antecedent part of the rule shows the positional values of the pixel '1' as depicted in Table 5.8.

TABLE 5.8 Positional representation of a pixel in a 4×4 table

1	2	3	4
5	6	7	8
9	10	11	12
13	14	15	16

If the support count of {4, 8, 12, 16} is 2 and the support count of digit '1' is 3, and the number of cases in D is 10, then the support of the rule {4, 8, 12, 16} $\Longrightarrow$ digit:'1' is 20% and the confidence is 66.7%. If the user-defined minimum support (σ) is 10%, then the rule is *frequent* and if the user-defined confidence (c) is 60%, then the rule is *accurate*.

5.7 FREQUENT ITEMSETS FOR CLASSIFICATION USING PC TREE

The PC tree can be used to solve pattern classification problems. The PC tree is constructed using labelled training data and then the test data is checked against each path of the PC tree to determine its class label. The example below shows how the training patterns can be stored in a PC tree.

EXAMPLE 8: Let us consider the patterns for digits '0' and '9' shown in Fig. 5.13 (a).

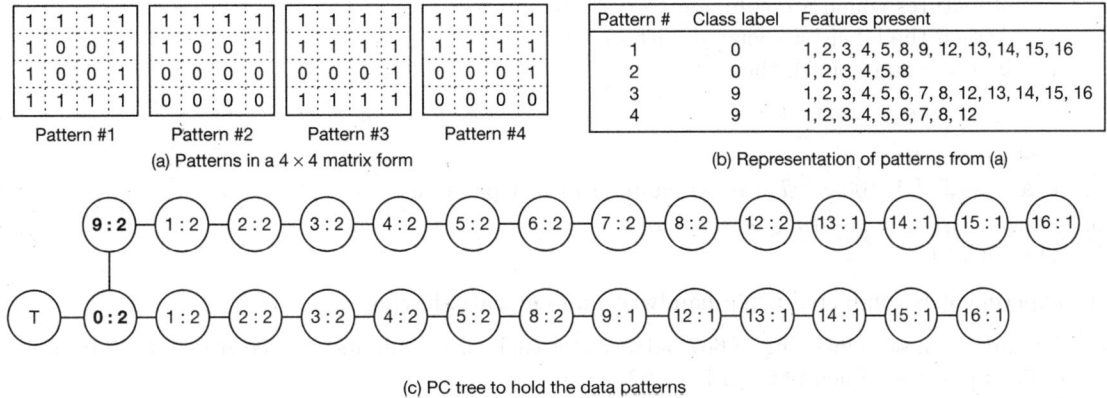

(a) Patterns in a 4 × 4 matrix form

Pattern #	Class label	Features present
1	0	1, 2, 3, 4, 5, 8, 9, 12, 13, 14, 15, 16
2	0	1, 2, 3, 4, 5, 8
3	9	1, 2, 3, 4, 5, 6, 7, 8, 12, 13, 14, 15, 16
4	9	1, 2, 3, 4, 5, 6, 7, 8, 12

(b) Representation of patterns from (a)

(c) PC tree to hold the data patterns

Note: In (c), the horizontal link represents the c-pointer and the vertical link represents the s-pointer.
Each node has two parts: item name and count value.

FIG. 5.13 Data patterns for '0' and '9'

Here, pattern 2 is a noisy version of '0' and pattern 4 is a noisy version of '9'. Each pattern is a matrix of size 4 × 4. A '1' in a location indicates the presence and a '0' indicates the absence of the feature. Figure 5.13 (b) gives an equivalent representation of the patterns using features that are present. Features are numbered row wise. Figure 5.13 (c) shows the PC tree for the corresponding data patterns based on class labels followed by the features present. Note that even though there are four tuples in Fig. 5.13 (b) the corresponding PC tree (Fig. 5.13 (c)) has only two branches. This property of the PC tree will be helpful in the following ways:

- It abstracts the number of training patterns to be considered for classifying the test patterns which in turn reduces the space and time requirements.
- It decreases improper classification of test patterns by not considering those like Pattern#2 and Pattern#4 which are completely subsumed by Pattern#1 and Pattern#3, respectively.

We give the classification algorithm based on the PC tree (PC classifier) below:

INPUTS:

1. T_S - Test data set (A collection of test patterns)
2. T_R - Training data set (A collection of training patterns)

OUTPUTS:

1. C_{time} - Classification time
2. CA - Classification accuracy

STEPS:

1. Generate a PC tree using T_R (refer Section 5.5.2 for details).
2. *start-time = time().*
3. For each $s_i \in T_S$

   a. Find n_i, the set of positions of non-zero values corresponding to s_i.
   b. Find the nearest neighbor branch e_k in the PC tree depending on the maximum number of features which are common to both e_k and n_i.
   c. Attach the label l associated with e_k to n_i.
   d. If ($l ==$ label of s_i), then

$$correct = correct + 1$$

4. *end-time = time().*
5. $CA = \frac{correct}{|T_S|} \times 100$ // $|T_S|$ is the number of test patterns.
6. Output $C_{time} = end\text{-}time - start\text{-}time$.
7. Output CA.

An experiment is conducted using handwritten digit data, having

- Training data set consisting of 667 patterns for each class of digits labelled from 0 to 9, totalling to 6670 patterns. Each pattern has 192 features.
- Testing data set consisting of 3333 patterns.

The classification accuracy using the PC classifier is 93.61%.

5.8 FREQUENT ITEMSETS FOR CLUSTERING USING THE PC TREE

The PC tree can be used for clustering data in a direct way. Here, a **cluster description** is the set of features along the path from the root to the leaf node of the PC tree. All the patterns that are mapped onto the path form the members of the cluster.

The PC tree requires less space to hold the clusters for two reasons: (i) if two or more patterns share a prefix, P, then P is stored *only once*. (ii) only non-zero positional values are stored in the PC tree.

A PC tree-based clustering algorithm (PC cluster) is given below:

STORE:

For each pattern, $P_i \in D$ (database):

If any prefix sub-pattern, SP_i, of P_i exists as a prefix in a branch e_b:

Put features in SP_i in e_b, by incrementing the corresponding *Count* field value of the nodes in the PC tree. Put the sub-patterns, if any, of P_i by appending additional nodes with the *Count* field value equal to 1 to the path in e_b.

Else Put P_i as a new branch of the PC tree.

RETRIEVE:

For each branch, $B_i \in PC$ (PC tree):

For each node, $N_j \in B_i$:

Output N_j.Feature.

The features corresponding to the nodes in the branch of the PC tree, from root to leaf, constitute the prototype.

EXAMPLE 9: Consider the patterns for the digits '1' and '7' as shown in Fig. 5.14 (a).

Pattern #1

0	0	1	0
0	0	1	0
0	0	1	0
0	0	1	0

Pattern #2

1	1	1	0
0	0	1	0
0	0	1	0
0	0	1	0

Pattern #3

0	0	1	0
0	0	1	0
0	0	1	0
0	1	1	1

Pattern #4

1	1	1	0
0	0	1	0
0	0	1	0
0	0	0	0

Pattern #5

0	0	1	0
0	0	1	0
0	0	1	0
0	1	1	0

Pattern #6

0	0	1	0
0	0	1	0
0	0	1	0
0	0	0	0

(a) Patterns in a 4 × 4 matrix form

Pattern #	Class label	Features present
1	1	3, 7, 11, 15
2	7	1, 2, 3, 7, 11, 15
3	1	3, 7, 11, 14, 15, 16
4	7	1, 2, 3, 7, 11
5	1	3, 7, 11, 14, 15
6	1	3, 7, 11

(b) Representation of patterns from (a)

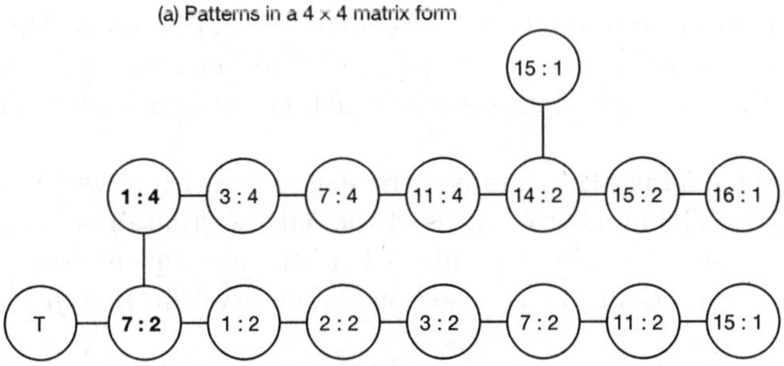

(c) PC tree to hold the data patterns

Note: In (c), the horizontal link represents the child-pointer and the vertical link represents the sibling-pointer. Each node has two parts: item name and count value.

FIG. 5.14 Data patterns for '1' and '7'

Each pattern is a matrix of size 4×4. A '1' in a location indicates the presence and a '0' indicates the absence of the feature. Figure 5.14 (b) gives an equivalent representation of the patterns using features that are present. Figure 5.14 (c) shows the PC tree for the corresponding data patterns. Here, there are three clusters and the cluster descriptions are {3,7,11,14,15,16,}, {3,7,11,15}, {1,2,3,7,11,15}. In the case of the PC tree, the cluster description itself forms the prototype. Members of the cluster, with description {3,7,11,14,15,16}, are {3,7,11,14,15,16}, {3,7,11,14,15} and {3,7,11}.

An experiment is conducted using handwritten digit data, having

- Training data set consisting of 667 patterns for each class of digit labelled from 0 to 9, totalling to 6670 patterns. Each pattern has 192 features.
- Testing data set consisting of 3333 patterns.

PC cluster took 62% of the time of Leader to construct the clusters and the clusters generated by PC cluster occupied 31% of the space occupied by the clusters generated by Leader.

SUMMARY

In this chapter, we discussed co-occurrences of items for finding interesting rules. Some important issues discussed are as follows:

- Co-occurrences of items play an important role in finding interesting patterns. Association rules are based on co-occurrences of items in a large data set, where the support and confidence are greater than the user-defined support and confidence values.
- Itemsets with support greater than the user-defined minimum support are frequent itemsets. Among the frequent itemsets, association rules are constructed whose confidence is greater than the user-defined confidence value.
- The Apriori algorithm is a classic frequent itemset mining algorithm. During its 'generate' phase, it generates all possible candidate k-itemsets from frequent $(k-1)$-itemsets. During the 'test' phase, the algorithm checks whether the generated candidate itemsets are frequent or not. The number of scans of the data set (database) is k if there are $(k-1)$-frequent itemsets.
- The FP tree-based frequent itemset mining algorithm is a tree-based algorithm, which does not generate any candidate itemsets. The number of scans of the data set (database) is 2.
- Classification association rules are special association rules where the consequent part contains a class label. Using frequent itemsets in ML is based on the observation that *frequent itemsets are discriminative*.
- The PC tree is an abstraction of the database, which is amenable to frequent itemset mining. Construction of the PC tree requires one database scan. Further, it can be used for generating classification rules and clustering.
- Dynamic mining is a notion of handling change in data, knowledge and user-defined parameters (like support value) for discovering frequent itemsets. The PC tree is an appropriate abstraction for performing dynamic mining.
- Frequent itemset-based ML models are highly amenable for explanation. This is because each item set will be dealing with a binary attribute and its value.

EXERCISES

1. Consider the transaction data set shown in the table below.

Transaction database, D

Transaction ID	IDs of the items
t_1	A, B, E
t_2	B, D
t_3	B, C
t_4	A, B, D
t_5	A, C
t_6	B, C
t_7	A, C
t_8	A, B, C, E
t_9	A, B, C

   a. With $\sigma = 2$ (here we are referring to absolute support, since we are using support count), what are the large 1-itemsets?
   b. Using the Apriori property, generate and prune candidate 2-itemsets from large 1-itemsets.
   c. With the items shown in the table, what are the maximum number of candidate itemsets possible?
   d. For the transaction database shown in the table, generate all frequent itemsets using the Apriori algorithm.

2. Consider the transaction database shown in the table in Q1 with the frequent itemset, $L_3 = \{A, B, E\}$. List all of its non-empty frequent itemsets. List all the corresponding association rules, where c is (i) 50% (ii) 33% (iii) 29% and (iii) 100%.

3. Consider the transaction data set shown in the table in Q1. Construct an FP tree with $\sigma = 2$.

4. For the FP tree generated in Q3, construct a conditional pattern tree, conditional FP tree and frequent patterns.

5. For the transaction data set shown in the table in Q1, construct a PC tree.

6. For the PC tree constructed in Q5, generate cluster descriptions for $\sigma = 2$.

7. For the transaction database D shown in the table in Q1, the conceptual hierarchy (is_a tree) is given in the figure below. Construct the generalized PC tree (GPC tree) using the PC tree constructed in Q5.

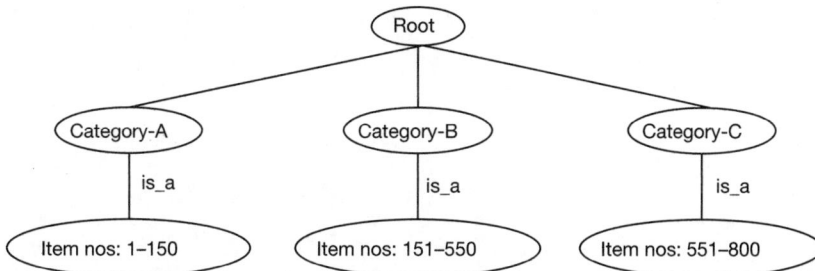

Conceptual hierarchy (is_a tree)

8. Which of the following activities requires the least number of database scans:

    a. Construction of an FP tree
    b. Finding the mean
    c. Perceptron learning algorithm
    d. k-means algorithm

9. Let c_1, c_2 and c_3 be the confidence values for the association rules $p \implies q$, $p \implies \{q, r\}$ and $\{p, r\} \implies q$, respectively. The possible relationships that may exist among c_1, c_2 and c_3 are:

    a. $c_1 == c_2$ and $c_1 == c_3$
    b. $c_1 < c_2$ and $c_2 < c_3$
    c. $c_2 \leq c_1$ and $c_2 \leq c_3$
    d. $c_3 \leq c_2$ and $c_2 \leq c_1$

10. Given that A is a frequent itemset, one of the following is false:

    a. Every $B(\subseteq A)$ is frequent
    b. There is a $B(\subseteq A)$ which is infrequent
    c. There is at least one $B(\supset A)$ which is frequent
    d. B is infrequent if it is a superset of an infrequent set

Bibliography

- Han, J, Kamber, M and Pei, J. 2012. *Data Mining Concepts and Techniques*, Morgan Kaufmann Publishers, Waltham.

- Murty, MN and Susheela Devi, V. 2015. *Introduction to Pattern Recognition and Machine Learning*, World Scientific Publishing Co. Pte. Ltd.: Singapore.

- Ananthanarayana, VS, Narasimha Murty, M and Subramanian, DK. 2011. An Incremental Data Mining Algorithm for Compact Realization of Prototypes, *Pattern Recognition Journal*, 34(11): 2249–2251. http://www.sciencedirect.com/science/article/pii/S0031320301000280

- Ananthanarayana, VS, Narasimha Murty, M and Subramanian, DK. 2001. Efficient Clustering of Large Data Sets, *Pattern Recognition Journal*, 34(12): 2561–2563. http://www.sciencedirect.com/science/article/pii/S0031320301000978

- Ananthanarayana, VS, Subramanian, DK and Narasimha Murty, M 2000. Scalable, Distributed and Dynamic Mining of Association Rules, *International Conference on High Performance Computing*, p. 559–566. https://link.springer.com/chapter/10.1007/3-540-44467-x_51

CHAPTER 6
Representation

Learning Objectives

At the end of this chapter, you will be able to:

- Explain how data items, classes and clusters are represented
- Describe schemes for dimensionality reduction
- Explain feature selection and feature extraction

6.1 INTRODUCTION TO REPRESENTATION

The most crucial step in ML is representation. We have seen in the earlier chapters how the performance of various classifiers can be affected when the dimensionality of the data is large. The problems faced by various classifiers are as follows:

- KNN and its variants require a large training data; typically, the number of training patterns is 5–10 times the dimensionality of the data. Even though fractional norms can be used to improve the situation, the computational burden can be huge.
- DTC is not suitable for high-dimensional data. Even DTs based on axis-parallel splits can face difficulties in dealing with high-dimensional data. Solutions based on random forest and gradient boosting can come to our rescue; however, performance may improve if large-scale training data is available.
- The Bayes classifier and its popular variant NBC can suffer if the dimensionality is large. For example, NBC needs to estimate L probabilities for each class, where L is the dimensionality of the data. Further, these estimates can be erroneous if the number of training data points is small.
- Building a PC tree or computing frequent itemsets using the Apriori algorithm can be time consuming when the dimensionality is large. The height of the PC tree can increase, leading to memory problems; the number of database scans used by the Apriori algorithm can increase and may lead to huge computational burden.
- Even the neural network-based schemes that we are going to study in later chapters need large-scale data to deal with high-dimensional applications. In the presence of large-scale data, these models may overfit and one needs to impose suitable constraints on the learning algorithms.

So, in general, dimensionality reduction seems to be an attractive solution to several of these problems in ML. These schemes may be divided as follows:

- **Feature selection:** Here, given a set $\mathcal{F}_L$ of L features $f_1, f_2, \ldots, f_L$, the problem is to select a subset $\mathcal{F}_l$ of l features. So, $\mathcal{F}_l \subset \mathcal{F}_L$, where $l < L$. So, we select these l features based on some criterion. Note that no new features are considered; $\mathcal{F}_l$ is a collection of some of the given features only.
- **Linear feature extraction:** This is a popular scheme for dimensionality reduction. Here, a feature g_j is extracted such that

$$g_j = \sum_{i=1}^{L} \alpha_{ij} f_i,$$

where α_{ij} is a real number and it indicates the contribution of feature f_i to the extracted feature g_j. Note that these g_j are typically different from f_i; they are extracted using a linear combination of the given features. Further, we extract l features $g_1, g_2, \ldots, g_l$ such that $l < L$.
- **Non-linear feature extraction:** Here, we extract new features $h_1, h_2, \ldots, h_l$ using features in $\mathcal{F}_L$ such that $l < L$. Typically $h_j = f(f_1, f_2, \ldots, f_L)$, for $j = 1, 2, \ldots, l$, where f is a non-linear function and $l < L$. A popular scheme of current interest is the **autoencoder** that is based on *neural networks*. We discuss this approach in a later chapter where we provide some background on neural networks.

6.2 FEATURE SELECTION

Feature selection schemes can be categorised as follows:

- **Filter Methods:** Here, we rank the L features using some criterion that may exploit the class labels, but which does not directly use any classifier in the selection process. Based on the ranking, the top l features are selected. Some of the fitness measures used to evaluate and rank the features are:

  - *Variance:* We prefer a feature that has high variance among its values. We illustrate the idea with the example data shown in Table 6.1.

TABLE 6.1 Example of data set to illustrate the filter method

Pattern	Feature1	Feature2	Feature3	Class
1	0	0	0	C_0
2	6	0	3	C_1
3	1	0	0	C_0
4	5	1	3	C_1
5	6	0	0	C_1
6	0	0	1	C_0

The variances of the three features are:

1. $Var[Feature1] = 7.333$
2. $Var[Feature2] = 0.139$
3. $Var[Feature3] = 1.805$

So, we rank the three features in the order *Feature1*, *Feature3* and *Feature2* based on decreasing variance. If we want to select two features, we select *Feature1* and *Feature3*. This is done with the hope that features with larger variance values are more discriminative.

This is the case with *Feature*1; it 0, 1, 0 for class C_0 and 6, 5, 6 for C_1; the values are well separated between C_0 and C_1.

Similarly, *Feature*3 has values 0, 0, 1 for C_0 and 3, 3, 0 for C_1; the values are not so well separated. However, the mean of C_0 values is $\frac{1}{3}$ and the mean of C_1 values is 2; the means differ by a larger value. So, features with larger variance are more discriminative in this example. This criterion can be used even when the patterns are unlabelled.

— *Dependency:* Here, the correlation between a pair of features is exploited in ranking. If the correlation coefficient is larger between two features, then one of the features is dropped; the feature that has high average correlation with the retained features is dropped. The correlation coefficient, $r_{p,q}$, between two features *Featurep* and *Featureq* is defined as follows:

$$r_{p,q} = \frac{\sum_{j=1}^{n}(Featurep_j - \bar{Featurep}) \times (Featureq_j - \bar{Featureq})}{\sum_{j=1}^{n}(Featurep_j - \bar{Featurep})^2 \times \sum_{j=1}^{n}(Featureq_j - \bar{Featureq})^2},$$

where n is the number of patterns, $Featurep_j$ is the value of *Featurep* for pattern j and $\bar{Featurep}$ is the mean value of *Featurep* across the n patterns.

EXAMPLE 1: Consider again the data in Table 6.1. The correlation coefficient values between the pairs of features are:

$r_{1,2} = 0.207$
$r_{1,3} = -0.517$
$r_{2,3} = 0.61$

If we want to select the best feature out of the three features, then we select the one that has maximum average correlation coefficient with the remaining two features.

*Feature*1: The average is $\frac{(0.207-0.517)}{2} = -0.155$.

*Fetaure*2: The average is $\frac{(0.207+0.610)}{2} = 0.409$

*Feature*3: The average is $\frac{(0.610-0.517)}{2} = 0.047$.

So, the best feature is *Feature*2. However, if we want to select two features, then we drop a feature that is highly correlated with the remaining two features. We have seen that *Feature*2 is highly correlated with *Feature*1 and *Feature*3. So, we drop *Feature*2.

— *Mutual information* (MI:) If a feature and a class have a larger MI value, then the feature is good. It is seen that MI-based feature selection works well on high-dimensional data sets. The MI between two RVs X and Y is defined as

$$MI(X,Y) = \sum_x \sum_y P_{X,Y}(x,y) log \left(\frac{P_{X,Y}(x,y)}{P_X(x) \times P_Y(y)} \right),$$

where $P_{X,Y}(x,y)$ is the joint PMF of X and Y and $P_X(x)$ and $P_Y(y)$ are the marginal PMFs.

It measures the information that X and Y have in common. It indicates by how much the uncertainty of a variable is reduced, after knowing the other variable.

If X and Y are independent, they are uncorrelated; then knowing one variable does not give any information about the other variable. In some cases, their mutual information is zero. We will illustrate this notion using an example.

EXAMPLE 2: Consider the data shown in Table 6.2. There are a total of 60,400 documents in a collection, out of which 400 are from Data Mining (*DM*) class and 60000 are from Computer Science (*CS*) class.

TABLE 6.2 Example of data set to illustrate mutual information

Feature value	Data Mining	Computer Science
Data is Present	60	44000
Data is Absent	340	16000
Mining is Present	350	500
Mining is Absent	50	59500

We have considered two words *Data* and *Mining* and indicated the number of documents in which each of these words is present and the number of documents that do not have each of these words.

For example, in Table 6.2, we have 4 rows. The first row shows that the word *Data* is present in 60 documents of the *DM* class and 44,000 documents of the *CS* class. Similarly, the fourth row indicates that the word *Mining* is absent in 50 documents of the *DM* class and 59,500 documents of the *CS* class. The other two rows can be interpreted similarly.

We can use MI between two features or between a feature and a class. Here, we calculate the MI values between each of the two features, *Data* and *Mining*, and the *DM* class. So, we consider the RV X to be associated with a word and the RV Y to be associated with the class *DM*. Note that both X and Y are binary variables taking one of two values.

In the case of X, it is either the presence (P) or absence (A) of *Data* while RV Y takes present (yes) in *DM* or not (no).

To calculate various probabilities, we will use the MLE scheme that employs ratios of frequency counts:

$$P(X = P, Y = yes) = \frac{60}{60400} = 0.001; \ P(X = P, Y = no) = \frac{44000}{60400} = 0.73$$

$$P(X = A, Y = yes) = \frac{340}{60400} = 0.0056; \ P(X = A, Y = no) = \frac{16000}{60400} = 0.265$$

$$P(X = P) = \frac{44060}{60400} = 0.73; \ P(X = A) = \frac{16340}{60400} = 0.27$$

$$P(Y = yes) = \frac{400}{60400} = 0.0066; \ P(Y = no) = \frac{60000}{60400} = 0.9934$$

When we compute MI, we need to sum over four terms as there are two variations in X (P or A) and two variations in Y (*yes* or *no*). So, when we consider X to be *Data* and Y to be the *DM* class, by referring to the formula for MI, we have

$$MI(Data, DM) = P(X = P, Y = yes) \times log\_2 \left(\frac{P(X = P, Y = yes)}{P(X = P) \times P(Y = yes)} \right)$$

$$+ P(X = A, Y = yes) \times log\_2 \left(\frac{P(X = A, Y = yes)}{P(X = A) \times P(Y = yes)} \right)$$

$$+ P(X = P, Y = no) log\_2 \left(\frac{P(X = P, Y = no)}{P(X = P) \times P(Y = no)} \right)$$

$$+ P(X = A, Y = no) \times log\_2 \left(\frac{P(X = A, Y = no)}{P(X = A) \times P(Y = no)} \right).$$

So,

$$MI(Data, DM) = 0.001 \times log\_2 \left(\frac{0.001}{0.73 \times 0.0066} \right) (= -0.0023)$$

$$+ 0.0056 \times log\_2 \left(\frac{0.0056}{0.27 \times 0.0066} \right) (= 0.0093)$$

$$+ 0.73 \times log\_2 \left(\frac{0.73}{0.73 \times 0.9934} \right) (= 0.0070)$$

$$0.265 \times log\_2 \left(\frac{0.265}{0.27 \times 0.9934} \right) (= -0.0046) = 0.0094.$$

Similarly, we can compute MI(*Mining*, DM); we leave the details to an exercise. So, if we were to select one feature, we would select the one that has a larger MI value.

- **Wrapper Methods:** In this case, the selection of a subset of features is based on the performance of a classifier on the selected subset. The subset of features which gives the maximum classification performance is selected. This subset of features is used in designing the ML model. These schemes can be more expensive compared to the filter-based schemes because we are considering subsets of $\mathcal{F}_L$ here; note that the number of subsets is much larger than the number of features. Some simplifications of this method are given below:

  - *Sequential methods:* Here, features are added sequentially. The steps in the sequential forward scheme (SFS) are as follows:

    a. Initially $\mathcal{F} = \mathcal{F}_L$, $\mathcal{F}_p = \{\}$ and $p = 0$.
    b. Find the best feature, based on some criterion, from $\mathcal{F}$ in conjunction with the existing p features in $\mathcal{F}_p$. Let it be f_j. Update as follows:

    $\mathcal{F} = \mathcal{F} - \{f_j\}$ (remove f_j from $\mathcal{F}$)

    $\mathcal{F}_p = \mathcal{F}_p \cup \{f_j\}$ (add f_j to $\mathcal{F}_p$)

    $p = p + 1$ (increment p)

    c. Repeat step b till $p = l$, where l is the required number of features.

    A major problem with SFS is that it is greedy; addition of the qth feature is based on already selected $q - 1$ features. There are several variations to this scheme. For example, the backward selection scheme starts with the entire set of L features and keeps deleting one feature at a time till we end up with a set of l features; the feature to be deleted is the one, among the existing features, that leads to a minimum reduction in classification accuracy. One of the most refined schemes in this category is sequential floating forward selection (SFFS).

  - *Sequential floating forward selection (SFFS):* This scheme permits both addition and deletion of features and may be summarized as follows:

    a. Set $p = 0$ and $\mathcal{F} = \mathcal{F}_L$, $\mathcal{F}_p = \{\}$.

    b. *Addition*: Find the best feature, in terms of classification accuracy on a validation set, from $\mathcal{F}$ in conjunction with the existing p features in $\mathcal{F}_p$. Let it be f_j. Update as follows:

$$\mathcal{F} = \mathcal{F} - \{f_j\} \text{ (remove } f_j \text{ from } \mathcal{F})$$

$$\mathcal{F}_p = \mathcal{F}_p \cup \{f_j\} \text{ (add } f_j \text{ to } \mathcal{F}_p)$$

$$p = p + 1 \text{ (increment } p)$$

    c. *Selective deletion*: Choose the appropriate feature, from $\mathcal{F}_p$, for possible deletion; this feature is the one, among the features in $\mathcal{F}_p$, that has a minimum impact on the classification accuracy of the validation set. Let the feature selected be f_k. Delete f_k from $\mathcal{F}_p$ if there is no reduction in classification accuracy; decrement the value of p and go to step c. If there is reduction in accuracy, then add the deleted feature back and go to step b.

EXAMPLE 3: Consider a classification problem with three features f_1, f_2 and f_3. Let f_1 be the best feature. So, we get $\mathcal{F}_1 = \{f_1\}$. Let $\{f_1, f_2\}$ be a better set compared to $\{f_1, f_3\}$ or $\{f_1\}$. So, f_2 is added to get $\mathcal{F}_2 = \{f_1, f_2\}$.

We try to delete a feature from $\mathcal{F}_2$. The weaker feature is f_2. However, we cannot delete it as there will be a decrease in classification accuracy as $\{f_1, f_2\}$ is a better set compared to $\{f_1\}$. So, we add it back and go for addition.

Let $\{f_1, f_2, f_3\}$ be better than $\{f_1, f_2\}$. So, we add f_3 to get $\mathcal{F}_3 = \{f_1, f_2, f_3\}$. We try to delete a feature. Let us say that $\{f_2, f_3\}$ is the optimal set; this can be reached by deleting f_1 now.

Observe that feature f_1 added initially can be deleted at a later point to reach the optimal set using SFFS; this is not possible using SFS.

- **Embedded Methods:** Here, an ML model is built using L features and the model directly selects/indicates the relevant subset of features. Several classifiers including ones based on decision trees, support vector machines, neural networks and Bayes' classifier can be exploited to realize such an embedded scheme. For example, refer to Chapter 3, Section 3.4 to learn how the DTC can be used for feature selection.

6.3 LINEAR FEATURE EXTRACTION

Here, given a set, $\mathcal{F}_L$, of L features, we extract new features that are linear combinations of the features in $\mathcal{F}_L$. A feature g_j is extracted such that

$$g_j = \sum_{i=1}^{L} \alpha_{ij} f_i,$$

where α_{ij} is a real number and it indicates the contribution of feature f_i to the extracted feature g_j. In order to appreciate the linear feature extraction schemes, we require knowledge of basic linear algebra. This is discussed next.

6.3.1 Vector Spaces

A vector space is a non-empty collection of vectors V satisfying some useful properties. This notion is important in ML because we represent any data point as a vector. Two popular operations on vectors are:

- **Vector addition**: We have the following properties:

  - The sum or addition of any two vectors $u, v \in V$ gives a vector, $u + v \in V$. This is called the **closure** property of addition. This operation of vector addition is *commutative*, that is, $u+v = v+u$ for any pair of vectors $u, v \in V$. It is *associative*, that is, $(u+v)+w = u+(v+w)$ for all vectors $u, v, w \in V$.
  - There is a zero vector $0 \in V$ such that $u + 0 = u$ for all $u \in V$. For each vector $u \in V$, there exists a vector $-u \in V$ such that $u + (-u) = 0$. We will illustrate with an example.

    EXAMPLE 4: Let us consider vectors in a two-dimensional real space denoted by $\Re^2$. Let $u = (-1, 2.9), v = (0, 3.6)$ and $w = (1, 0.5)$ be three vectors from $\Re^2$.
    $u + v = v + u = (-1, 6.5) \in \Re^2$. Note that $u + w = w + u = (0, 3.4) \in \Re^2$ and $v + w = w + v = (1, 4.1) \in \Re^2$.
    $(u + v) + w = (-1, 6.5) + (1, 0.5) = (0, 7.0)$ and $u + (v + w) = (-1, 2.9) + (1, 4.1) = (0, 7.0)$. Both are equal.
    $-u = (1, -2.9)$. So, $u + (-u) = (0, 0)$.

- **Scalar multiplication**: We have the following properties:

  - Let c be a scalar (real number), that is, $c \in \Re$ and a vector $u \in V$. Then the scalar multiple cu is in V and $c(u + v) = cu + cv$.
  - For all scalars $c, d \in \Re$ and $u \in V$, $(c + d)u = cu + du$ and $(cd)u = c(du)$.
  - There is a scalar (real number) 1 such that $1u = u$ for all $u \in V$.

    EXAMPLE 5: Let us consider again vectors in a two-dimensional real space denoted by $\Re^2$. Let $u = (-1, 2.9), v = (0, 3.6)$ and $w = (1, 0.5)$ be three vectors from $\Re^2$. Let $c = 2$ and $d = 6.5$.
    $cu = 2(-1, 2.9) = (-2, 5.8) \in \Re^2$. Note that $c(u + w) = 2(0, 3.4) = (0, 6.8) \in \Re^2$.
    $(c + d)u = 8.5(-1, 2.9) = (-8.5, 24.65)$. $cu + du = (-2, 5.8) + (-6.5, 18.85) = (-8.5, 24.65)$. Both are equal.
    $(cd)u = 13(-1, 2.9) = (-13, 37.7)$. $c(du) = 2(-6.5, 18.85) = (-13, 37.7)$. Both are equal.

Some important observations are given below:

- We have considered scalars to be real numbers and vector space $\Re^2$ defined over real numbers in the examples. In general, it is possible to have other vector spaces. For example, we can have scalars to be complex numbers and vector spaces over complex numbers. However, for the kind of ML applications we typically encounter, vector spaces over reals are adequate.
- We have considered only two operations: vector addition and scalar multiplication on the vectors. It is possible to have other operations like vector multiplication, specifically inner product operation. Spaces that permit such operation are called inner product spaces.
- The notion of vector is more generic than the conventional vectors we have used in the examples. For example, we can have a vector space of 2×2 matrices. Now these matrices are the elements (vectors) of the vector space. Note that

  - If $c \in \Re$ is a scalar and $u = \begin{bmatrix} u_1 & u_2 \\ u_3 & u_4 \end{bmatrix}$ and $v = \begin{bmatrix} v_1 & v_2 \\ v_3 & v_4 \end{bmatrix}$ are two 2×2 matrices, we can have vector addition and scalar multiplication on the matrices.

$$u + v = \begin{bmatrix} u_1 + v_1 & u_2 + v_2 \\ u_3 + v_3 & u_4 + v_4 \end{bmatrix}$$

$$cv = \begin{bmatrix} cv_1 & cv_2 \\ cv_3 & cv_4 \end{bmatrix}$$

– It is possible to show that the set of all matrices in $\Re^{2\times 2}$ form a vector space. We leave it as an exercise.

Subspace of a Vector Space

In most ML applications, we may not be interested in the entire vector space. A subset of the vector space might be adequate. For example, in dimensionality reduction, we represent the vectors given in some L-dimensional space in an l-dimensional subspace. So, the notion of subspace is important in ML. The **subspace** of a vector space is a subset S of V having the following requirements:

- The zero vector $0 \in V$ is present in S.
- S is closed under vector addition. That means if u and v are in S, then $u + v$ is in S.
- S is closed under multiplication by scalars. That is, for each $u \in S$ and for each scalar c, $cu \in S$.

The above properties are adequate to show that every subspace is a vector space under the operations of vector addition and scalar multiplication defined on V. We illustrate this using an example.

EXAMPLE 6: Consider the vector space $\Re^2$. Each vector is based on two features, $Feature1$ and $Feature2$. Now using a feature selection scheme, we have selected $Feature1$; the value of $Feature2$ is made as 0 for all the vectors in the subspace. So, vectors in the subspace S will be of the form $u = (\alpha, 0)$. We can show that the set of all such u values form a subspace.

Consider the zero vector of V ($= (0, 0)$). It is present in S as the vector of the form $(\alpha, 0) = (0, 0)$ when $\alpha = 0$.

Consider two vectors $u = (\alpha, 0)$ and $v = (\beta, 0)$, where α and β are real numbers. So, $u + v = (\alpha + \beta, 0)$ which belongs to S.

Consider a scalar c and $u = (\alpha, 0)$. Now $cu = c(\alpha, 0) = (c\alpha, 0)$; this vector automatically belongs to S as the second component is 0. So, S is a subspace of V.

6.3.2 Basis of a Vector Space

We start with some important notions to gain a better understanding of vector spaces:

- We say that a vector $u \in V$ is a **linear combination** of vectors $u_1, u_2, \ldots, u_n \in V$ if there exist scalars $c_1, c_2, \ldots, c_n \in \Re$ such that

$$u = c_1 u_1 + c_2 u_2 + \cdots + c_n u_n$$

EXAMPLE 7: Consider the vector space $\Re^2$. The vector $(1, 2) \in \Re^2$ is a linear combination of $(-1,0)$ and $(1,1)$ in $\Re^2$ such that

$$(1, 2) = 1(-1, 0) + 2(1, 1),$$

where $(-1,0)$ and $(1,1)$ are vectors in $\Re^2$ and 1 and 2 are the scalars that were used.

- A set of vectors $u_1, u_2, \ldots, u_n$ is **linearly independent** if no vector in the set can be represented as a linear combination of the remaining $n-1$ vectors in the set. In other words, the only solution of

$$c_1 u_1 + c_2 u_2 + \cdots + c_n u_n = 0$$

is $c_1 = c_2 = \cdots = c_n = 0$.

EXAMPLE 8: The vectors $(1,2)$, $(-1,0)$ and $(1,1)$ are not linearly independent. This is because

$$c_1(1,2) + c_2(-1,0) + c_3(1,1) = 0 \Rightarrow c_1 = 1, c_2 = -1, c_3 = -2$$

is one possible solution where c_i are non-zero. However, the vectors $(-1,0)$ and $(1,1)$ are linearly independent as

$$c_1(-1,0) + c_2(1,1) = 0 \Rightarrow c_2 = 0 \text{ and } c_1 = 0$$

is the only possible solution.

- A set of vectors S in V is called a **basis** if the vectors are linearly independent and any vector in V can be represented as a linear combination of vectors in S.

EXAMPLE 9: Consider the vector space $\Re^2$. The set of vectors $\{(-1,0), (1,1)\}$ is a basis for $\Re^2$. This is because these two vectors are linearly independent and any vector (x,y) in $\Re^2$ can be represented as a linear combination of these two vectors as follows:

$$(x,y) = (y - x)(-1,0) + y(1,1)$$

A simpler basis is the set $\{(1,0), (0,1)\}$. A vector $(x,y) \in \Re^2$ can be represented as a linear combination of these two vectors that is given by

$$(x,y) = x(1,0) + y(0,1)$$

This is called the **standard basis** for $\Re^2$.

6.3.3 Row Vectors and Column Vectors

Thus far, we have been representing pattern or data points as row vectors. However, a representation scheme employs column vectors to represent data points. So, the standard basis in the previous subsection is represented as

$$\left\{ \begin{pmatrix} 1 \\ 0 \end{pmatrix}, \begin{pmatrix} 0 \\ 1 \end{pmatrix} \right\}$$

The **dot product** between two vectors u_i and u_j in an l-dimensional space is

$$u_i^t u_j = u_{i1} u_{j1} + u_{i2} u_{j2} + \cdots + u_{il} u_{jl} = \sum_{p=1}^{l} u_{ip} u_{jp}$$

A set of vectors is said to be **orthonormal** if the dot product between u_i and u_j for all $i \neq j$ is 0, that is, $u_i^t u_j = 0$ for $i \neq j$ (orthogonal) and $u_i^t u_i = \| u_i \|^2 = 1$ (unit vectors). So, in simple terms, a pair of vectors is orthonormal if it is orthogonal and each vector in the pair is a unit vector.

EXAMPLE 10: The standard basis for $\Re^2$ is an orthonormal basis. This is because the two vectors are orthogonal as

$$\begin{pmatrix} 1 \\ 0 \end{pmatrix}^t \begin{pmatrix} 0 \\ 1 \end{pmatrix} = \begin{pmatrix} 0 \\ 1 \end{pmatrix}^t \begin{pmatrix} 1 \\ 0 \end{pmatrix} = 0$$

and

$$\begin{pmatrix} 1 \\ 0 \end{pmatrix}^t \begin{pmatrix} 1 \\ 0 \end{pmatrix} = \begin{pmatrix} 0 \\ 1 \end{pmatrix}^t \begin{pmatrix} 0 \\ 1 \end{pmatrix} = 1$$

We can represent any vector in an l-dimensional space as a linear combination of basis vectors. It is possible to show that l linearly independent vectors will form a basis for an l-dimensional space. If the basis vectors are orthonormal, then the coefficients associated with the basis vectors in the linear combination can be obtained easily.

EXAMPLE 11: Let a vector z be represented using basis vectors $u_1, u_2, \ldots, u_l$ as

$$z = c_1 u_1 + c_2 u_2 + \cdots + c_l u_l$$

Taking the dot product with u_1 on both sides, we get

$$u_1^t z = c_1 u_1^t u_1 + c_2 u_1^t u_2 + \cdots + c_l u_1^t u_l = c_1 \times 1 + c_2 \times 0 + \cdots + c_l \times 0 = c_1$$

So,

$$c_1 = u_1^t z; c_2 = u_2^t z; \cdots ; c_l = u_l^t z$$

So, the linear combination coefficients can be found using dot product computations when the basis vectors are orthonormal.

6.3.4 Linear Transformations

A linear transformation, in general, is a mapping from a vector space V to a vector space W. However, as our ML applications deal with real spaces, we restrict our definition as follows. A **linear transformation** T is a mapping from $\Re^L$ to $\Re^l$ satisfying

- $T(u+v) = T(u) + T(v)$
- $T(cu) = c\, T(u)$

for all vectors $u, v \in \Re^L$ and scalars $c \in \Re$. We are interested in matrix transforms in ML and we observe that they are linear transforms.

If we have n data points, each being a vector in an L-dimensional space, we can represent the n data points using a matrix, A, of size $n \times L$ that has n rows and L columns. So, there will be nL elements and each element is a real number in our applications. For example, the data matrix will have a row for each of the n patterns and a column for each of the L features.

In linear feature extraction, we transform the L-dimensional vectors into l-dimensional vectors. This could be viewed as a matrix transform as follows:

- Consider the transpose A^t of the data matrix A; the matrix A^t is an $L \times n$ matrix.
- Let B be a matrix of size $l \times L$ which can be called the transform matrix.
- Let $C = BA^t$, which means C is an $l \times n$ matrix. Each column of C is a pattern in the low-dimensional (l-dimensional) space; it is obtained by transforming each column of A^t using B.

We illustrate this with an example.

EXAMPLE 12: Consider the variance-based feature selection scheme discussed in Section 6.2 and the data shown in Table 6.1. The corresponding data matrix and its transpose are:

$$A = \begin{bmatrix} 0 & 0 & 0 \\ 6 & 0 & 3 \\ 1 & 0 & 0 \\ 5 & 1 & 3 \\ 6 & 0 & 0 \\ 0 & 0 & 1 \end{bmatrix} \text{ and } A^t = \begin{bmatrix} 0 & 6 & 1 & 5 & 6 & 0 \\ 0 & 0 & 0 & 1 & 0 & 0 \\ 0 & 3 & 0 & 3 & 0 & 1 \end{bmatrix}$$

Now, feature selection based on variance of the features retains only *Feature*1 and *Feature*3 and discards *Feature*2, as discussed in Section 6.2. So, considering the transform matrix $B = \begin{bmatrix} 1 & 0 & 0 \\ 0 & 0 & 1 \end{bmatrix}$, we get

$$BA^t = \begin{bmatrix} 1 & 0 & 0 \\ 0 & 0 & 1 \end{bmatrix} \times \begin{bmatrix} 0 & 6 & 1 & 5 & 6 & 0 \\ 0 & 0 & 0 & 1 & 0 & 0 \\ 0 & 3 & 0 & 3 & 0 & 1 \end{bmatrix} = \begin{bmatrix} 0 & 6 & 1 & 5 & 6 & 0 \\ 0 & 3 & 0 & 3 & 0 & 1 \end{bmatrix}$$

Note that each column of A^t is the given three-dimensional vector corresponding to a data point and it is transformed into a two-dimensional vector using B; the matrix B selects the first and third rows of A^t to get the resulting two-dimensional vectors. So, this example illustrates that feature selection can be viewed as a matrix transformation.

- We can easily see that it is matrix transformation because:
  - If B is an $l \times L$ matrix and u, v are L-dimensional vectors and $c \in \Re$ is a scalar, then $B(u + v) = Bu + Bv$, and
  - $B(cu) = cBu$.

  Using the above properties, it is possible to show that $B0 = 0$ and $B(u_1 + u_2 + \cdots + u_p) = Bu_1 + Bu_2 + \cdots + Bu_p$.

- There is a special and simple matrix of the form $I_2 = \begin{bmatrix} 1 & 0 \\ 0 & 1 \end{bmatrix}$. Note that $Ix = x$ for any $x \in \Re^2$. That is why it is called an **identity matrix** and the corresponding transformation is an **identity transformation**. For $\Re^n$, we have the identity matrix to be $I_n = \begin{bmatrix} 1 & 0 & \cdots & 0 \\ 0 & 1 & \cdots & 0 \\ \vdots & \vdots & \vdots & \vdots \\ 0 & 0 & \cdots & 1 \end{bmatrix}$,

  which has n rows and n columns.

6.4   EIGENVALUES AND EIGENVECTORS

Consider the vector and matrix shown in Fig. 6.1. Let matrix $A = \begin{bmatrix} 1 & 2 \\ 2 & 4 \end{bmatrix}$ and vector $x = \begin{pmatrix} 1 \\ 2 \end{pmatrix}$. Then

$$Ax = \begin{pmatrix} 5 \\ 10 \end{pmatrix} = 5 \begin{pmatrix} 1 \\ 2 \end{pmatrix} = 5x$$

Such non-zero vectors that satisfy the equality $Ax = \lambda x$ are called eigenvectors of A and λ is the eigenvalue of A.

EXAMPLE 13:   Consider the matrix $A = \begin{bmatrix} 1 & 2 \\ 2 & 4 \end{bmatrix}$. We have see that $\begin{pmatrix} 1 \\ 2 \end{pmatrix}$ is an eigenvector and the corresponding eigenvalue is 5 as

$$Ax = \begin{pmatrix} 5 \\ 10 \end{pmatrix} = 5 \begin{pmatrix} 1 \\ 2 \end{pmatrix} = 5x$$

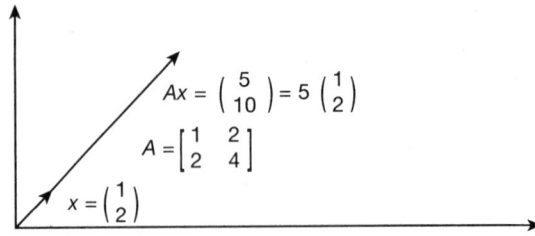

Fig. 6.1 Eigenvalue and eigenvector

Note that vector $x = \begin{pmatrix} -2 \\ 1 \end{pmatrix}$ also satisfies the equality of the form $Ax = 0x$, where 0 is an eigenvalue and the corresponding eigenvector is $\begin{pmatrix} -2 \\ 1 \end{pmatrix}$. An eigenvalue and the corresponding eigenvector are together treated as an eigenpair. So, for matrix A, the two eigenpairs we have seen are $\left\langle 5, \begin{pmatrix} 1 \\ 2 \end{pmatrix} \right\rangle$ and $\left\langle 0, \begin{pmatrix} -2 \\ 1 \end{pmatrix} \right\rangle$

So, the eigenvector equation is $Ax = \lambda x$, where A is a matrix, x is a non-zero vector and $\lambda \in \Re$ is a scalar. Some important observations are given below:

- The equation $Ax = \lambda x$ has a trivial solution $x = 0$. But it is important to note that an **eigenvector is a non-zero vector.**
- It is possible that one or more eigenvalues of a matrix are 0. For example, one of the eigenvalues in Example 11 is 0, but the corresponding eigenvector is non-zero and it is $\begin{pmatrix} -2 \\ 1 \end{pmatrix}$.
- The matrix A is a square matrix of size $n \times n$, where n is a positive integer greater than or equal to 2. If A is an $m \times n$ matrix $(m \neq n)$, then if x in the LHS of $Ax = \lambda x$ is an $n \times 1$ vector, then x in the RHS will be an $m \times 1$ matrix. But we are interested in an x that is the same in both LHS and RHS. This is possible only when $m = n$. So, we have eigenvalues only for square matrices.
- Consider

$$Ax = \lambda x \Rightarrow Ax = \lambda I x \Rightarrow (A - \lambda I)x = 0$$

If $(A - \lambda I)$ is an invertible matrix, then the solution of $(A - \lambda I)x = 0$ by multiplying both sides of the equation by $(A - \lambda I)^{-1}$, which gives the solution to be $x = 0$. However, for x to be an eigenvector, we want a non-zero x. It is possible to obtain a non-zero x if $(A - \lambda I)$ is not invertible or singular. A singular matrix will have its determinant as 0. So, in order to get eigenvalues and from them the eigenvectors of a matrix A, we need the determinant of $A - \lambda I$ to be zero.

- It is possible that the matrix A has only real entries, but its eigenvalues are complex.

EXAMPLE 14: Consider the matrix $A = \begin{bmatrix} 0 & -1 \\ 1 & 0 \end{bmatrix}$. The determinant of the matrix $A - \lambda I = $ $Determinant\left(\begin{bmatrix} -\lambda & -1 \\ 1 & -\lambda \end{bmatrix}\right) = \lambda^2 + 1$. Eigenvalues are the roots of the equation $\lambda^2 + 1 = 0$; so, the eigenvalues are $+i$ and $-i$, where $i = \sqrt{-1}$ is complex. The corresponding eigenvectors are

$\begin{pmatrix} -1 \\ i \end{pmatrix}$ and $\begin{pmatrix} 1 \\ i \end{pmatrix}$. However, if we want real eigenvalues and eigenvectors for a real matrix, it is sufficient that the matrix is symmetric.

- The **equation** $determinant(A - \lambda I) = 0$ **is called the characteristic equation** of A as the eigenvalues and eigenvectors are also called characteristic values and characteristic vectors, respectively.

6.4.1 Symmetric Matrices

A matrix A is symmetric if $A = A^t$, where A^t contains the rows of A as its columns and the columns of A as its rows; equivalently a symmetric matrix will have $A_{i,j} = A_{j,i}$, that is, the jth element in the ith row is equal to the ith element in the jth row.

Symmetric matrices have very good properties and a variety of ML applications deal with them. These include linear feature extraction using principal components or singular value decomposition and spectral clustering. Their properties are as follows:

- Eigenvalues of a real symmetric matrix are real.

EXAMPLE 15: Consider the matrix A in Example 13. The matrix is

$$A = \begin{bmatrix} 1 & 2 \\ 2 & 4 \end{bmatrix} = A^t$$

We have observed its eigenvalues to be 5 and 0. We can also compute the roots of the equation

$$determinant(A - \lambda I) = 0 \Rightarrow (1 - \lambda)(4 - \lambda) - 4 = 0 \Rightarrow \lambda^2 - 5\lambda = 0$$

The roots of $\lambda^2 - 5\lambda = 0$ are 0 and 5. So, in this example the eigenvalues of a symmetric matrix are real.

It is possible that a non-symmetric real matrix has real eigenvalues. We will explore it in an exercise.

- A **diagonal matrix** has non-zero elements only on its diagonal. All the off-diagonal entries are 0. The diagonal elements of a diagonal matrix are its eigenvalues.

EXAMPLE 16: Consider the diagonal matrix $\begin{bmatrix} 2 & 0 & 0 \\ 0 & 0 & 0 \\ 0 & 0 & 1 \end{bmatrix}$. Its characteristic equation is $(\lambda - 2)(\lambda)(\lambda - 1) = 0$. So, the roots of this equation, the eigenvalues of the matrix, are 2, 0 and 1.

- A **triangular matrix** has non-zero elements only in the **upper triangular part** of the matrix, that is, $A_{i,j} = 0$ if $j < i$, or in the **lower triangualr part**, that is, $A_{i,j} = 0$ if $j > i$.

EXAMPLE 17: Consider the matrix

$$A = \begin{bmatrix} 2 & 0 & 3 \\ 0 & 0 & 2 \\ 0 & 0 & 1 \end{bmatrix}$$

It is an upper triangular matrix. Its eigenvalues are 2, 0 and 1.

We do not show this property. We leave it as an exercise.

- It is possible to select the eigenvectors of a symmetric matrix such that they are orthonormal.

EXAMPLE 18: Consider the matrix $A = \begin{bmatrix} 1 & 2 \\ 2 & 4 \end{bmatrix}$ seen in Example 13. Its eigenpairs are:

$$\left\langle 5, \begin{pmatrix} 1 \\ 2 \end{pmatrix} \right\rangle \text{ and } \left\langle 0, \begin{pmatrix} -2 \\ 1 \end{pmatrix} \right\rangle$$

So, the two eigenvectors are orthogonal as $\begin{pmatrix} 1 \\ 2 \end{pmatrix}^t \begin{pmatrix} -2 \\ 1 \end{pmatrix} = 0$. We can make them orthonormal by normalizing (dividing) each component of each vector by its respective norm.

The norm of $\begin{pmatrix} 1 \\ 2 \end{pmatrix}$ is $\sqrt{1^2 + 2^2} = \sqrt{5}$. So, unit norm vector is $\begin{pmatrix} \frac{1}{\sqrt{5}} \\ \frac{2}{\sqrt{5}} \end{pmatrix}$.

Similarly, the norm of $\begin{pmatrix} -2 \\ 1 \end{pmatrix}$ is $\sqrt{(-2)^2 + 1^2} = \sqrt{5}$. So, the corresponding unit norm vector is $\begin{pmatrix} \frac{-2}{\sqrt{5}} \\ \frac{1}{\sqrt{5}} \end{pmatrix}$.

6.4.2 Rank of a Matrix

The rank of an $m \times n$ matrix is the number of linearly independent rows (*row rank*); it also equals the number of linearly independent columns (*column rank*).

EXAMPLE 19: Consider a 3×2 matrix $A = \begin{bmatrix} 2 & 0 \\ 1 & 2 \\ 2 & 3 \end{bmatrix}$. Note that the two columns are linearly independent as

$$c \begin{pmatrix} 2 \\ 1 \\ 2 \end{pmatrix} + d \begin{pmatrix} 0 \\ 2 \\ 3 \end{pmatrix} = 0 \Rightarrow c = 0 \text{ and } d = 0$$

The three rows are linearly dependent as, for example, the third is a linear combination of the first two rows:

$$(2,3) = \frac{1}{4} \times (2,0) + \frac{3}{2} \times (1,2)$$

However, the row vectors (2,0) and (1,2) are linearly independent because

$$c(2,0) + d(1,2) = (0,0) \Rightarrow d = 0 \text{ and } c = 0$$

So, rank(A) = 2 = row rank(A) = column rank(A).

It is possible to show that if an $n \times n$ matrix A has rank $r(< n)$, then A will have r 0 eigenvalues.

EXAMPLE 20: Consider $A = \begin{bmatrix} 1 & 2 \\ 2 & 4 \end{bmatrix}$. Its two rows and two columns are linearly dependent on each other and any row or column is linearly independent. So, rank(A) = 1. So, one eigenvalue is 0 and the other is 5, as seen in Example 13.

6.5 PRINCIPAL COMPONENT ANALYSIS

Principal components (PCs) are the most popular directions employed in ML for linear feature extraction. PCs are the eigenvectors of the covariance matrix of the data.

Eigenvalues are considered in decreasing (non-increasing) order of their value and the eigenvectors corresponding to the l-largest eigenvalues are considered for low-dimensional representation. Eigenvalues capture the variance, in different directions, that is present in the data.

The covariance matrix of the data is symmetric; so, it is possible to select the leading l orthonormal eigenvectors. The value of l is chosen based on total variance, in the data, explained by the selected l eigenvalues. Each new feature is a linear combination of the original features. We will examine the role played by eigenvectors in PCs. Let $x_1, x_2, \ldots, x_n$ be n training vectors in an L-dimensional space. Any such vector x can be represented as a linear combination of L orthonormal basis vectors as $x = \sum_{i=1}^{L} \alpha_i u_i$, where $\alpha_i \in \Re$ and u_i is the ith basis vector for $i = 1, 2, \ldots, L$.

Let us approximate x by an l-dimensional representation y for $l < L$ as $y = \sum_{i=1}^{l} \alpha_i u_i$. So, the difference between x and its approximation y is $x - y = \sum_{i=l+1}^{L} \alpha_i u_i$.

We consider the quantity given by

$$(x - y)^t (x - y) = \left(\sum_{j=l+1}^{L} \alpha_j u_j^t \right) \left(\sum_{j=l+1}^{L} \alpha_j u_j \right)$$

We choose the error to be

$$E[(x - y)^t (x - y)] = E\left[\sum_{j=l+1}^{L} \alpha_j^2 \right]$$

because u_i are orthonormal basis vectors. It is the expected value of $(x - y)^t (x - y)$.

We have

$$u_j^t x = u_j^t \sum_{i=1}^{L} \alpha_i u_i - \alpha_j$$

because $u_j^t u_i = 0$ for $i \neq j$ and $u_j^t u_j = 1$. So, $\alpha_j = u_j^t x$. We will substitute this in the error equation to obtain $Error = E[\sum_{j=l+1}^{L} \alpha_j^2] = E[\sum_{j=l+1}^{L} \alpha_j \alpha_j^t] = E[\sum_{j=l+1}^{L} u_j^t x x^t u_j]$. Note that $\alpha_j = \alpha_j^t$ because α_j is a scalar.

So, error is $\sum_{j=l+1}^{L} u_j^t \Sigma u_j$, where Σ is the covariance matrix if the data is zero-mean data. The covariance matrix $\Sigma = E[x x^t]$. The covariance matrix is a symmetric matrix. The eigenvector equation is $\Sigma u_j = \lambda_j u_j$. So, $Error = \sum_{j=l+1}^{L} u_j^t \Sigma u_j = \sum_{j=l+1}^{L} u_j^t u_j \lambda_j = \sum_{j=l+1}^{L} \lambda_j$.

We want to minimize the error, that is, minimize the sum of the eigenvalues $\lambda_{l+1} \cdots \lambda_L$. This can be achieved by selecting the first l eigenvalues to be larger. That is, we arrange the eigenvalues of Σ in descending order as $\lambda_1 \geq \lambda_2 \geq \cdots \geq \lambda_l \geq \lambda_{l+1} \geq \cdots \geq \lambda_L$ and use the corresponding (leading) l eigenvectors $u_1, u_2, \ldots, u_l$ in representing the approximate vector y of the given x.

Note that if xs are zero-mean vectors, then α are also zero mean because $E[\alpha_i] = E[u_i^t x] = u_i^t E[x] = 0$. Further, when we use zero-mean data,

$$covariance(\alpha_i, \alpha_j) = E[\alpha_i \alpha_j]$$

The **projections on to the PCs are uncorrelated.** We know that $\alpha_i = u_i^t x$ and $\alpha_j = u_j^t x$. So, covariance between α_i and α_j is

$$E[\alpha_i \alpha_j] = E[u_i^t x x^t u_j] = u_i^t \Sigma u_j = \lambda_j u_i^t u_j = 0$$

Note that we have used the fact that $\alpha_j = \alpha_j^t$ as α_j is a scalar and u_i and u_j are orthonormal. So, the covariance between the projected values is 0. So, they are uncorrelated.

EXAMPLE 21: Consider the two-dimensional data matrix X of 6 patterns.

$$X = \begin{bmatrix} 1 & 2 \\ 2 & 1 \\ 2 & 2 \\ 6 & 7 \\ 7 & 6 \\ 6 & 6 \end{bmatrix} \quad X_{zm} = \begin{bmatrix} -3 & -2 \\ -2 & -3 \\ -2 & -2 \\ 2 & 3 \\ 3 & 2 \\ 2 & 2 \end{bmatrix}$$

The sample mean of the data is $\frac{1}{6}(24, 24) = (4, 4)$. After subtracting the mean from each vector in X, we get X_{zm} which is the zero-mean data. We can get the covariance matrix from the zero-mean data by computing $\frac{1}{6} X_{zm}^t X_{zm}$. It is

$$\Sigma = \begin{bmatrix} \frac{17}{3} & \frac{16}{3} \\ \frac{16}{3} & \frac{17}{3} \end{bmatrix}$$

It is a symmetric matrix with the characteristic equation $(\lambda - \frac{17}{3})^2 - (\frac{16}{3})^2$; so the eigenvalues are 11 and $\frac{1}{3}$. By considering the equation

$$\Sigma x = \lambda x$$

and solving for x using λ to be 11 and $\frac{1}{3}$, we get the respective eigenvectors to be $\begin{pmatrix} 1 \\ 1 \end{pmatrix}$ and $\begin{pmatrix} 1 \\ -1 \end{pmatrix}$.

These eigenvectors are orthogonal but not orthonormal; we can make them orthonormal by dividing each vector by its norm to make it a unit norm vector. The norm is $\sqrt{2}$ for both the vectors. So, the orthonormal vectors are $\begin{pmatrix} \frac{1}{\sqrt{2}} \\ \frac{1}{\sqrt{2}} \end{pmatrix}$ and $\begin{pmatrix} \frac{1}{\sqrt{2}} \\ -\frac{1}{\sqrt{2}} \end{pmatrix}$.

We show the two eigenvector directions in Fig. 6.2. They are orthogonal to each other as the covariance matrix is symmetric.

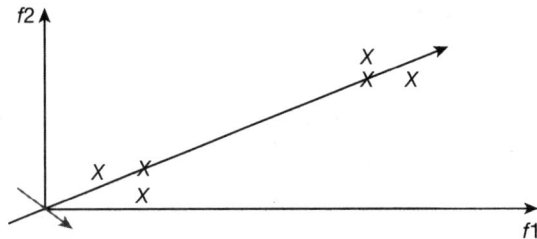

FIG. 6.2 The two PCs for the data set of six patterns

When we project the data points in X_{zm} on to the new feature directions, that is, the orthonormal eigenvector directions, we get the projected matrix X_{proj}.

$$X_{proj} = \begin{bmatrix} \frac{-5}{\sqrt{2}} & \frac{-1}{\sqrt{2}} \\ \frac{-5}{\sqrt{2}} & \frac{1}{\sqrt{2}} \\ \frac{-4}{\sqrt{2}} & 0 \\ \frac{5}{\sqrt{2}} & \frac{-1}{\sqrt{2}} \\ \frac{5}{\sqrt{2}} & \frac{1}{\sqrt{2}} \\ \frac{4}{\sqrt{2}} & 0 \end{bmatrix}$$

Note that the first element of the first row of X_{proj} is obtained by considering the dot product of $\begin{pmatrix} \frac{1}{\sqrt{2}} \\ \frac{1}{\sqrt{2}} \end{pmatrix}$ and the first vector in X_{zm}, which is $(-3, -2)$; the second element of the first row of X_{proj} is obtained by considering the dot product of $\begin{pmatrix} \frac{1}{\sqrt{2}} \\ -\frac{1}{\sqrt{2}} \end{pmatrix}$ and the first vector in X_{zm}, which is $(-3, -2)$. Other entries are computed in a similar manner by taking each row of X_{zm}. Note that the data in X_{proj} has zero mean.

6.5.1 Experimental Results on Olivetti Face Data

Let us consider the Olivetti Face data set. The dimensionality of the data set is 4096 (64×64 pixels in each image). The accuracy of the KNN classifier for k value in the range [1,7] is computed using distances based on L_2, L_1, $L_{0.8}$, $L_{0.5}$ and $L_{0.1}$. The results are shown in Fig. 6.3 (a). The X-axis depicts the value of k and the number of neighbors used and the Y-axis shows the test accuracy using KNN.

From the covariance matrix of the training data patterns, 120 eigenvalues and their eigenvectors are calculated. The eigenvectors are sorted in descending order based on their corresponding eigenvalues. The test accuracy of KNN with k in the range [1,7] is computed using distances based on L_2, L_1, $L_{0.8}$, $L_{0.5}$ and $L_{0.1}$. The results are shown in Fig. 6.3 (b). From Fig. 6.3, we observe the following:

- Feature extraction based on the top 120 PCs gave better test accuracy, on an average, than using all the 4096 pixels.
- $L_{0.5}$ and $L_{0.1}$ norms are not so important in the reduced dimensional space.
- The results based on L_1 and $L_{0.8}$ are superior to those based on Euclidean distance as the value of k is increased.

6.6 SINGULAR VALUE DECOMPOSITION

It is possible to factorize a matrix if its eigenvalues are distinct. It is even better if the matrix is symmetric irrespective of whether the eigenvalues are distinct or not. We will illustrate with an example.

EXAMPLE 22: Consider matrix $A = \begin{bmatrix} 1 & 2 \\ 2 & 4 \end{bmatrix}$. We have seen in Example 13 that the eigenpairs of this matrix are $\left\langle 5, \begin{pmatrix} 1 \\ 2 \end{pmatrix} \right\rangle$ and $\left\langle 0, \begin{pmatrix} -2 \\ 1 \end{pmatrix} \right\rangle$.

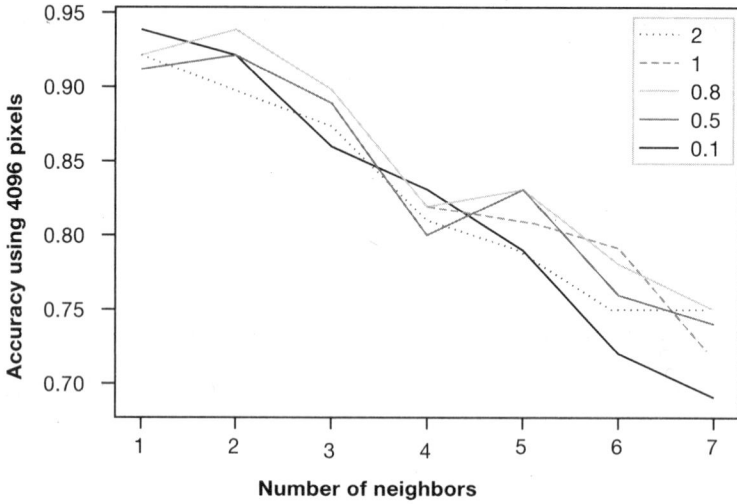

(a)

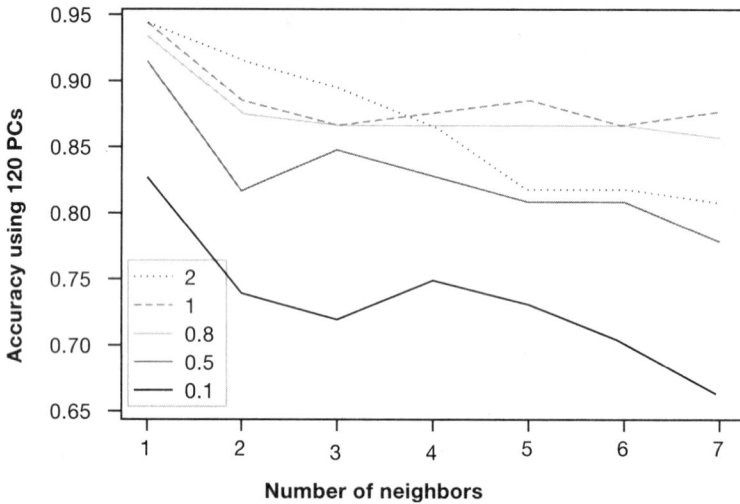

(b)

FIG. 6.3 Results on the Olivetti Face data set (for colour figure, please see Colour Plate 2)

The eigenvectors are orthogonal. We can make them orthonormal by scaling the components of each vector by its norm. The norm of both the vectors is $\sqrt{5}$. By dividing all the entries by $\sqrt{5}$, we get the orthonormal eigenvectors to be $u_1 = \begin{pmatrix} \frac{1}{\sqrt{5}} \\ \frac{2}{\sqrt{5}} \end{pmatrix}$ and $u_2 = \begin{pmatrix} \frac{-2}{\sqrt{5}} \\ \frac{1}{\sqrt{5}} \end{pmatrix}$.

If we keep these two orthonormal eigenvectors as the columns of a matrix U, then $AU = [Au_1 Au_2] = [5u_1 0u_2] = U \begin{bmatrix} 5 & 0 \\ 0 & 0 \end{bmatrix}$. Because the columns of U are orthonormal, U is an **orthogonal matrix**. An important property is that if U is orthogonal, then $U^{-1} = U^t$.

In this example, we have $AU = U\Omega$, where Ω is a diagonal matrix and $\Omega = \begin{bmatrix} 5 & 0 \\ 0 & 0 \end{bmatrix}$. Post-multiplying both sides of $AU = U\Omega$ by $U^{-1}(= U^t$ in this case), we get $A = U\Omega U^t$. So, the given matrix A is factorized to have U, Ω and U^t as the three factors.

Matrix factorization is very useful in ML. It is possible to see that clustering and feature extraction/selection may be viewed as matrix factorization schemes. It is possible to factorize matrices that are not square using SVD.

Typically, the input data is a matrix, X, of size $n \times L$, where n is the number of data points and L is the number of features or dimensions.

Singular value decomposition (SVD) is a matrix factorization given by

$$X_{n \times L} = U_{n \times n} \Sigma_{n \times L} V_{L \times L}^T$$

In general, X is not a square matrix (n may not be equal to L). However, XX^t is a square matrix of size $n \times n$ and $X^t X$ is a square matrix of size $L \times L$.

Another important property is that both XX^t and $X^t X$ are symmetric matrices. SVD exploits this symmetry.

Let $\{v_1, v_2, \ldots, v_L\}$ be the orthonormal basis for $\Re^L$ consisting of the eigenvectors of $X^t X$ and let $\lambda_1, \lambda_2, \ldots, \lambda_L$ be the corresponding eigenvalues. Then $||Xv_i||^2 = (Xv_i)^t Xv_i = v_i^t (X^t Xv_i) = v_i^t (\lambda_i v_i) = \lambda_i$ for $i = 1, 2, \ldots, L$ using the facts that v_i is an eigenvector of $X^t X$ and the corresponding eigenvalue is λ_i and v_is are orthonormal.

So, $||Xv_i||^2 = \lambda_i$, where the LHS is non-negative for a real matrix X. So, λ_i is non-negative for $i = 1, 2, \ldots, L$.

The **singular values** of X are the square roots of the eigenvalues, which are non-negative, of $X^t X$; they are denoted by $\sigma_i = \sqrt{\lambda_i}$. Note that $\sigma_i = ||Xv_i||$ as $\lambda_i = ||Xv_i||^2$.

If $v_1, v_2, \ldots, v_L$ are the orthonormal eigenvectors of $X^t X$, then $X^t Xv_i = \lambda_i v_i$. Pre-multiplying both sides of this equality by X, we get $XX^t(Xv_i) = \lambda_i(Xv_i)$, which means Xv_i are eigenvectors of XX^t with the same eigenvalue λ_i.

Let $w_i = Xv_i$ for $i = 1, 2, \ldots, L$. Because v_i are orthonormal, Xv_is are orthogonal. We can normalize them by dividing by $||Xv_i||$ to get $u_i = \frac{1}{||Xv_i||}w_i = \frac{1}{||Xv_i||}Xv_i$.

We know that $\frac{1}{||Xv_i||} = \frac{1}{\sigma_i}$ as $||Xv_i|| = \sigma_i$. So, $u_i = \frac{1}{\sigma_i}Xv_i \Rightarrow Xv_i = \sigma_i u_i$ for $i = 1, 2, \ldots, L$. So, by placing u_i as columns of U and v_i as columns of V, we have $XV = U\Sigma$, where Σ is a diagonal matrix made up of σ_i.

Since V is an orthogonal matrix, $XV = U\Sigma \Rightarrow XVV^t = U\Sigma V^t \Rightarrow X = U\Sigma V^t$ as $XVV^t = X$ $(V^t = V^{-1})$ Note that if the rank of X is r, where $r \le min(n, L)$, then we will have $\sigma_{r+1} = \sigma_{r+2} = \cdots = \sigma_L = 0$.

EXAMPLE 23: Let us consider three data points in a two-dimensional space given by the data matrix X:

$$X = \begin{bmatrix} 1 & -1 \\ -2 & 2 \\ 2 & -2 \end{bmatrix}$$

$$X^t X = \begin{bmatrix} 9 & -9 \\ -9 & 9 \end{bmatrix}$$

The eigenvalues of $X^t X$ are 18 and 0.

The corresponding orthonormal eigenvectors are $v_1 = \begin{pmatrix} \frac{1}{\sqrt{2}} \\ \frac{1}{\sqrt{2}} \end{pmatrix}$ and $v_2 = \begin{pmatrix} \frac{1}{\sqrt{2}} \\ \frac{1}{\sqrt{2}} \end{pmatrix}$. So,

$V = \begin{bmatrix} \frac{1}{\sqrt{2}} & \frac{1}{\sqrt{2}} \\ \frac{-1}{\sqrt{2}} & \frac{1}{\sqrt{2}} \end{bmatrix}$ and the singular values are $\sigma_1 = \sqrt{18} = 3\sqrt{2}$ and $\sigma_2 = 0$. So, $\Sigma = \begin{bmatrix} 3\sqrt{2} & 0 \\ 0 & 0 \\ 0 & 0 \end{bmatrix}$.

To construct U, find Av_1 and Av_2 and normalize. $Av_1 = \begin{pmatrix} \frac{2}{\sqrt{2}} \\ \frac{-4}{\sqrt{2}} \\ \frac{4}{\sqrt{2}} \end{pmatrix}$ and $Av_2 = \begin{pmatrix} 0 \\ 0 \\ 0 \end{pmatrix}$. The

normalized version of Av_1 is obtained by dividing with $3\sqrt{2}$ which gives us $u_1 = \begin{pmatrix} \frac{1}{3} \\ -\frac{2}{3} \\ \frac{2}{3} \end{pmatrix}$.

The other two vectors in U are to be chosen so that they form an orthonormal basis for $\Re^3$. So, they are of the form p such that $u_1^t p = 0 \Rightarrow p_1 - 2p_2 + 2p_3 = 0$.

So, one such vector is $\begin{pmatrix} 2 \\ 1 \\ 0 \end{pmatrix}$. We need to normalize it to get $u_2 = \begin{pmatrix} \frac{2}{\sqrt{5}} \\ \frac{1}{\sqrt{5}} \\ 0 \end{pmatrix}$. Similarly, a vector

that is orthogonal to both u_1 and u_2 is $\begin{pmatrix} -2 \\ 4 \\ 5 \end{pmatrix}$ and its normalized version $u_3 = \begin{pmatrix} \frac{-2}{\sqrt{45}} \\ \frac{4}{\sqrt{45}} \\ \frac{5}{\sqrt{45}} \end{pmatrix}$.

So, we have

$$X = \begin{bmatrix} \frac{1}{3} & \frac{2}{\sqrt{5}} & \frac{-2}{\sqrt{45}} \\ -\frac{2}{3} & \frac{1}{\sqrt{5}} & \frac{4}{\sqrt{45}} \\ \frac{1}{3} & 0 & \frac{5}{\sqrt{45}} \end{bmatrix} \begin{bmatrix} 3\sqrt{2} & 0 \\ 0 & 0 \\ 0 & 0 \end{bmatrix} \begin{bmatrix} \frac{1}{\sqrt{2}} & \frac{-1}{\sqrt{2}} \\ \frac{1}{\sqrt{2}} & \frac{1}{\sqrt{2}} \end{bmatrix}$$

Note that there is no need to compute u_2 and u_3 in this example. In fact, we can simply write

$$X = \begin{bmatrix} \frac{1}{3} \\ -\frac{2}{3} \\ \frac{1}{3} \end{bmatrix} \begin{bmatrix} 3\sqrt{2} & 0 \end{bmatrix} \begin{bmatrix} \frac{1}{\sqrt{2}} & \frac{-1}{\sqrt{2}} \\ \frac{1}{\sqrt{2}} & \frac{1}{\sqrt{2}} \end{bmatrix},$$

because only one singular value is non-zero.

6.6.1 PCA and SVD

Both PCs and singular vectors can be used for dimensionality reduction. The difference between them is that in the case of PCA, we are using the covariance matrix by converting the data to have zero mean. In the case of SVD, we are using the data matrix X directly. So, we are using the eigenvectors of the correlation matrix.

We compare the performance of KNN based on 200 PCs and 200 SVs on the Olivetti Face data set. Figure 6.4 (a) shows the results based on PCs while Fig. 6.4 (b) shows the results based on singular vectors.

- In both the cases, the results based on L_2 and L_1 norms are better than those based on fractional norms. This is anticipated as the dimensionality is reduced.
- On an average, the performance based on PCs is a shade better than that using SVs.

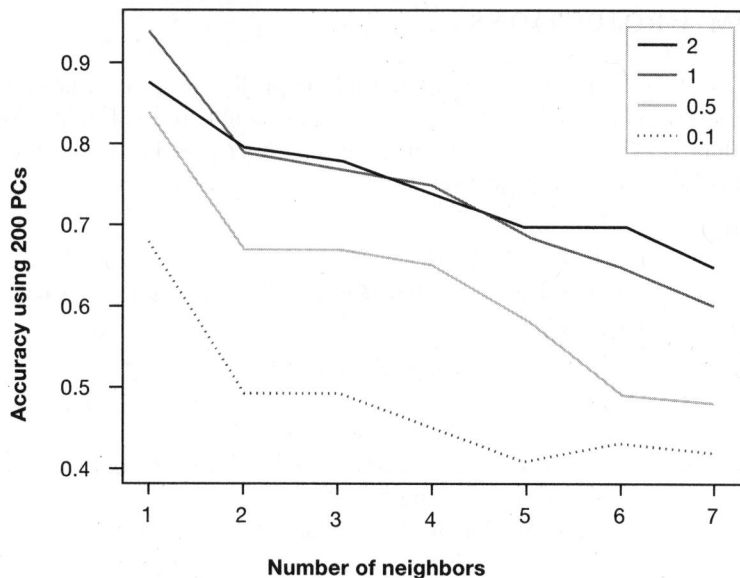

(a)

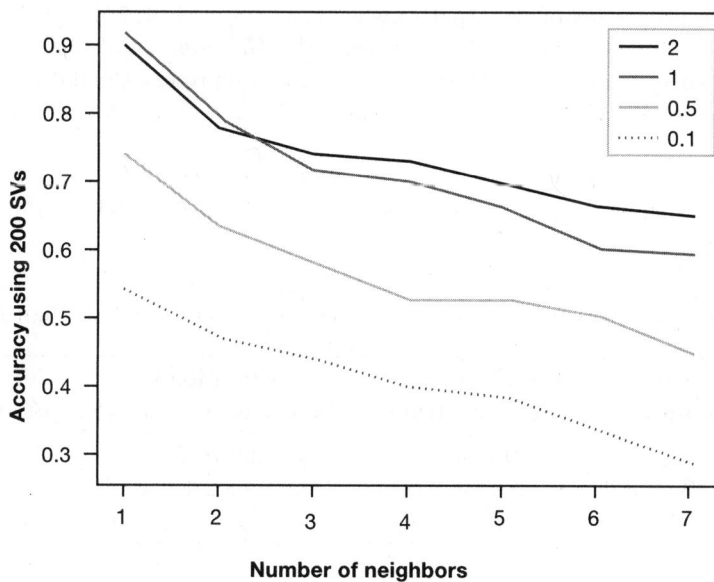

(b)

FIG. 6.4 Results on the Olivetti Face data set: 200 PCs and 200 SVs (for colour figure, please see Colour Plate 2)

6.7 RANDOM PROJECTIONS

We have discussed in Chapter 2 that in the case of random projections, we reduce the dimensionality of the data from the L-dimensional space to a lower-dimensional space of dimension l, where l can be much smaller than L. This is achieved by using a matrix R that has random entries. Some of the properties of random projections are:

- $P_{n \times l} = X_{n \times L} R_{L \times l}$
- The value of l can be much smaller than L.
- If the entries in R are independently selected from a zero mean and unit variance distribution, then it is possible to show that

$$E[||p_i||^2] = ||x_i||^2,$$

where p_i is the ith row of P and x_i is the ith row of X.
- Further, by selecting the value of l appropriately, we can preserve pairwise distances up to a factor of $(1 \pm \epsilon)$, where ϵ is a small real number.
- Similarly, it is possible to preserve dot products. That is

$$E[p_i^t p_j] = x_i^t x_j$$

Let us consider the MNIST data set, the subset of patterns belonging to class labels $\{7, 9\}$ from both the training and test sets. The number of training patterns is 12,107 and the number of test patterns is 2010. Each of these patterns is of size 28×28 ($= 784$) pixels.
The dimensionality of the data is reduced using the *Gaussian random projection* model from the Python sklearn package. The features extracted to represent the data points are based on 10 PCs, 10 SVs or 10 RPs. The KNN classifier with values of $k = 1, 11, 21, 31, 41$ is used for classification.
- The average, over different values of k, test set accuracies based on L_1 norm (city-block distance) and L_2 norm (Euclidean distance), and the time taken by each scheme for classification are shown in Table 6.3.

TABLE 6.3 Accuracies of schemes using PCs, SVs and RPs on MNIST Digit Classes 7 and 9

Features (scheme)	City-Block (average accuracy)	Euclidean (average accuracy)	Time (in secs)
10 PCs	70.15%	69.76%	4.51
10 SVs	69.3%	68.2%	4.16
10 RPs	83%	84%	1.72

We make the following observations from Table 6.3:

- Results based on PCs and SVs are almost similar. Accuracy of the scheme based on PCs is a shade better than the scheme based on SVs.
- Time taken based on SVs is marginally less than the one based on PCs.
- Results based on RPs are superior both in terms of accuracy and time compared to those based on PCs and SVs. However, we cannot generalize by saying that RPs are superior always.

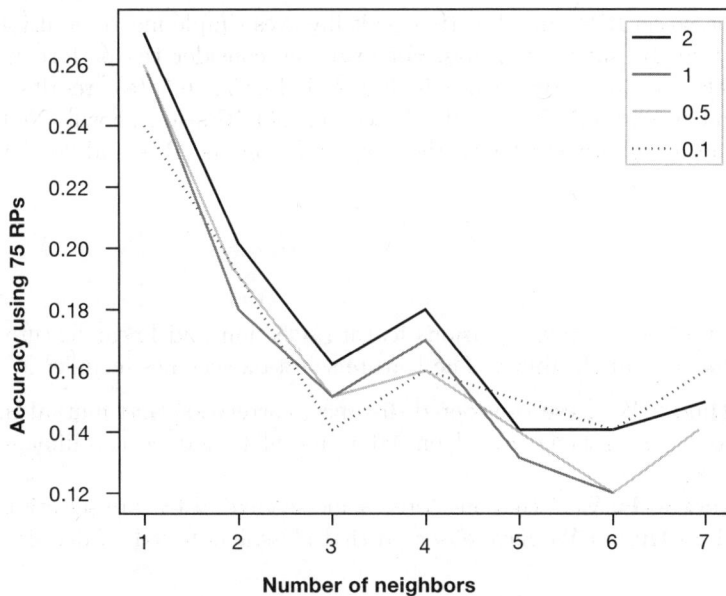

(a)

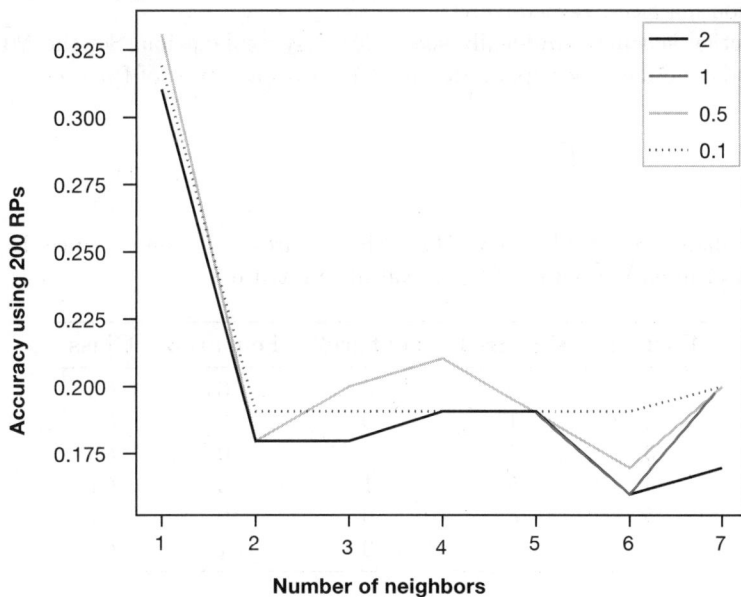

(b)

FIG. 6.5 Results on the Olivetti Face data set: 75 RPs and 200 RPs (for colour figure, please see Colour Plate 2)

- The scheme based on RPs takes less time as it involves simple matrix multiplication. However, accuracy may not be superior always. For example, consider the performance based on RPs on the Olivetti Face data set shown in Fig. 6.5. In Fig. 6.5 (a), results based on 75 RPs are shown and in Fig. 6.5 (b), results based on 200 RPs are shown. Neither of these two cases has an accuracy comparable to the ones based on 200 PCs and 200 SVs that are shown in Fig. 6.4.

SUMMARY

In this chapter, we have examined various feature selection and linear feature extraction techniques in terms of their applicability. The feature selection schemes include:

- **Filter methods:** We have considered distance, correlation and mutual information (MI)-based methods. The scheme based on MI is useful to reduce the dimensionality of large dimensional data sets.
- **Wrapper methods:** We have considered some sequential feature selection schemes.
- **Embedded methods:** We have observed that classifiers based on decision trees are of this category.

Subsequently, we have considered linear feature extraction schemes. These are based on principal components, singular vectors and random projections.

- Random projections provide a good alternative to deal with high-dimensional data sets; they take a smaller amount of time to extract compared to the other schemes.
- Schemes based on PCs and SVs are comparable as they deal with eigenvectors of covariance and correlation matrices, respectively.
- Feature selection schemes are ideally suited for easy explanation. So, the MI-based scheme is a very good candidate for explanation as it involves a subset of features.

EXERCISES

1. Consider the data in the table below. Rank the features based on variance. Does this ranking reflect the discriminative power of larger variance features?

Pattern	Feature1	Feature2	Feature3	Class
1	6	0	0	C_0
2	1	0	3	C_1
3	5	0	0	C_0
4	5	1	3	C_1
5	6	0	0	C_1
6	1	0	1	C_0

2. What are the maximum and minimum possible values of $r_{i,j}$, the correlation coefficient between two features $Feature i$ and $Feature j$? How do you justify?

3. Consider the data in the table below. Compute the correlation coefficients $r_{1,2}$, $r_{1,3}$ and $r_{2,3}$. If you need to select two out of the three features, then which feature is dropped?

Pattern	Feature1	Feature2	Feature3	Class
1	6	0	7	C_0
2	1	0	-3	C_1
3	5	0	5	C_0
4	5	1	5	C_1
5	6	0	7	C_1
6	1	0	-3	C_0

4. Use the details provided in Table 6.2. Compute the MI value of the word *Mining* with respect to the class DM. Consider the MI($Data$, DM) obtained in Example 2. Which of the two features is better?

5. Show that the set of all three-dimensional vectors, that is, $\Re^3$, is a vector space by examining all the properties of a vector space with the operations vector addition and scalar multiplication.

6. Show that the set of all 2×2 matrices from $\Re^{2\times 2}$ forms a vector space.

7. Consider the vector space $\Re^2$. Let S be a subset of $\Re^2$ such that the vectors in S are of the form (α, α), where $\alpha \in \Re$. Show that S is a subspace of the vector space $\Re^2$.

8. Consider a subset T of S in Q6. T has vectors of the form (α, α), where α is a non-negative real number, that is, $\alpha \geq 0$. Is T a subspace of $\Re^2$? Justify.

9. Consider the following sets of vectors. Does each of them form a basis for $\Re^2$? Justify.

   a. $\{(1,-1), (-1,1)\}$
   b. $\{(1,-1), (0,1)\}$

10. Consider the following transformations from $\Re^2$ to $\Re^2$. Are they linear transformations? Is it possible to view them as matrix transformations? What are the matrices?

    a. $T((a,b)) = (b, a)$
    b. $T((a,b)) = (b - a, b)$

11. Consider the matrix $A = \begin{bmatrix} 1 & 2 \\ 0 & 3 \end{bmatrix}$. This is not a symmetric matrix. Are its eigenvalues real? Justify your answer.

12. Let A be an upper triangular matrix of size $n \times n$, that is, $A_{i,j} = 0$ for $j < i$. Show that its diagonal entries form the eigenvalues.

13. Consider the two-dimensional data matrix X of 4 patterns.

$$X = \begin{bmatrix} 1 & 2 \\ 2 & 1 \\ 8 & 9 \\ 9 & 8 \end{bmatrix}.$$

Find

   a. The zero-mean data matrix X_{zm}.
   b. Obtain the covariance matrix.
   c. Compute the eigenvalues and orthonormal eigenvectors (PCs) of the covariance matrix.
   d. Project the data onto the PC directions and obtain the projection matrix X_{proj}.

14. Consider the matrix

$$X = \begin{bmatrix} 1 & -1 \\ 0 & 1 \\ 1 & 0 \end{bmatrix}.$$

Obtain XX^t and X^tX. What are the ranks of matrices X, XX^t and X^tX. How many non-zero eigenvalues are there for either XX^t or X^tX?

15. Consider the matrix

$$X = \begin{bmatrix} 1 & -1 \\ 0 & 1 \\ 1 & 0 \end{bmatrix}.$$

Obtain the SVD of X.

PRACTICAL EXERCISES

1. Download the Wisconsin Breast Cancer data set from sklearn. There are 569 samples corresponding to two classes. Each is a 30-dimensional vector. For more details, visit `https://scikit-learn.org/stable/modules/generated/sklearn.datasets.load\_breast\_cancer.html`.

   Your Tasks: There are three tasks. For all the tasks, split the data set into train and test parts using the default option. Split the data randomly 10 times and report the average accuracy. The tasks are:

   a. **Task 1:** In this task, you are supposed to reduce the dimensionality using 12 random projections. Use KNN with values of $K = 1, 3, 5, 7, \ldots, 25$. For each value of K, use KNN based on Minkowski distance with $r = 1$ and 2. Also consider fractional norms with $r = 0.5$ and 0.1. Compute the percentage accuracy.

   b. **Task 2:** In this task, you are supposed to reduce the dimensionality using 12 singular vectors. Use KNN with values of $K = 1, 3, 5, \ldots, 25$. For each value of K, use KNN based on Minkowski distance with $r = 1$ and 2. Also consider fractional norms with $r = 0.5$ and 0.1. Compute the percentage accuracy.

   c. **Task 3:** In this task, you are supposed to reduce the dimensionality using 12 principal components. Compute the KNN accuracy for different values of K and different distance measures as in Task 2.

Bibliography

- Murty, MN and Susheela Devi, V. 2015. *Introduction to Pattern Recognition and Machine Learning*, World Scientific Publishing Co. Pte. Ltd.: Singapore.

- Schölkopf, B and Smola, AJ. 2001. *Learning with Kernels*, MIT Press.

- Rifkin, RM. 2008. *Multiclass Classification*, Lecture Notes, Spring08, MIT, USA.

- Witten, IH, Frank, E and Hall, MA. 2011. *Data Mining: Practical Machine Learning Tools and Techniques*, Third Edition, Morgan Kaufmann Publishers, Burlington.

- Prakash, M and Murty, MN. 1995. A genetic approach for selection of (near-) optimal subsets of principal components for discrimination, *Pattern Recognition Letters* 16: 781–787, Elsevier.

- Lee, DD and Seung, HS. 2000. Algorithms for non-negative matrix factorization, NIPS'00: *Proceedings of the 13th International Conference on Neural Information Processing Systems*, 535–541.

- MNIST data set: `https://www.tensorflow.org/datasets/catalog/mnist`

- ORL Face data set: `https://www.kaggle.com/datasets/tavarez/the-orl-database-for-training-and-testing`

- Scikit-Machine Learning in Python: `https://scikit-learn.org/stable/`

CHAPTER 7
Clustering

Learning Objectives

At the end of this chapter, you will be able to:

- Define clustering
- Describe the commonly used clustering algorithms

7.1 INTRODUCTION TO CLUSTERING

Clustering refers to the process of arranging or organizing objects according to specific criteria. It plays a crucial role in uncovering concealed knowledge in large data sets. Clustering involves dividing or grouping the data into smaller data sets based on similarities/dissimilarities. Depending on the requirements, this grouping can lead to various outcomes, such as partitioning of data, data re-organization, compression of data and data summarization.

7.1.1 Partitioning of Data

Clustering of data is a crucial aspect of efficient data access in database-based applications.

EXAMPLE 1: To illustrate, let us take the example of an EMPLOYEE data table that contains 30,000 records of fixed length. We assume that there are 1000 distinct values of DEPT_CODE and that the employee records are evenly distributed among these values.

If this data table is clustered by DEPT_CODE, accessing all the employees of the department with DEPT_CODE = "15" requires $\log_2(1000) + 30$ accesses. The first term involves accessing the index table that is constructed using DEPT_CODE, which is achieved through binary search. The second term involves fetching 30 (that is, 30000/1000) records from the clustered (that is, grouped based on DEPT_CODE) employee table, which is indicated by the fetched entry in the index table.

Without clustering, accessing 30 employee records from a department would require, on average, $30 \times 30000/2$ accesses.

7.1.2 Data Re-organization

EXAMPLE 2: We can demonstrate the concept of data re-organization using the binary pattern matrix displayed in Table 7.1.

TABLE 7.1 Row data

Pattern	f1	f2	f3	f4	f5	f6
1	1	0	1	0	1	0
2	0	1	0	1	0	1
3	1	0	1	0	1	0
4	0	1	0	1	0	1

This data set comprises four binary patterns, each with six dimensions or features labelled f1 to f6 and assigned a unique ID ranging from 1 to 4. By clustering the rows, we can obtain the re-organized data set shown in Table 7.2. This process groups similar patterns together based on their row-wise similarities.

TABLE 7.2 Data re-organization using row as the criterion

Pattern	f1	f2	f3	f4	f5	f6
1	1	0	1	0	1	0
3	1	0	1	0	1	0
2	0	1	0	1	0	1
4	0	1	0	1	0	1

Further, by clustering the re-organized data set from Table 7.2 by column and grouping similar columns together, we obtain the data set displayed in Table 7.3.

TABLE 7.3 Data re-organization using rows and columns

Pattern	f1	f2	f3	f4	f5	f6
1	1	1	1	0	0	0
3	1	1	1	0	0	0
2	0	0	0	1	1	1
4	0	0	0	1	1	1

This table reveals the hidden structure in the data. The cluster comprising patterns 1 and 3 can be described using the conjunction $f1 \wedge f3 \wedge f5$. Likewise, the second cluster comprising patterns 2 and 4 can be described using $f2 \wedge f4 \wedge f6$.

7.1.3 Data Compression

Clustering can also assist in data compression by reducing the time complexity of accessing the data features and minimizing the space required to store them.

EXAMPLE 3:   To demonstrate, consider the data set shown in Table 7.1. Through clustering, we can compress the data set, resulting in the table shown in Table 7.4. Note that the frequent pattern (FP) tree and pattern count (PC) tree structures are some examples for achieving such compression of data.

TABLE 7.4 Compressed data

Pattern	f1	f2	f3	f4	f5	f6	Count
1, 3	1	0	1	0	1	0	2
2, 4	0	1	0	1	0	1	2

7.1.4 Summarization

The objective of data summarization is to extract a representative subset of samples from a large data set. The purpose is to simplify and expedite analysis on a smaller data set.

EXAMPLE 4:   For instance, when dealing with a data set of the marks of 100 students in a subject, statistical measures like mean (the numerical average of the marks), mode (the most frequently repeated marks) and median (the value in the middle of all the marks when the marks are ranked in order) can provide summarized information. Such summarized information can be derived through clustering.

7.1.5 Matrix Factorization

It is possible to view clustering as matrix factorization.

Let there be n data points in an l-dimensional space. We can represent it as a matrix $X_{n \times l}$. It is possible to approximate X as a product of two matrices $B_{n \times K}$ and $C_{K \times l}$. So, $X \approx BC$, where B is the cluster assignment matrix and C is the representatives matrix.

EXAMPLE 5:   Consider the data matrix

$$X = \begin{bmatrix} 6 & 6 & 6 \\ 6 & 6 & 8 \\ 2 & 4 & 2 \\ 2 & 2 & 2 \end{bmatrix}$$

There are four patterns in a three-dimensional space. If we use the leader algorithm with a threshold of 3 units, we get two clusters: (6,6,6) and (6,6,8) that belong to cluster 1 with (6,6,6) as its leader and (2,4,2) and (2,2,2) that belong to cluster 2 with (2,4,2) as its leader.

So, we have $B = \begin{bmatrix} 1 & 0 \\ 1 & 0 \\ 0 & 1 \\ 0 & 1 \end{bmatrix}$. It indicates that the first pattern, (6,6,6), is assigned to cluster 1.

Correspondingly in B we have a 1 in the first row and first column. Matrix B has four rows corresponding to four patterns in X; B has two columns corresponding to two clusters. Matrix C will have K rows if there are K clusters and the ith row is the representative of the ith cluster.

In the current example, $C = \begin{bmatrix} 6 & 6 & 6 \\ 2 & 4 & 2 \end{bmatrix}$. So the resulting matrix factorization is given by

$$\begin{bmatrix} 6 & 6 & 6 \\ 6 & 6 & 8 \\ 2 & 4 & 2 \\ 2 & 2 & 2 \end{bmatrix} \approx \begin{bmatrix} 1 & 0 \\ 1 & 0 \\ 0 & 1 \\ 0 & 1 \end{bmatrix} \times \begin{bmatrix} 6 & 6 & 6 \\ 2 & 4 & 2 \end{bmatrix}$$

Leader is a hard clustering algorithm where each pattern is assigned fully, that is, with a value 1, to a single cluster. So, each row of the B matrix will have one 1 and the rest as 0s.

It is possible to have soft clustering. In this case, we can have multiple non-zero entries in the range [0,1] in each row, indicating that a pattern is assigned to more than one cluster. The sum of the entries in a row is 1.

Instead of the leader, if we use the centroid of points in the cluster as its representative, then $C = \begin{bmatrix} 6 & 6 & 7 \\ 2 & 3 & 2 \end{bmatrix}$. There is no change in B. Note that the centroid of $\{(6,6,6), (6,6,8)\}$ is $(6,6,7)$ and the centroid of the second cluster $\{(2,4,2), (2,2,2)\}$ is $(2,3,2)$.

7.2 CLUSTERING OF PATTERNS

Clustering is a technique that involves grouping a set of patterns, resulting in the creation of cohesive clusters or groups from a given collection of patterns. The process of clustering aims to group similar patterns together while keeping dissimilar patterns separate. Figure 7.1 illustrates the clustering of a two-dimensional data set (represented by f1 and f2), where three clusters are visually identifiable.

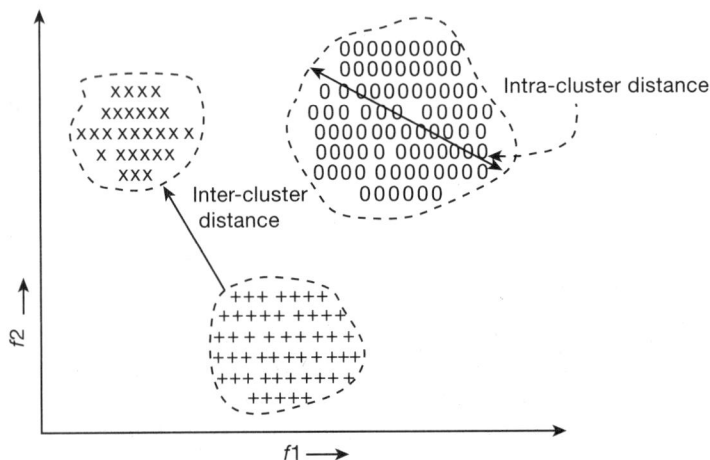

FIG. 7.1 Inter- and intra-cluster distances

The clustering process may ensure that the distance between any two points within the same cluster (intra-cluster distance), as measured by a dissimilarity measure such as Euclidean distance, is smaller than the distance between any two points belonging to different clusters (inter-cluster distance).

This indicates that the similarity between points within a cluster is higher than the similarity between clusters. We illustrate this idea with the following example.

EXAMPLE 6:   Let us consider a two-dimensional data set with 11 data points having two features f1 and f2 as listed in Table 7.5.

TABLE 7.5 Data set with two features

Data point	f1	f2
x_1	1	1
x_2	1	3
x_3	2	2
x_4	3	1
x_5	3	3
x_6	4	7
x_7	5	7
x_8	6	9
x_9	7	1
x_{10}	7	3
x_{11}	9	1

Any two points are placed in the same cluster if the distance between them is lower than a certain threshold. In this example, the squared Euclidean distance is used to measure the distance between the points, and a threshold of 10 units is set to cluster them. The squared Euclidean distance (d) between two points, x_i and x_j, is calculated as follows:

$$d(x_i, x_j) = \sum_{k=1}^{l} (x_i(k) - x_j(k))^2,$$

where l represents the dimensionality of the points.

In this example, since we are dealing with two-dimensional data points, l equals 2. Using this formula, the squared Euclidean distance between all pairs of points can be found and is presented in Table 7.6.

In Table 7.6, the clusters are clearly visible within the matrix itself. The table contains three sub-matrices of sizes 5×5, 3×3 and 3×3 that meet the condition of having a maximum value of 10 in any entry. For instance, the sub-matrix of size 5×5, corresponding to the first five patterns (say, Cluster_A = $\{x_1, x_2, x_3, x_4, x_5\}$) has values ranging from 0 to 8, with every other entry in the first five rows exceeding 10. Similarly, it can be deduced that patterns x_6, x_7 and x_8 belong to the second cluster (Cluster_B = $\{x_6, x_7, x_8\}$) while patterns x_9, x_{10} and x_{11} belong to the third cluster (Cluster_C = $\{x_9, x_{10}, x_{11}\}$).

One can observe from Table 7.6 that none of the data points is present in more than one cluster. Such a clustering is called *hard clustering*, otherwise it is known as *soft or overlapping clustering*. We discuss soft clustering in Section 7.6.

TABLE 7.6 Squared Euclidean distance between pairs of data points listed in Table 7.4

Data point	x_1	x_2	x_3	x_4	x_5	x_6	x_7	x_8	x_9	x_{10}	x_{11}
x_1	0	4	2	4	8	45	52	89	36	40	64
x_2	4	0	2	8	4	25	32	61	40	36	68
x_3	2	2	0	2	2	29	34	65	26	26	50
x_4	4	8	2	0	4	37	40	73	16	20	36
x_5	8	4	2	4	0	17	20	45	20	16	40
x_6	45	25	29	37	17	0	1	8	45	25	61
x_7	52	32	34	40	20	1	0	5	40	20	52
x_8	89	61	65	73	45	8	5	0	65	37	73
x_9	36	40	26	16	20	45	40	65	0	4	4
x_{10}	40	36	26	20	16	25	20	37	4	0	8
x_{11}	64	68	50	36	40	61	52	73	4	8	0

Let $\mathfrak{X} = \{x_1, x_2, \ldots, x_n\}$ be n data points which are mapped to C clusters, $\mathfrak{C} = \{X_1, X_2, \ldots, X_c\}$ such that

- $\cup_{i=1}^{C} X_i = \mathfrak{X}$
- $X_i \neq \phi, \forall i \in \{1, 2, \ldots, c\}$

If $\forall i, j, i \neq j, x_i \cap x_j = \phi$, then the clustering is **hard**, else it is **soft**.

7.2.1 Data Abstraction

Clustering is a useful method for data abstraction, and it can be applied to generate clusters of data points that can be represented by their centroid, medoid, leader or some other suitable entity.

The centroid is computed as the sample mean of the data points in a cluster, and it is given by $\frac{1}{n_C} \times \sum(x_i \in C)$, where n_C is the number of patterns in the cluster C. Note that it may not coincide with any specific data point, as shown in Fig. 7.2.

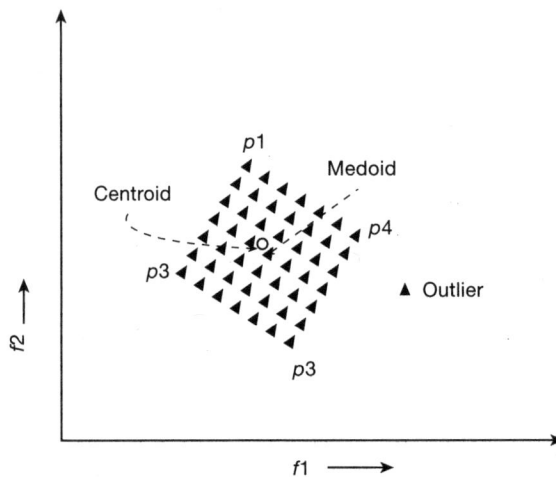

FIG. 7.2 Visualization of centroid and medoid

The medoid is the point that minimizes the sum of distances to all other points in the cluster. Figure 7.2 illustrates this process, where an outlier that is located far away from the data points in the cluster is also present. The centroid can shift dramatically based on the position of the outlier, while the medoid remains stable within the boundaries of the original cluster. Therefore, clustering algorithms that utilize medoids are more resilient to noisy patterns or outliers.

EXAMPLE 7: Consider the data set shown in Table 7.5 and the cluster thus generated given in Table 7.6. The centroid of the clusters, Cluster_A (that is, μ_A), Cluster_B (that is, μ_B) and Cluster_C (that is, μ_C), are given below:

$$\mu_A = \frac{1}{5}[(1,1) + (1,3) + (2,2) + (3,1) + (3,3)] = \left(\frac{1+1+2+3+3}{5}, \frac{1+3+2+1+3}{5}\right) = (2,2)$$

$$\mu_B = \frac{1}{3}[(4,7) + (5,7) + (6,9)] = \left(\frac{4+5+6}{3}, \frac{7+7+9}{3}\right) = (5, 7.67)$$

$$\mu_C = \frac{1}{3}[(7,1) + (7,3) + (9,1)] = \left(\frac{7+7+9}{3}, \frac{1+3+1}{3}\right) = (7.67, 1.67)$$

For the same clusters, the medoids are given below:

$$m_A = (2,2), m_B = (5,7) \text{ and } m_C = (7,3)$$

We know that the medoid of a group of points is defined as a data point within the cluster that is located at the centre, where the total distance from the points in the cluster is at its minimum. Using the squared Euclidean distance as the metric for calculating distance, the distance of the points from m_A, m_B and m_C can be determined as follows:

For Cluster_A: Considering X_5 as the medoid (m_A):

$d(x_1, m_A) = 2$
$d(x_2, m_A) = 2$
$d(x_3, m_A) = 0$
$d(x_4, m_A) = 2$
$d(x_5, m_A) = 2$

The sum of the distances $= 8$. Note that if you choose any other data point, say x_1 as m_A, then the squared Euclidean distance is:

$d(x_1, m_A) = 0$
$d(x_2, m_A) = 4$
$d(x_3, m_A) = 2$
$d(x_4, m_A) = 4$
$d(x_5, m_A) = 8$

The sum of the distances $= 18$. Similarly, one can find that on choosing any point other than x_3 as the medoid, the corresponding sum of the distances is greater than 8.

For Cluster_B: Considering x_7 as medoid (m_B):

$d(x_6, m_B) = 1$
$d(x_7, m_B) = 0$
$d(x_8, m_B) = 5$

The sum of the distances $= 6$. Note that if you choose any other data point as m_B, then the squared Euclidean distance is greater than 6. So x_7 is the medoid of Cluster_B.

For Cluster_C: Considering x_9 as medoid (m_C):

$d(x_9, m_C) = 0$
$d(x_{10}, m_C) = 4$
$d(x_{11}, m_C) = 4$

The sum of the distances $= 8$. Note that if you choose any other data point as m_C, then the squared Euclidean distance is greater than 8. So X_9 is the medoid of Cluster_C.

In the above example, even though centroid and medoid are the same for Cluster_A, they may not be the same as shown for Cluster_B and Cluster_C.

The following example shows that the medoid of a cluster is more robust to the addition of outliers.

EXAMPLE 8: Consider the cluster of three data points given below:

$$x_1 = (4, 1) \quad x_2 = (5, 1) \quad x_3 = (6, 3)$$

The centroid μ for this cluster is

$$\mu = \left(\frac{4 + 5 + 6}{3}, \frac{1 + 1 + 3}{3} \right) = (5, 1.67)$$

and the medoid m for this cluster can be obtained from Table 7.7.

TABLE 7.7 Squared Euclidean distance between the data points

	x_1	x_2	x_3	$\sum$
x_1	0	1	8	9
x_2	1	0	5	6
x_3	8	5	0	13

In Table 7.7, the data point x_2 has the minimum squared Euclidean distance with all other data points (refer the last column of the table). So, x_2 (that is, $(5,1)$) is the medoid.

If we add an outlier point, $x_4 = (20,1)$, then the new centroid is

$$\dot{\mu} = \left(\frac{4 + 5 + 6 + 20}{4}, \frac{1 + 1 + 3 + 1}{4} \right) = (8.75, 1.5)$$

and the new medoid $\dot{m}$ for this cluster can be obtained from Table 7.8. In Table 7.8, the data point x_3 has the minimum squared Euclidean distance with all other data points (refer the last column of the table). So, x_3 (that is, $(5,1)$) is the medoid.

One can observe that because of x_4 (an outlier in this case), the centroid is shifted from $\mu = (5,1.67)$ to $\dot{\mu} = (8.75, 1.5)$ and the medoid is shifted from $m = (5,1)$ to $\dot{m}=(6,3)$. But the change from m to $\dot{m}$ is very small when compared to the change from μ to $\dot{\mu}$. So, the medoid of the cluster is more robust to the addition of outliers.

TABLE 7.8 Squared Euclidean distance between the data points

	x_1	x_2	x_3	x_4	$\sum$
x_1	0	1	8	256	265
x_2	1	0	5	225	231
x_3	8	5	0	200	213
x_4	256	225	200	0	681

Note that in the above examples, we have used centroid and medoid as the representatives or descriptions of the cluster. However, it is feasible to have more than one or multiple representative elements per cluster.

As an instance, a cluster can be represented by four extreme points, namely, p1, p2, p3 and p4, as illustrated in Fig. 7.2. Additionally, it is possible to use a logical description to depict a cluster. For instance, a cluster can be characterised by (dept_name = "Computer Sc" ∨ dept_name = "Mech Engg" ∧ Salary > 200,000).

Clusters are useful in several decision-making situations like classification, prediction, etc. However, when the number of cluster representatives obtained is smaller than the number of input patterns, there is a reduction in computation while performing classification or prediction activity. The following example demonstrates this.

EXAMPLE 9: Consider the data set shown in Table 7.9. Let (2, 4) be a test pattern which needs to be classified using the nearest neighbor classifier on the 11 labelled patterns of Table 7.9.

TABLE 7.9 Data set with two features and corresponding labels

Data point	f1	f2	Class label
x_1	1	1	A
x_2	1	3	A
x_3	2	2	A
x_4	3	1	A
x_5	3	3	A
x_6	4	7	B
x_7	5	7	B
x_8	6	9	B
x_9	7	1	B
x_{10}	7	3	B
x_{11}	9	1	B

Note that (2, 4) is the nearest neighbor of (3, 3) with a squared Euclidean distance of 2 units between them. This can be obtained by computing 11 distance values – calculating the squared Euclidean distance between the test pattern (2, 4) and each of the 11 labelled patterns. So, (2, 4) is classified as belonging to class "A" because its nearest neighbor (3, 3) belongs to "A".

However, by clustering the 11 patterns using a suitable clustering algorithm and representing each resulting cluster by its centroid, we can reduce the number of distance values to be computed, from the number of labelled patterns to the number of cluster representatives. This may be illustrated as follows.

Let the 11 patterns be clustered using the same criterion as the one used in Example 7, that is, the squared Euclidean distance between any two patterns placed in the same cluster should be within a threshold of 10 units. This results in three clusters: one from class "A" and two from class "B". The cluster corresponding to class "A" is:

$$Cluster\_A = \{(1,1), (1,3), (2,2), (3,1), (3,3)\} \text{ with centroid } \mu_A = (2,2)$$

The two clusters corresponding to class "B" are:

$$Cluster\_B = \{(4,7), (5,7), (6,9)\} \text{ with centroid } \mu_B = (5, 7.67)$$

$$Cluster\_C = \{(7,1), (7,3), (9,1)\} \text{ with centroid } \mu_C = (7.67, 1.67)$$

So, instead of using the 11 labelled patterns, if one uses the centroids of the three clusters as the labelled data, then there is a reduction in the data to be analysed by a factor of 3. The reduced labelled data is given in Table 7.10.

TABLE 7.10 Reduced data set with only cluster representatives

Data point	f1	f2	Class label
μ_A	2	2	A
μ_B	5	7.67	B
μ_C	7.67	1.67	B

By referring to Table 7.10, it is possible to determine that the nearest neighbor of the point (2, 4) belongs to class "A", specifically the point (2, 2). As a result, the test pattern is classified as belonging to class "A" by making only three distance computations between the test pattern and each of the three centroids. Moreover, it should be noted that for the distance computations, only the centroids of the three clusters need to be stored, leading to a reduction in both time and space.

In this example, since we need to compute only three distances instead of 11 and need to store only three centroids out of 11 patterns, there is a 73% reduction in both time and space.

It could be argued that clustering the labelled data and computing the centroids may require a significant amount of time and space. However, it is possible to perform the clustering process as soon as the labelled data becomes available, prior to the classification stage. Furthermore, computation of the cluster centroid is a one-time affair. After obtaining the representative centroids, we can use them as labelled data to classify new test patterns, resulting in a significant reduction in both time and space requirements.

This clustering-based classification approach is particularly valuable for solving large-scale pattern classification problems that commonly arise in machine learning and also in dealing with *class imbalance*, where patterns in the majority class are clustered to have a smaller number of representatives from the majority class.

7.2.2 Clustering Algorithms

A wide variety of clustering algorithms are in use today. Figure 7.3 shows the taxonomy of clustering algorithms.

Clustering algorithms can be classified into hard and soft clustering, depending on whether clusters share data points or not. Hard clustering algorithms generate a partition of the given data set, while hierarchical algorithms generate a nested sequence of partitions. On the other hand, soft clustering algorithms utilize fuzzy sets, rough sets or evolutionary algorithms.

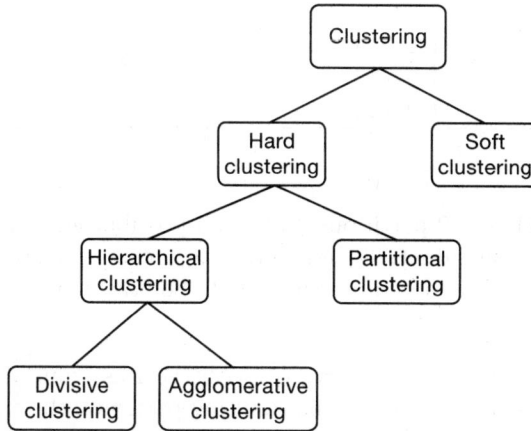

FIG. 7.3 Classification of clustering algorithms

Hierachical Algorithms

The process of generating a sequence of data partitions using hierarchical algorithms can be represented using a tree structure called a **dendrogram**. There are two types of hierarchical algorithms – divisive and agglomerative.

Divisive algorithms follow a top-down approach, where a single cluster with all the patterns is split into smaller clusters at each step until there is only one pattern in each cluster or a collection of singleton clusters. On the other hand, agglomerative algorithms follow a bottom-up approach, where n singleton clusters are created for n input patterns. At each level, the two most similar clusters are merged to reduce the size of the partition by 1.

Once two patterns are placed in the same cluster in agglomerative algorithms, they remain in the same cluster in subsequent levels. Similarly, once two patterns are placed in different clusters in divisive algorithms, they remain in different clusters in subsequent levels.

In the next sections, we discuss divisive clustering, agglomerative clustering and partitional clustering in detail.

7.3 DIVISIVE CLUSTERING

Divisive algorithms are either polythetic, where the division is based on more than one feature, or monothetic, when only one feature is considered at a time. The **polythetic scheme** is based on finding all possible 2-partitions of the data and choosing the best among them. If there are n patterns, the number of distinct 2-partions is given by $\frac{2^n - 2}{2} = 2^{n-1}$ - 1.

EXAMPLE 10: For example, if the data set contains three patterns, x_1, x_2 and x_3, the possible 2-partitions among these patterns are given in Table 7.11.

TABLE 7.11 All possible 2-partition sets for the patterns x_1, x_2 and x_3

Possibilities	Non-empty subset	Its complement	2-partition
1	$\{x_1\}$	$\{x_2, x_3\}$	$\{x_1, \{x_2, x_3\}$
2	$\{x_2\}$	$\{x_1, x_3\}$	$\{x_2, \{x_1, x_3\}$
3	$\{x_3\}$	$\{x_1, x_2\}$	$\{x_3, \{x_1, x_2\}$
4	$\{x_2, x_3\}$	$\{x_1\}$	$\{\{x_2, x_3\}, x_1\}$
5	$\{x_1, x_3\}$	$\{x_2\}$	$\{\{x_1, x_3\}, x_2\}$
6	$\{x_1, x_2\}$	$\{x_3\}$	$\{\{x_1, x_2\}, x_3\}$

Note that in Table 7.11, the 2-partitions under the possibilities 1 and 4, 2 and 5 and 3 and 6 are repetitions. Further, the subsets ϕ and $\{x_1, x_2, x_3\}$ can be ignored. This is because ϕ is an empty set and $\{x_1, x_2, x_3\}$ is an improper subset. So, the number of possible 2-partitions is $\frac{2^3 - 2}{2} = 3$.

Among all possible 2-partitions, the partition with the least sum of the sample variances of the two clusters is chosen as the best. From the resulting partition, the cluster with the maximum sample variance is selected and is split into an optimal 2-partition. This process is repeated till we get singleton clusters. If a collection of patterns (data points) is split into two clusters with p patterns $x_1, \ldots, x_p$ in one cluster and q patterns $y_1, \ldots, y_q$ in the other cluster, with the centroids of the two clusters being C1 and C2, respectively, then the sum of the sample variances will be

$$\sum_{i=1}^{p}(x_i - C1)^2 + \sum_{j=1}^{q}(y_j - C2)^2$$

EXAMPLE 11: Consider the data set containing eight points (patterns) with two features as shown in Table 7.12. Figure 7.4 shows the visual representation of these eight data patterns.

TABLE 7.12 Example of data patterns with two features

Patterns	f1	f2
x_1	5.5	5.5
x_2	7	6.5
x_3	7	5.5
x_4	10	6
x_5	10.75	6
x_6	10	8
x_7	10.5	8
x_8	7	8

Figure 7.5 shows the dendrogram corresponding to the divisive clustering using the procedure discussed above. The top of the dendrogram shows a single cluster consisting of all eight data points. By evaluating all possible 2-partitions (out of $2^7 - 1 = 127$), the best 2-partition $\{\{x_1, x_2, x_3, x_8\}, \{x_4, x_5, x_6, x_7\}\}$ is obtained and shown in the dendrogram.

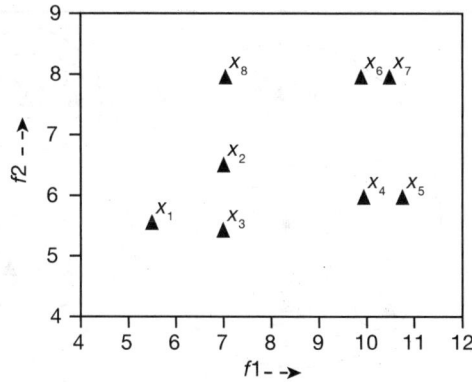

FIG. 7.4 Visual representation of data patterns listed in Table 7.12

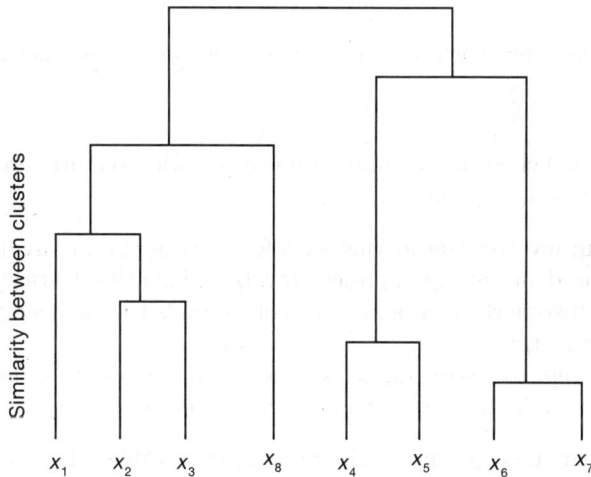

FIG. 7.5 Divisive clustering for polythetic clustering for the data points listed in Table 7.12

The cluster $\{x_4, x_5, x_6, x_7\}$ is then selected to split into two clusters, $\{x_4, x_5\}$ and $\{x_6, x_7\}$. After the split, the three clusters at that level of the dendrogram are $\{x_1, x_2, x_3, x_8\}$, $\{x_4, x_5\}$ and $\{x_6, x_7\}$.

At the subsequent levels, the cluster $\{x_1, x_2, x_3, x_8\}$ is split into $\{x_1, x_2, x_3\}$ and $\{x_8\}$, and $\{x_1, x_2, x_3\}$ is further divided into $\{x_1\}$ and $\{x_2, x_3\}$. This results in five clusters: $\{x_1\}$, $\{x_2, x_3\}$, $\{x_8\}$, $\{x_4, x_5\}$ and $\{x_6, x_7\}$.

The dendrogram in Fig. 7.5 illustrates these clusters, as well as the partitions with 6, 7 and 8 clusters at successive levels. The same is represented in an onion layer diagram (Fig. 7.6), which shows the cluster generation order numerically. Observe that at the final level, each cluster has only one point; such clusters are called singleton clusters.

It is important to highlight that in order to find the optimal 2-partition for a cluster of size m, it is necessary to consider $2^m - 1$ possible 2-partitions and select the best one. Therefore, generating

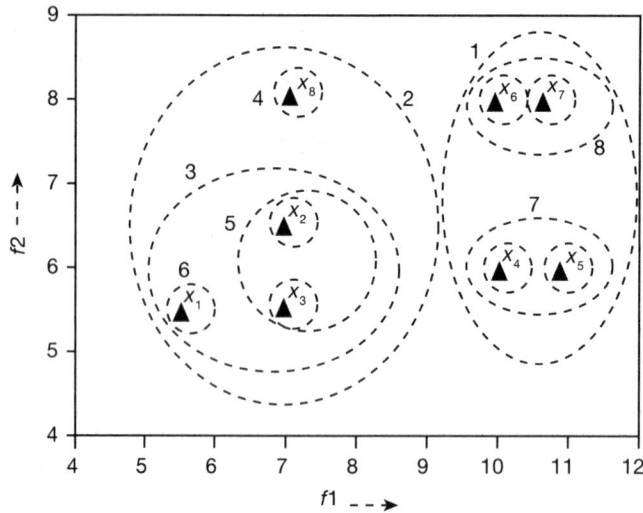

FIG. 7.6 Divisive clustering with cluster generation order for the data points listed in Table 7.12

all possible 2-partitions and choosing the most suitable partition requires an effort of $\mathcal{O}(2^n)$, where n represents the number of data points.

Monothetic clustering involves considering each feature direction individually and dividing the data into two clusters based on the gap in projected values along that feature direction. Specifically, the data set is split into two parts at a point that corresponds to the mean value of the maximum gap observed among the feature values.

This process is then repeated sequentially for the remaining features, further partitioning each cluster. The following example illustrates monothetic clustering.

EXAMPLE 12: Consider the eight data points provided in Table 7.12, each having two features, f1 and f2. The first step is to split the data based on the maximum inter-pattern gap observed in the f1 direction. Sorting the f1 values in ascending order yields $x_1 : 5.5$, $x_2 : 7$, $x_3 : 7$, $x_8 : 7$, $x_4 : 10$, $x_6 : 10$, $x_7 : 10.5$ and $x_5 : 10.75$. The largest gap of 3 units is between x_8 and x_4. We select the mid-point between 7 and 10, which is 8.5 (that is, $7 + 1.5$), and use it to split the data into two clusters: $C1 = \{x_1, x_2, x_3, x_8\}$ and $C2 = \{x_4, x_5, x_6, x_7\}$.

Next, each of these clusters is further divided based on the f2 values. Sorting the patterns in C1 by their f2 values gives $x_1 : 5.5$, $x_3 : 5.5$, $x_2 : 6.5$ and $x_8 : 8$. The largest gap of 1.5 units ($8 - 6.5 = 1.5$ units) occurs between x_2 and x_8. We split C1 at the midpoint 7.25 (that is, $6.5 + 0.75$) along the f2 direction, resulting in two clusters: $C11 = \{x_8\}$ and $C12 = \{x_1, x_2, x_3\}$. Similarly, by splitting C2 using the value 7 for f2, we obtain $C21 = \{x_6, x_7\}$ and $C22 = \{x_4, x_5\}$; by splitting C12 using the value 6.25 for f1, we obtain $C121 = \{x_1\}$ and $C122 = \{x_2, x_3\}$. Figure 7.7 shows this monothetic clustering.

However, a challenge with this approach is that in the worst case, the initial data set may be divided into 2^l clusters when considering all l features. Having such a large number of clusters might not be practical, necessitating an additional merging phase. In this phase, pairs of clusters are selected based on their proximity and merged into a single cluster. This process is repeated until the desired number of clusters is achieved.

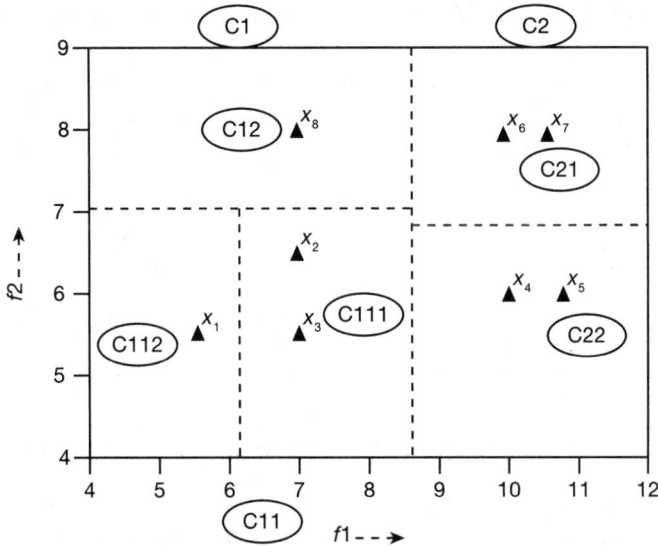

FIG. 7.7 Monothetic clustering for the data points listed in Table 7.12

One commonly used criterion for proximity is the distance between the centroids of the clusters. The time complexity of sorting the n elements and finding the maximum inter-pattern gap in each feature direction is $\mathcal{O}(n \log n)$. Considering l features, the overall effort is $\mathcal{O}(ln \log n)$. However, this approach may become infeasible for large values of l and n.

7.4 AGGLOMERATIVE CLUSTERING

An agglomerative clustering algorithm generally uses the following steps:

1. Compute the proximity matrix for all pairs of patterns in the data set.
2. Find the closest pair of clusters based on the computed proximity measure and merge them into a single cluster. Update the proximity matrix to reflect the merge, adjusting the distances between the newly formed cluster and the remaining clusters.
3. If all the patterns belong to a single cluster, terminate the algorithm. Otherwise, go back to Step 2 and repeat the process until all the patterns are in one cluster.

By iteratively merging the closest clusters, the algorithm gradually builds a hierarchy of clusters, with each iteration reducing the number of clusters until a stopping criterion is met. The following example illustrates agglomerative clustering.

EXAMPLE 13: Consider the eight data points specified in Table 7.12. To start with, each data point is a singleton cluster. As per the method, the first step is to compute the proximity between the data points. Table 7.13 shows the proximity between the data points using city-block distance. Note that the city-block (Manhattan) distance between any two data points, x_1 and x_2, is given by $M(x_1, x_2) = \sum_{i=1}^{l}(\mid x_1(i) - x_2(i) \mid)$, where l is the number of features in the data point.

TABLE 7.13 City-block distance between pairs of data points

	x_1	x_2	x_3	x_4	x_5	x_6	x_7	x_8
x_1	0	2.5	1.5	5	5.5	7	7.5	4
x_2	2.5	0	1	3.5	4.25	4.5	5	1.5
x_3	1.5	1	0	3.5	4.25	5.5	6	2.5
x_4	5	3.5	3.5	0	0.75	2	2.5	5
x_5	5.75	4.25	4.25	0.75	0	2.75	2.25	5.75
x_6	7	4.5	5.5	2	2.75	0	0.5	3
x_7	7.5	5	6	2.5	2.25	0.5	0	3.5
x_8	4	1.5	2.5	5	5.75	3	3.5	0

The second step is to find the closest pair of clusters, whose city-block distance is minimum. Since the clusters $\{x_6\}$ and $\{x_7\}$ are the closest to each other with a distance of 0.5 units, they are merged as C1 to realise a partition of 7 clusters.

The proximity matrix after the merger is given in Table 7.14. The merging uses a *single-link* strategy. That is, the distance between any pair of clusters Cp and Cq is $min_{x_i \in Cp \text{ and } x_j \in Cq}(d(x_i, x_j))$.

TABLE 7.14 City-block distance among the data points after merging x_6 and x_7 as one cluster

	x_1	x_2	x_3	x_4	x_5	$C1=\{x_6, x_7\}$	x_8
x_1	0	2.5	1.5	5	5.5	7	4
x_2	2.5	0	1	3.5	4.25	4.5	1.5
x_3	1.5	1	0	3.5	4.25	5.5	2.5
x_4	5	3.5	3.5	0	0.75	2	5
x_5	5.75	4.25	4.25	0.75	0	2.75	5.75
$C1 = \{x_6, x_7\}$	7	4.5	5.5	2	2.25	0	3
x_8	4	1.5	2.5	5	5.75	3	0

In the third step, we need to repeat Step 2 till we reach a single cluster of the required number of clusters. In Table 7.14, one can observe that the pair of clusters $\{x_4\}$ and $\{x_5\}$ has a minimum distance (that is, 0.75 units) and hence the clusters are merged next to form C2 = $\{x_4, x_5\}$. Similarly, the clusters $\{x_2\}$ and $\{x_3\}$ are merged to create C3 = $\{x_2, x_3\}$. Then, C3 is merged with cluster $\{x_1\}$, resulting in C4 = $\{x_1, x_2, x_3\}$. Subsequently, C4 is merged with cluster $\{x_8\}$, leading to C5 = $\{x_1, x_2, x_3, x_8\}$. Clusters C2 and C1 are then merged to form C6 = $\{x_4, x_5, x_6, x_7\}$.

At this point, there are two clusters remaining, namely, C5 and C6. The process can be terminated once the desired number of clusters is achieved. The dendrogram in Fig. 7.8 illustrates the merging of clusters at different levels.

The time complexity of agglomerative clustering is $\mathcal{O}(n^2)$, where n is the number of data points. This is to compute the proximity between all pairs of data points in the data set. Since the computed proximity matrix needs to be stored, the space complexity is also $\mathcal{O}(n^2)$.

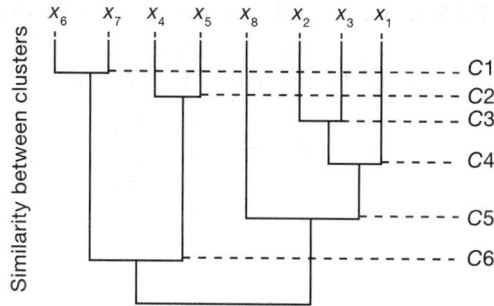

FIG. 7.8 Agglomerative clustering for the data points listed in Table 7.12

7.5 PARTITIONAL CLUSTERING

We have discussed a partitional clustering algorithm, the Leader clustering algorithm, in an earlier chapter. One of the most popular algorithms under this category is the k-means clustering algorithm.

7.5.1 *K*-Means Clustering

The k-means algorithm can be described in the following steps:

1. Select k initial cluster centres from the given n data points. The remaining $(n - k)$ data points are then assigned to one of the k clusters based on their proximity to the closest cluster centre.
2. Compute the new cluster centres based on the current assignment of data points. This is done by calculating the mean or centroid of all the data points belonging to each cluster.
3. Assign each of the n data points to the cluster centre that is closest in proximity to it.
4. Check if there have been any changes in the assignment of data points to clusters. If no changes have occurred, terminate the algorithm; otherwise, go back to Step 2 and repeat the process.

By iteratively updating the cluster centres and reassigning data points until convergence, the k-means algorithm aims to find a partition that minimizes the overall within-cluster variance or distance.

We illustrate the algorithm using the following example.

EXAMPLE 14: Consider the data set containing eight points with two features as shown in Table 7.15.

Let us assume that the number of clusters, $k = 3$. If we select the initial clusters as x_1, x_4 and x_7, then Cluster 1 has $(1, 1)$ as its cluster centre, Cluster 2 has $(7, 2)$ as its cluster centre and Cluster 3 has $(7, 9)$ as its cluster centre.

Table 7.16 shows the Euclidean distance (a proximity measure) between the cluster centre and the other data points. The patterns x_2, x_3, x_5, x_6 and x_8 are assigned to their respective clusters as shown in the last column of Table 7.16.

TABLE 7.15 Example of data patterns with two features

Data points	f1	f2
x_1	1	1
x_2	1	3
x_3	3	3
x_4	7	2
x_5	8	2
x_6	9	3
x_7	7	9
x_8	8	9

TABLE 7.16 Euclidean distances with cluster assignments

Data points	Euclidean distance from C1	Euclidean distance from C2	Euclidean distance from C3	Assigned cluster
x_1	0	$\sqrt{37}$	10	C1
x_2	2	$\sqrt{37}$	$\sqrt{72}$	C1
x_3	$\sqrt{8}$	$\sqrt{17}$	$\sqrt{52}$	C1
x_4	$\sqrt{37}$	0	7	C2
x_5	$\sqrt{50}$	1	$\sqrt{50}$	C2
x_6	$\sqrt{68}$	$\sqrt{5}$	$\sqrt{40}$	C2
x_7	10	7	0	C3
x_8	$\sqrt{113}$	$\sqrt{50}$	1	C3

The next step involves computing the new cluster centres. The new cluster centre of C1 will be the mean of the patterns in Cluster 1(mean of x_1, x_2 and x_3), which will be (1.67, 2.33). The cluster centre of C2 will be (8,2.33) and the cluster centre of $C3$ will be (7.5, 9).

The patterns are again assigned to the closest cluster depending on the distance from the cluster centres. Now, x_1, x_2 and x_3 are assigned to Cluster 1, x_4, x_5 and x_6 are assigned to Cluster 2 and x_7 and x_8 are assigned to Cluster 3. Table 7.17 shows the Euclidean distance from the new cluster centre to the data points with their new assigned clusters.

Since there is no change in the clusters formed, this is the final set of clusters. The visual representation of the clusters is given in Fig. 7.9.

One of the important points to be noted in k-means clustering is that the algorithm is *sensitive to the selection of initial centroids*. That is, cluster formation changes with the initially chosen cluster centres. For example, if we select the initial cluster centre as x_1, x_2 and x_3, the clustering generated is as shown in Fig. 7.10. (Cluster formation using k-means clustering with $k = 3$ is left as an exercise to the reader.)

The goodness of the k-means clustering algorithm is based on the sum of the squared deviations of data points in a cluster from its centre. More formally, if C_i is the ith cluster and μ_i is its centre,

TABLE 7.17 Euclidean distances with cluster assignments

Data points	Euclidean Euclidean from C1	Euclidean distance from C2	Euclidean distance from C3	Assigned cluster
x_1	**1.49**	7.13	10.3	C1
x_2	**0.95**	7	8.85	C1
x_3	**1.49**	5.24	7.5	C1
x_4	5.34	**1.05**	7.02	C2
x_5	6.34	**0.33**	7.02	C2
x_6	7.36	**1.2**	6.18	C2
x_7	8.54	6.74	**0.5**	C3
x_8	9.2	6.67	**0.5**	C3

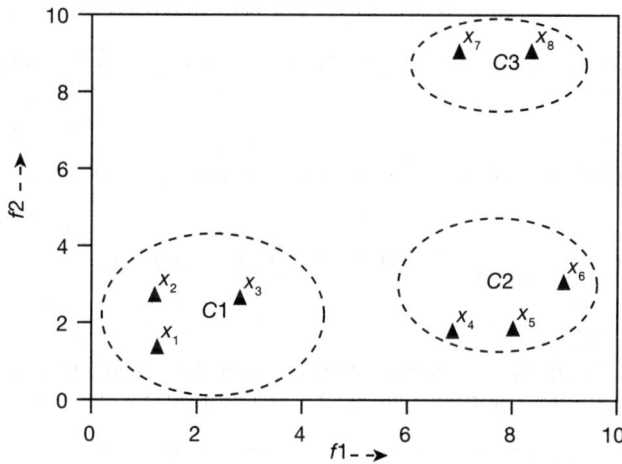

FIG. 7.9 K-means clustering (with $k = 3$) for the data points listed in Table 7.15

then the criterion function minimised by the algorithm is

$$\sum_{i=1}^{k} \sum_{x \in C_i} (x - \mu_i)^t (x - \mu_i)$$

This is called the **sum-of-squared-error criterion**. This should be minimal for optimal clustering.

Choosing a suitable value for k is a practical challenge in k-means clustering. A commonly employed strategy is to initialize the cluster centres for k in a way that maximizes their distance from each other, which has shown to be effective in real-world scenarios. Alternatively, the **elbow method** is a popular approach to determine the value of k. The steps used in the elbow method are as follows:

1. Perform the k-means clustering algorithm (as discussed earlier) for different values of k.

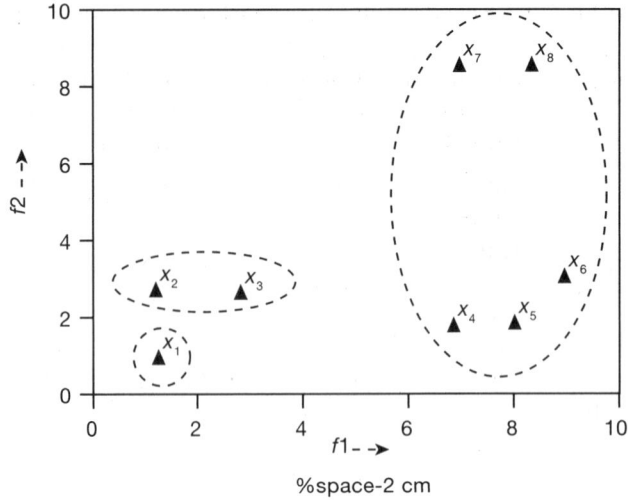

FIG. 7.10 K-means clustering (with $k = 3$) for the data points listed in Table 7.15 but with initial cluster centre, x_1, x_2 and x_3

2. For each k, calculate the total in-cluster sum of squared error (IC-SSE):

$$\text{IC-SSE} = \sum_{i=1}^{k} \sum_{x \in C_i} (x - \mu_i)^2$$

3. Plot the curve, IC-SSE verses k.
4. The location of a bend (knee) in the plot is generally considered as an indicator of the appropriate value of k. Figure 7.11 is an indicative plot showing the elbow method.

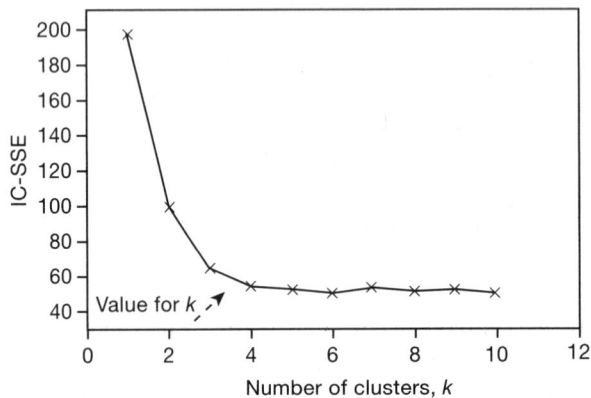

FIG. 7.11 Elbow method for approximating the k value

The k-means clustering algorithm has a time complexity of $\mathcal{O}(nlkp)$, where n represents the number of data points, l denotes the number of features in each data point, k specifies the number of clusters and p indicates the number of iterations. As for space requirements, it is $\mathcal{O}(kn)$. These characteristics of the algorithm make it more practical and hence it is considered as one of the popular clustering algorithms.

7.5.2 K-Means++ Clustering

In the case of k-means clustering, we randomly select the initial k cluster centres (centroids). We know that in k-means clustering, cluster formation depends on the initial selection of the centroids. To obtain good clusters, we need to select the initial cluster centroids smartly. The k-means++ clustering algorithm is mainly used for identifying the initial cluster centres. The idea here is to choose the initial k centroids away from each other, in a probabilistic sense, so that they become good representatives of the clusters formed.

The steps to identify the initial k centroids are as follows:

1. Initially, select a random data point as the first centroid.
2. Compute the distance (say Euclidean distance) between each centroid and the other data points. Record the minimum distances between any centroid and the data point.
3. Compute the probabilities among the computed minimum distances.
4. Choose the next centroid with maximum probability among all probabilities computed in Step 3.
5. Repeat steps 2 to 4 till we get k number of centroids.

We illustrate the k-means++ clustering for initialization using the following example.

EXAMPLE 15: Consider the data set with following data points:

$$x_1 = (8,5), x_2 = (9,4), x_3 = (6,10)$$
$$x_4 = (4,4), x_5 = (2,4), x_6 = (11,2)$$

Figure 7.12 gives the visual representation of the data set.

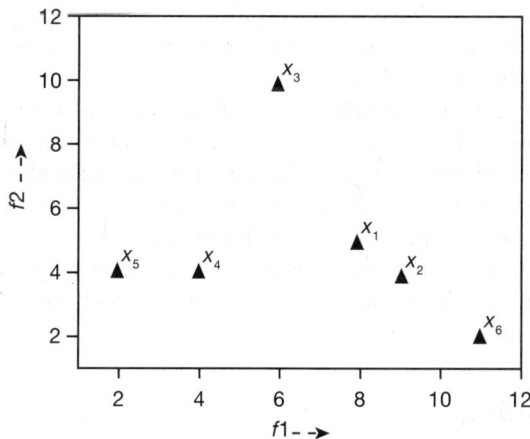

FIG. 7.12 Visualization of the data set

Let x_1 be the randomly selected first cluster centre. Table 7.18 gives the Euclidean distances and probabilities between x_1 and the other data points. Choose x_5 as the second centroid, as its probability in the list is maximum.

TABLE 7.18 Euclidean distances and probabilities between x_1 and other data points

Data point	$d(x_1, x_i)$	$Min(d(x_1, x_i))$	$Prob_i = \frac{d(x_1, x_i)}{\sum d(x_1, x_i)}$
x_1	-	-	-
x_2	1.41	1.41	0.066
x_3	5.39	5.39	0.254
x_4	4.12	4.12	0.194
x_5	6.08	6.08	**0.286**
x_6	4.24	4.24	0.2

Table 7.19 gives the Euclidean distances and probabilities between x_1, x_5 and other data points.

TABLE 7.19 Euclidean distances and probabilities between x_1, x_5 and other data points

Data point	$d(x_1, x_i), d(x_5, x_i)$	$Min(d(x_1, x_i), d(x_5, x_i))$	$Prob_i = \frac{Min(d(x_1, x_i), d(x_5, x_i))}{\sum Min(d(x_1, x_i), d(x_5, x_i))}$
x_1	-	-	-
x_2	1.41, 7.0	1.41	0.09
x_3	5.39, 7.2	5.39	**0.36**
x_4	·4.12, 2	4.12	0.27
x_5	-	-	-
x_6	4.24, 9.22	4.24	0.28

Choose x_3 as the third centroid, as its probability in the list is maximum. We now run the k-means clustering algorithm with initial centroids, x_1, x_2 and x_3.

Note that in k-means++, we have some initialization (set-up) cost when compared to the conventional k-means algorithm. But convergence tends to be faster and better with lower heterogeneity. One can visualize this in Fig. 7.13.

Figure 7.13 represents cluster formation using random initial centroids and selection of initial centroids by k-means++. Figure 7.13 (a) shows the randomly selected initial three centroids (circled), Fig. 7.13 (b) shows three possible clusters after executing the k-means algorithm. Figure 7.13 (c) shows the initial 3-centroids (circled) by k-means++ and Fig. 7.13 (d) shows cluster formation after executing the k-means clustering algorithm.

Visually it is clear that (i) the amount of centroid movement is less if we use the k-means++ initialization step (the grey arrows in both the cases shows this) and (ii) the cluster formed has lower heterogeneity with k-means++ initialization.

7.5.3   Soft Partitioning

The conventional k-means algorithm assigns data points exclusively to the cluster which is nearest to it, and this leads to a hard partition. Due to inadequate selection of the initial cluster centre, the algorithm may result in a locally optimal solution. Various solutions were proposed in the literature to address this issue.

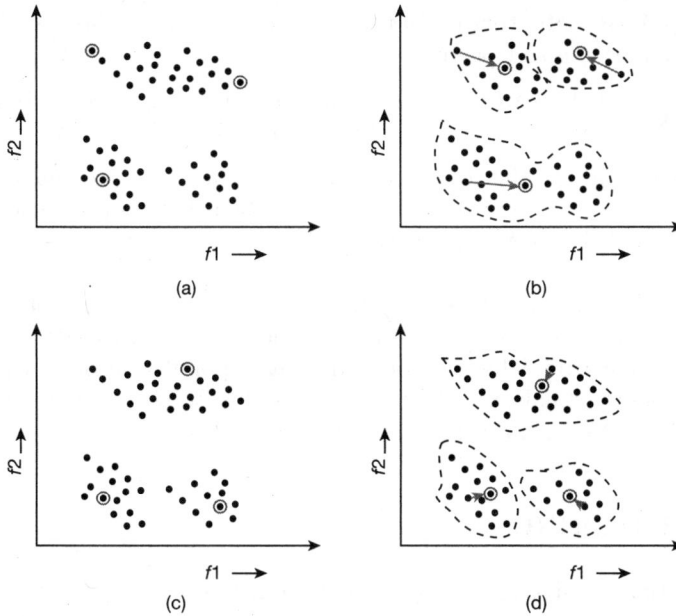

FIG. 7.13 Visualization of goodness of using k-means clustering for initialization (for colour figure, please see Colour Plate 2)

Split and merge strategy: Here, clusters generated by the k-means algorithm are evaluated for potential splitting and merging.

If the sample variance of a cluster exceeds a specified threshold value (τ_v), the cluster is divided into two clusters by selecting the most dissimilar pair of patterns as initial seeds. Likewise, if the distance between the centres of two clusters is below a distance threshold (τ_d), they are merged into a single cluster. By appropriately setting τ_v and τ_d, it becomes possible to obtain an optimal partition. For example, the cluster $\{x_4, x_5, x_6, x_7, x_8\}$ in Fig. 7.10 is selected for splitting based on the variance surpassing the user-defined threshold, τ_v. By choosing either x_4, x_5 or x_6 as initial centroids for one cluster and either x_7 or x_8 for the other, we obtain the clusters $\{x_4, x_5, x_6\}$ and $\{x_7, x_8\}$.

Additionally, by merging the clusters $\{x_1\}$ and $\{x_2, x_3\}$ because the distance between x_1 and the centroid of $\{x_2, x_3\}$ falls below the user-defined threshold τ_d, the partition shown in Fig. 7.9 can be achieved. However, implementing this strategy is challenging due to the difficulty of estimating τ_v and τ_d.

Multiple membership approach: This approach involves the influence of not only the closest cluster centre to a given data point but also neighboring cluster centres. These cluster centroids are affected to varying degrees by the data point, with the closest centre being more influenced than others. Several existing methods for this approach include:

- **Fuzzy clustering:** Each data point is assigned to multiple clusters, typically more than one, based on a membership value. The value is computed using the data point and the corresponding cluster centroid.
- **Rough clustering:** Each cluster is assumed to have both a non-overlapping part and an overlapping part. Data points in the non-overlapping portion belong exclusively to that cluster, while data points in the overlapping part may belong to multiple clusters.

- **Neural network-based clustering:** In this method, varying weights associated with the data points are used to obtain a soft partition.

Stochastic approach: In this category, the focus is on achieving the global optimal solution for cluster formation through probabilistic methods. Some existing methods for this approach include:

- **Simulated annealing:** In this case, the current solution is randomly updated, and the resulting solution is accepted with a certain probability. If the resulting solution is better than the current solution, it is accepted; otherwise, it is accepted with a probability ranging from 0 to 1.
- **Tabu search:** Unlike simulated annealing, multiple solutions are stored, and the current solution is modified in various ways to determine the next configuration.
- **Evolutionary algorithms:** This method maintains a population of solutions. In addition to the fitness values of individuals, a random search based on the interaction among solutions with mutation is employed to generate the next population.

7.6 SOFT CLUSTERING

In numerous applications, assigning an object or pattern (data point) to a single cluster is often not feasible.

For instance, when assigning a document in text mining, it is possible for the same document to fall under both the "politics" and "sports" categories. Similarly, a person in a social network can belong to multiple subject areas within social media. This means that an object or pattern can be assigned to more than one category. This type of categorization, where an object or pattern can be associated with multiple groups, is known as soft partitioning or soft clustering.

The general form of soft clustering involves considering n data points, denoted as $x_1, x_2, \ldots, x_n$, and k clusters, denoted as $C_1, C_2, \ldots, C_k$. In soft clustering, a data point can belong to one or more clusters. This relationship can be represented by an $n \times k$ matrix, where a non-zero entry in the cell (i, j) indicates that the data point x_i belongs to the cluster C_j, while $(i, j) = 0$ indicates otherwise.

In each row, the number of non-zero entries can range from 1 to k. Excluding the all-zero vector, which implies that x_i must be assigned to at least one of the k clusters, there are $(2^k - 1)$ possible choices for each row in terms of the non-zero entries. Therefore, the total number of possibilities for all n rows is $(2^k - 1)^n$. If we store one of p possible values in each cell to indicate the extent to which x_i belongs to C_j, then the number of possibilities is bounded by $(p^k - 1)^n$, considering the number of distinct values each cell can assume as p.

7.6.1 Fuzzy C-Means Clustering

In this type, each data point x_i can be assigned to a cluster C_j with a membership value μ_{ij}, which represents the degree to which x_i belongs to C_j. The membership value μ_{ij} is assumed to range between 0 and 1, and for each data point x_i, the sum of all μ_{ij} values across clusters is equal to 1.

The fuzzy c-means clustering algorithm follows an iterative scheme similar to the k-means algorithm. It begins by identifying c initial cluster centres from the data set or membership values between each data point and the cluster. The following steps are then performed iteratively:

1. Compute membership values:

$$\mu_{ij} = \frac{\mathrm{d}(x_j, C_i)^{-1/(M-1)}}{\sum_{l=1}^{c} \mathrm{d}(x_j, C_l)^{-1/(M-1)}}$$

2. Compute the fuzzy cluster centroid:

$$C_i = \frac{\sum_{j=1}^{m} (\mu_{ij})^M \times x_j}{\sum_{j=1}^{m} (\mu_{ij})^M}$$

In the above expressions, μ_{ij} represents the membership value of data point x_j in cluster C_i, where $j = 1, 2, \ldots, n$ and $i = 1, 2, \ldots, c$; $\mathrm{d}(x_j, C_i)$ denotes the Euclidean distance between x_j and C_i; C_i represents the centroid of the ith cluster and M is the fuzzyifying constant which determines the behaviour of the algorithm. When $M = 1$, the fuzzy algorithm functions similar to the hard k-means algorithm, where $\mu_{ij} = 1$ when X_j is assigned to C_i. As M increases or tends to infinity, μ_{ij} approaches $\frac{1}{c}$, as the exponents in both the numerator and denominator tend towards 0.

3. Repeat the above steps, and the algorithm terminates when there is no significant change in the computed values for membership and cluster centroid.

EXAMPLE 16: Consider the data points with two features shown in Table 7.21.

TABLE 7.20 Example of a data set with two features

Data points	f1	f2
x_1	1	1
x_2	2	2
x_3	4	3
x_4	5	3

Let the number of clusters, $c = 2$ and $M = 2$. To start with, assume the following membership values between the data points and the clusters($\mu^{(0)}$):

	x_1	x_2	x_3	x_4
C1	1	1	0	0
C2	0	0	1	1

The cluster centres are calculated as follows:

$$C1_{f1} = \frac{\mu_{11}^2 * x_1(f1) + \mu_{12}^2 * x_2(f1) + \mu_{13}^2 * x_3(f1) + \mu_{14}^2 * x_4(f1)}{\mu_{11}^2 + \mu_{12}^2 + \mu_{13}^2 + \mu_{14}^2}$$

$$C1(f1) = \frac{1^2 * 1 + 1^2 * 2 + 0^2 * 4 + 0^2 * 5}{1^2 + 1^2 + 0^2 + 0^2}$$

$$C1(f1) = \frac{3}{2} = 1.5$$

$$C1(f2) = \frac{\mu_{11}^2 * X1_{f2} + \mu_{12}^2 * X2_{f2} + \mu_{13}^2 * X3_{f2} + \mu_{14}^2 * X4_{f2}}{\mu_{11}^2 + \mu_{12}^2 + \mu_{13}^2 + \mu_{14}^2}$$

$$C1(f2) = \frac{1^2 * 1 + 1^2 * 2 + 0^2 * 3 + 0^2 * 3}{1^2 + 1^2 + 0^2 + 0^2}$$

$$C1(f2) = \frac{3}{2} = 1.5$$

So, the updated cluster centre for $C1 = (1.5, 1.5)$.

Similarly,

$$C2(f1) = \frac{\mu_{21}^2 * x_1(f1) + \mu_{22}^2 * x_2(f1) + \mu_{23}^2 * x_3(f1) + \mu_{24}^2 * x_4(f1)}{\mu_{21}^2 + \mu_{22}^2 + \mu_{23}^2 + \mu_{24}^2}$$

$$C2(f1) = \frac{0^2 * 1 + 0^2 * 2 + 1^2 * 4 + 1^2 * 5}{0^2 + 0^2 + 1^2 + 1^2}$$

$$C2(f1) = \frac{9}{2} = 4.5$$

$$C2(f2) = \frac{\mu_{21}^2 * X1_{f2} + \mu_{22}^2 * X2_{f2} + \mu_{23}^2 * X3_{f2} + \mu_{24}^2 * X4_{f2}}{\mu_{21}^2 + \mu_{22}^2 + \mu_{23}^2 + \mu_{24}^2}$$

$$C2(f2) = \frac{0^2 * 1 + 0^2 * 2 + 1^2 * 3 + 1^2 * 3}{0^2 + 0^2 + 1^2 + 1^2}$$

$$C2(f2) = \frac{6}{2} = 3$$

So, the updated cluster centre for $C2 = (4.5, 3)$.

The next step is recalculation of membership values. As a part of this step, let us first calculate the Euclidean distance between the cluster centres and the data points. This is done as follows:

$$d(C1, x_1) = d((1.5, 1.5), (1, 1)) = \sqrt{1.5 - 1)^2 + (1.5 - 1)^2} = 0.71$$

$$d(C1, x_2) = d((1.5, 1.5), (2, 2)) = \sqrt{(1.5 - 2)^2 + (1.5 - 2)^2} = 0.71$$

$$d(C1, x_3) = d((1.5, 1.5), (4, 3)) = \sqrt{(1.5 - 4)^2 + (1.5 - 3)^2} = 2.92$$

$$d(C1, x_4) = d((1.5, 1.5), (5, 3)) = \sqrt{(1.5 - 5)^2 + (1.5 - 3)^2} = 3.81$$

$$d(C2, x_1) = d((4.5, 3), (1, 1)) = \sqrt{(4.5 - 1)^2 + (3 - 1)^2} = 4.03$$

$$d(C2, x_2) = d((4.5, 3), (2, 2)) = \sqrt{(4.5 - 2)^2 + (3 - 2)^2} = 2.69$$

$$d(C2, x_3) = d((4.5, 3), (4, 3)) = \sqrt{(4.5 - 4)^2 + (3 - 3)^2} = 0.25$$

$$d(C2, x_4) = d((4.5, 3), (5, 3)) = \sqrt{(4.5 - 5)^2 + (3 - 3)^2} = 0.25$$

The membership values are:

$$\mu_{11} = \frac{\frac{1}{d(C1,x_1)}}{\frac{1}{d(C1,x_1)} + \frac{1}{d(C2,x_1)}}$$

$$\mu_{11} = \frac{\frac{1}{0.71}}{\frac{1}{0.71} + \frac{1}{4.03}} = 0.85$$

$$\mu_{12} = \frac{\frac{1}{d(C1,x_2)}}{\frac{1}{d(C1,x_2)} + \frac{1}{d(C2,x_2)}}$$

$$\mu_{12} = \frac{\frac{1}{0.71}}{\frac{1}{0.71} + \frac{1}{2.69}} = 0.79$$

$$\mu_{13} = \frac{\frac{1}{d(C1,x_3)}}{\frac{1}{d(C1,x_3)} + \frac{1}{d(C2,x_3)}}$$

$$\mu_{13} = \frac{\frac{1}{2.92}}{\frac{1}{2.92} + \frac{1}{0.25}} = 0.08$$

$$\mu_{14} = \frac{\frac{1}{d(C1,x_4)}}{\frac{1}{d(C1,x_4)} + \frac{1}{d(C2,x_4)}}$$

$$\mu_{14} = \frac{\frac{1}{3.81}}{\frac{1}{3.81} + \frac{1}{0.25}} = 0.06$$

Similarly, μ_{21}, μ_{22}, μ_{23}, μ_{24} can be computed and their values are 0.15, 0.21, 0.92 and 0.94, respectively. Note that for a given data point x_j, $\sum_{i=1}^{2}(\mu_{ij}) = 1$.

Now the updated membership values $\mu^{(1)}$ are given below:

	x_1	x_2	x_3	x_4
$C1$	0.85	0.79	0.08	0.06
$C2$	0.15	0.21	0.92	0.94

Since we do not end up with the same cluster centres or the same membership values, we iterate for the next step. [Note: To converge in a finite number of iterations, some threshold values, either for the difference between the current and previous membership values or for the difference between the current and the previous cluster centre values, can be specified.]

The updated cluster centres for the updated membership values ($\mu^{(1)}$) are as follows:

$$C1(f1) = \frac{0.85^2 * 1 + 0.79^2 * 2 + 0.08^2 * 4 + 0.06^2 * 5}{0.85^2 + 0.79^2 + 0.08^2 + 0.06^2}$$

$$C1(f1) = \frac{2.04}{1.36} = 1.5$$

$$C1_{f2} = \frac{0.85^2 * 1 + 0.79^2 * 2 + 0.08^2 * 3 + 0.06^2 * 3}{0.85^2 + 0.79^2 + 0.08^2 + 0.06^2}$$

$$C1(f2) = \frac{2.01}{1.36} = 1.48$$

So, the updated cluster centre for $C1 = (1.5, 1.48)$.

$$C2(f1) = \frac{0.15^2 * 1 + 0.21^2 * 2 + 0.92^2 * 4 + 0.94^2 * 5}{0.15^2 + 0.21^2 + 0.92^2 + 0.94^2}$$

$$C2(f1) = \frac{7.9143}{1.797} = 4.4$$

$$C2(f2) = \frac{0.15^2 * 1 + 0.21^2 * 2 + 0.92^2 * 3 + 0.94^2 * 3}{0.15^2 + 0.21^2 + 0.92^2 + 0.94^2}$$

$$C2(f2) = \frac{5.307}{1.7966} = 2.95$$

So, the updated cluster centre for $C2 = (4.4, 2.95)$.

The next step is recalculation of membership values. As a part of this step, let us first calculate the Euclidean distance between the cluster centres and the data points. This is done as follows:

$$d(C1, x_1) = d((1.5, 1.48), (1, 1)) = \sqrt{(1.5 - 1)^2 + (1.48 - 1)^2} = 0.693$$

$$d(C1, x_2) = d((1.5, 1.48), (2, 2)) = \sqrt{(1.5 - 2)^2 + (1.48 - 2)^2} = 0.721$$

$$d(C1, x_3) = d((1.5, 1.48), (4, 3)) = \sqrt{(1.5 - 4)^2 + (1.48 - 3)^2} = 2.925$$

$$d(C1, x_4) = d((1.5, 1.48), (5, 3)) = \sqrt{(1.5 - 5)^2 + (1.48 - 3)^2} = 3.815$$

$$d(C2, x_1) = d((4.4, 2.95), (1, 1)) = \sqrt{(4.4 - 1)^2 + (2.95 - 1)^2} = 3.919$$

$$d(C2, x_2) = d((4.4, 2.95), (2, 2)) = \sqrt{(4.4 - 2)^2 + (2.95 - 2)^2} = 2.58$$

$$d(C2, x_3) = d((4.4, 2.95), (4, 3)) = \sqrt{(4.4 - 4)^2 + (2.95 - 3)^2} = 0.40$$

$$d(C2, x_4) = d((4.4, 2.95), (5, 3)) = \sqrt{(4.4 - 5)^2 + (2.95 - 3)^2} = 0.60$$

The updated membership values are:

$$\mu_{11} = \frac{\frac{1}{0.693}}{\frac{1}{0.693} + \frac{1}{3.919}} = 0.85$$

$$\mu_{12} = \frac{\frac{1}{0.721}}{\frac{1}{0.721} + \frac{1}{2.58}} = 0.79$$

$$\mu_{13} = \frac{\frac{1}{2.925}}{\frac{1}{2.925} + \frac{1}{0.40}} = 0.12$$

$$\mu_{14} = \frac{\frac{1}{3.815}}{\frac{1}{3.815} + \frac{1}{0.60}} = 0.14$$

Similarly, $\mu_{21}, \mu_{22}, \mu_{23}, \mu_{24}$ can be computed and their values are 0.15, 0.21, 0.88 and 0.86, respectively.

Now the updated membership values $\mu^{(2)}$ are given below:

	x_1	x_2	x_3	x_4
$C1$	0.85	0.79	0.12	0.14
$C2$	0.15	0.21	0.88	0.86

Since we do not end up with the same cluster point or the same membership values, we iterate for the next step.

The updated cluster centres for the updated membership values $(\mu^{(2)})$ are as follows:

$$C1(f1) = \frac{0.85^2 * 1 + 0.79^2 * 2 + 0.12^2 * 4 + 0.14^2 * 5}{0.85^2 + 0.79^2 + 0.12^2 + 0.14^2}$$

$$C1(f1) = \frac{2.04}{1.355} = 1.5$$

$$C1(f2) = \frac{0.85^2 * 1 + 0.79^2 * 2 + 0.12^2 * 3 + 0.14^2 * 3}{0.85^2 + 0.79^2 + 0.12^2 + 0.14^2}$$

$$C1(f2) = \frac{2.01}{1.355} = 1.48$$

The updated cluster centre for $C1 = (1.5, 1.48)$.

$$C2(f1) = \frac{0.15^2 * 1 + 0.21^2 * 2 + 0.92^2 * 4 + 0.94^2 * 5}{0.15^2 + 0.21^2 + 0.92^2 + 0.94^2}$$

$$C2(f1) = \frac{7.9143}{1.797} = 4.4$$

$$C2(f2) = \frac{0.15^2 * 1 + 0.21^2 * 2 + 0.92^2 * 3 + 0.94^2 * 3}{0.15^2 + 0.21^2 + 0.92^2 + 0.94^2}$$

$$C2(f2) = \frac{5.307}{1.797} = 2.95$$

The updated cluster centre for $C2 = (4.4, 2.95)$.

Since there is no change in the cluster centre, the algorithm stops. The final membership value and cluster allocation for the data points is given in Table 7.21.

TABLE 7.21 Cluster formation using fuzzy c-means algorithm

Data point	μ^{C1}	μ^{C2}	Allocated cluster
x_1	0.85	0.15	$C1$
x_2	0.79	0.21	$C1$
x_3	0.12	0.88	$C2$
x_4	0.14	0.86	$C2$

7.7 ROUGH CLUSTERING

It is possible that some of the data points clearly belong to only one cluster while the other data points may belong to two or more clusters. This is abstracted by a rough set which is represented using two sets; these sets are called **lower approximation** and **upper approximation.** A data point can belong to at most one lower approximation. If a data point does not belong to any lower approximation, then it belongs to two or more upper approximations.

We can define a rough set using the lower approximation ($\underline{R}(S)$) and the upper approximation ($\overline{R}(S)$), where $S \subseteq \mathcal{X}$, a set of data points, is as follows:

$$\underline{R}(S) = \bigcup_i G_i, \text{ where } \quad G_i \subseteq S$$

$$\overline{R}(S) = \bigcup_i G_i, \text{ where } \quad G_i \cap S \neq \emptyset$$

In the lower approximation, each equivalence class must be entirely contained within S. On the other hand, in the upper approximation, equivalence classes that partially overlap with S are included. Consequently, the lower approximation comprises patterns that definitely belong to S, whereas the upper approximation contains patterns that may potentially belong to S.

Further note that if $\overline{R}(S) - \underline{R}(S) = \emptyset$, then there is no roughness in S with respect to R. We discuss one of the popular clustering algorithms, rough k-means clustering, in the next subsection which is based on these concepts.

7.7.1 Rough K-Means Clustering Algorithm

Let n be the number of data points, $\mathcal{X} = \{x_1, x_2, \ldots, x_n\}$, having l features, k be the number of clusters, w_l and w_u be the weights associated with the lower and upper approximations and ϵ be the threshold value.

1. Randomly assign each data object to exactly one lower approximation $\underline{R}(k)$. Note that by definition, the data object also belongs to the upper approximation $\overline{R}(k)$ of the same cluster.
2. Compute the cluster centroid C_j as follows:

   If $\underline{R}(k) \neq \emptyset$ and $\overline{R}(k) - \underline{R}(k) = \emptyset$

$$C_j = \sum_{x_i \in \underline{R}(k)} \frac{x_i}{|\underline{R}(k)|}$$

   Else if $\underline{R}(k) = \emptyset$ and $\overline{R}(k) - \underline{R}(k) \neq \emptyset$

$$C_j = \sum_{x_i \in \overline{R}(k) - \underline{R}(k)} \frac{x_i}{|\overline{R}(k) - \underline{R}(k)|}$$

   Else

$$C_j = w_l \times \sum_{x_i \in \underline{R}(k)} \frac{x_i}{|\underline{R}(k)|} + w_u \times \sum_{x_i \in \overline{R}(k) - \underline{R}(k)} \frac{x_i}{|\overline{R}(k) - \underline{R}(k)|}$$

3. For each data point in $\mathcal{X}$, compute the Euclidean distance with each cluster, C_j, $1 \leq j \leq k$ [that is, $d(x_i, C_j)$, $\forall i \in m$, $\forall j \in k$].
   Let $d(X, C_p) = min_{1 \leq j \leq k}(x, Cj)$.

4. Use the ratio, $\frac{d(x,C_j)}{d(x,C_p)}$, where $1 \leq i, p \leq k$, to determine the membership of X as follows:

   - If the ratio $\not\leq \epsilon$, then the corresponding data point x will not be part of the lower approximation.
   - Else, the corresponding data point x will be part of the lower approximation

5. Repeat from Step 2 until convergence (that is, Old centroid = New centroid)

EXAMPLE 17: Consider the data set shown in Table 7.22 with $k = 2$, $w_l = 0.7$, $w_u = 0.3$ and $\epsilon = 2$.

TABLE 7.22 Example of data set with two features

Data points	f1	f2
x_1	1	1
x_2	2	2
x_3	4	1
x_4	5	2
x_5	7	0
x_6	8	0

1. Randomly assign each data point to exactly one lower approximation (two clusters).

$$\underline{R}(k_1) = \{(1,1), (2,2), (4,1)\}$$
$$\underline{R}(k_2) = \{(5,2), (7,0), (8,0)\}$$

2. Since (i) $\underline{R}(k_1) \neq \emptyset$ and $\overline{R}(k_1) - \underline{R}(k_1) = \emptyset$; (ii) $\underline{R}(k_2) \neq \emptyset$ and $\overline{R}(k_2) - \underline{R}(k_2) = \emptyset$, the centroid is calculated using $C_j = \sum_{x_i \in \underline{R}(k)} \frac{x_i}{|\underline{R}(k)|}$.

$$C_1 = \left(\frac{1+2+4}{3}, \frac{1+2+1}{3} \right) = (2.33, 1.33)$$

$$C_2 = \left(\frac{5+7+8}{3}, \frac{2=0+0}{3} \right) = (6.67, 0.67)$$

3. Find the Euclidean distances between each data point and the cluster centre.
   - With reference to C_1

$$d((1,1),(2.33,1.33)) = \sqrt{(1-2.33)^2 + (1-1.33)^2} = 1.37$$

$$d((2,2),(2.33,1.33)) = \sqrt{(2-2.33)^2 + (2-1.33)^2} = 0.75$$

$$d((4,1),(2.33,1.33)) = \sqrt{(4-2.33)^2 + (1-1.33)^2} = 1.70$$

$$d((5,2),(2.33,1.33)) = \sqrt{(5-2.33)^2 + (2-1.33)^2} = 2.75$$

$$d((7,0),(2.33,1.33)) = \sqrt{(7-2.33)^2 + (0-1.33)^2} = 4.86$$

$$d((8,0),(2.33,1.33)) = \sqrt{(8-2.33)^2 + (0-1.33)^2} = 5.82$$

- With reference to C_2

$$d((1,1),(6.67,0.67)) = \sqrt{(1-6.67)^2 + (1-0.67)^2} = 5.68$$

$$d((2,2),(6.67,0.67)) = \sqrt{(2-6.67)^2 + (2-0.67)^2} = 4.86$$

$$d((4,1),(6.67,0.67)) = \sqrt{(4-6.67)^2 + (1-0.67)^2} = 2.69$$

$$d((5,2),(6.67,0.67)) = \sqrt{(5-6.67)^2 + (2-0.67)^2} = 2.14$$

$$d((7,0),(6.67,0.67)) = \sqrt{(7-6.67)^2 + (0-0.67)^2} = 0.75$$

$$d((8,0),(6.67,0.67)) = \sqrt{(8-6.67)^2 + (0-0.67)^2} = 1.49$$

4. Use the ratio, $\frac{d(x,C_j)}{d(x,C_p)}$, where $1 \le i, p \le k$, to determine the membership.

$(1,1) \Longrightarrow \dfrac{d((1,1),(6.67,0.67))}{d((1,1),(2.33,1.33))} = \dfrac{5.68}{1.37} = 4.14 \not\le 2$. So, x_1 will be part of $\underline{R}(k_1)$

$(2,2) \Longrightarrow \dfrac{d((2,2),(6.67,0.67))}{d((2,2),(2.33,1.33))} = \dfrac{4.86}{0.75} = 6.48 \not\le 2$. So, x_2 will be part of $\underline{R}(k_1)$

$(4,1) \Longrightarrow \dfrac{d((4,1),(6.67,0.67))}{d((4,1),(2.33,1.33))} = \dfrac{2.69}{1.7} = 1.58 < 2$. So, x_3 will not be part of $\underline{R}(k_1)$ and $\underline{R}(k_2)$

$(5,2) \Longrightarrow \dfrac{d((5,2),(2.33,1.33))}{d((5,2),(6.67,0.67))} = \dfrac{2.75}{2.14} = 1.28 < 2$. So, x_4 will not be part of $\underline{R}(k_1)$ and $\underline{R}(k_2)$

$(7,0) \Longrightarrow \dfrac{d((7,0),(2.33,1.33))}{d((7,0),(6.67,0.67))} = \dfrac{4.86}{0.75} = 6.48 \not\le 2$. So, x_5 will be part of $\underline{R}(k_2)$

$(8,0) \Longrightarrow \dfrac{d((8,0),(2.33,1.33))}{d((8,0),(6.67,0.67))} = \dfrac{5.82}{1.49} = 3.91 \not\le 2$. So, x_4 will be part of $\underline{R}(k_2)$

Now we have clusters

$$\underline{R}(k_1) = \{(1,1),(2,2)\} \qquad \overline{R}(k_1) = \{(1,1),(2,2),(4,1),(5,2)\}$$
$$\underline{R}(k_2) = \{(7,0),(8,0)\} \qquad \overline{R}(k_2) = \{(4,1),(5,2),(7,0),(8,0)\}$$

Here, (i) $\underline{R}(k_1) \ne \emptyset$ and $\overline{R}(k_1) - \underline{R}(k_1) \ne \emptyset$; (ii) $\underline{R}(k_2) \ne \emptyset$ and $\overline{R}(k_2) - \underline{R}(k_2) \ne \emptyset$. So, the centroid is calculated using

$$C_j = w_l \times \sum_{x \in \underline{R}(k)} \frac{x_i}{|\underline{R}(k)|} + w_u \times \sum_{x_i \in \overline{R}(k) - \underline{R}(k)} \frac{x_i}{|\overline{R}(k) - \underline{R}(k)|}$$

$$C_1 = 0.7 \times \left(\frac{1+2}{2}, \frac{1+2}{2} \right) + 0.3 \times \left(\frac{4+5}{2}, \frac{1+2}{2} \right) = (2.4, 1.5)$$

$$C_2 = 0.7 \times \left(\frac{7+8}{2}, \frac{0+0}{2} \right) + 0.3 \times \left(\frac{4+5}{2}, \frac{1+2}{2} \right) = (6.6, 0.45)$$

5. Repeat from Step 3 until convergence.

7.8 EXPECTATION MAXIMIZATION-BASED CLUSTERING

One of the approaches to mathematically represent clusters is through parametric probability distributions. In this method, the data set is regarded as a combination of these distributions, and clustering involves using a finite mixture density model consisting of k probability distributions, where each distribution represents a cluster. Hence, the clustering problem entails estimating the parameters of the probability distributions to best fit the data.

The EM (expectation maximization) algorithm is a suitable method for parameter estimation. In EM, a data point is assigned to a cluster based on its membership probability. The basic concept of the EM algorithm is as follows: it begins with an initial estimation of the mixture model parameters. Subsequently, it iteratively recalculates the probabilities for the data points based on the mixture density generated using these parameters. The parameters are then updated using these recalculated probabilities.

Let $\{x_1, x_2, \ldots, x_n\}$ be the data set to be clustered. Let Z_{ij} be the probability that $x_i \in c_j$ (a cluster).

1. Computing the probability that a data point belongs to a cluster:

$$Prob(x_i \in c_j) = Z_{ij} = \frac{exp[-\frac{1}{2\sigma^2}(x_i - \mu_j)^2]}{\sum_{l=1}^{k} exp[-\frac{1}{2\sigma^2}(x_i - \mu_l)^2]} \tag{7.1}$$

2. The μ values are updated using Z_{ij} as follows:

$$\mu_j = \frac{\sum_{i=1}^{n} Z_{ij} x_i}{\sum_{i=1}^{n} Z_{ij}} \tag{7.2}$$

Notice the similarity between the EM algorithm and the k-means algorithm of choosing the centroids (μ_j) and assigning data points to the nearest cluster means to get Z_{ij}.
3. Repeat the above two steps until there is no change in μ values.

The following example illustrate the above steps.

EXAMPLE 18: Let the data points be $x_1 = 1$, $x_2 = 6$, $x_3 = 6$, $x_4 = 7$. Let us consider two clusters, c_1 and c_2, and $\sigma = 1$. Let the two extreme points $x_1 = 1$ and $x_4 = 7$ be the initial cluster centres of c_1 and c_2, respectively. The values of Z_{ij} are calculated using Equation 7.1 and is shown in Table 7.23. For example,

$$Z_{11} = \frac{e^{-\frac{1}{2}(1-1)^2}}{e^{-\frac{1}{2}(1-1)^2} + e^{-\frac{1}{2}(1-7)^2}} \approx 0.9999$$

The updated μ values using Equation 7.2 for c_1 and c_2 are $\mu_1 = 1.5$ and $\mu_2 = 6.5$, respectively.

TABLE 7.23 Z_{ij} values

Data point i	Z_{i1}	Z_{i2}
$x_1 = 1$	0.9999	0.0001
$x_2 = 2$	0.9999	0.0001
$x_3 = 6$	0.0001	0.9999
$x_4 = 7$	0.0001	0.999

Let us add the data item $x_5 = 4$. Starting with the same initial conditions as earlier, we get $Z_{51} = 0.5$ and $Z_{52} = 0.5$. The updated $\mu_1 = 2$ and $\mu_2 = 6$. The Z_{ij} values are given in Table 7.24. The updated $\mu_1 = 2$ and $\mu_2 = 6$ and are the same as the previous values, so the algorithm terminates.

TABLE 7.24 Z_{ij} values

Data points	Z_{i1}	Z_{i2}
$x_1 = 1$	0.9999	0.0001
$x_2 = 2$	0.9996	0.0004
$x_3 = 6$	0.0004	0.9996
$x_4 = 7$	0.0001	0.999
$x_5 = 4$	0.5	0.5

7.9 SPECTRAL CLUSTERING

Clustering algorithms like k-means are effective for data with isotropic or spherical clusters; they are not suitable for non-isotropic clusters. For instance, when clusters are elongated in a specific direction or when they are concentric with similar centroids (similar to two circles or spheres with varying radii and points distributed along each circle or sphere).

Spectral clustering algorithms are well-suited for data sets that contain non-isotropic clusters, such as chain-like or concentric clusters. The spectral clustering algorithm assumes that the data set is in the form of a graph which can be represented by triplets, $< V, E, S >$. Here,

- $V = \{x_1, x_2, \ldots, x_n\}$, where each x_i in V corresponds to a data point in the collection.
- $E\{< x_i, x_j >: x_i \in V$ and $x_j \in V \}$ for $i, j = 1, 2, \cdots, n$. So, each element of E characterizes an edge between a pair of vertices.
- $S = \{A_{ij} : x_i, x_j \in V\}$. Each element of A characterizes similarity between a pair of nodes. $A_{ij} = 0$ if x_i and x_j are not similar (or not connected); and $A_{ij} = 1$ if x_i and x_j are similar (or connected). We assume that the graph is undirected; so, $A_{ij} = A_{ji}$. Further, we are assuming that the similarity values are binary, either 0 or 1.

Consider the graph shown in Fig. 7.14.

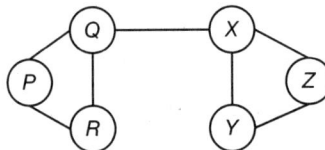

FIG. 7.14 Graphical representation of a data set with six vertices

The corresponding S matrix is given by

$$S = \begin{bmatrix} 1 & 1 & 1 & 0 & 0 & 0 \\ 1 & 1 & 1 & 1 & 0 & 0 \\ 1 & 1 & 1 & 0 & 0 & 0 \\ 0 & 1 & 0 & 1 & 1 & 1 \\ 0 & 0 & 0 & 1 & 1 & 1 \\ 0 & 0 & 0 & 1 & 1 & 1 \end{bmatrix}$$

where we are assuming that a node is similar to itself and so the diagonal entries are all 1.

Weight (Degree) matrix (D) is a diagonal matrix and $D_{ii} = \sum S_{ij}$ and $D_{ij} = 0$ if $i \neq j; i, j \in V$. That is, the ith diagonal element in D is the sum of the elements in the ith row of A. Matrix D for the graph (Fig. 7.14) is

$$D = \begin{bmatrix} 3 & 0 & 0 & 0 & 0 & 0 \\ 0 & 4 & 0 & 0 & 0 & 0 \\ 0 & 0 & 3 & 0 & 0 & 0 \\ 0 & 0 & 0 & 4 & 0 & 0 \\ 0 & 0 & 0 & 0 & 3 & 0 \\ 0 & 0 & 0 & 0 & 0 & 3 \end{bmatrix}$$

Let C_1 (a cluster) be a subset of V and C_2 (another cluster), the complement of C_1, be $V - C_1$. $Cut(C_1, C_2) = \sum_{X_i \in C_1, X_j \in C_2} A_{ij}$. C_1 and C_2 are said to be best partitions, provided $Cut(C_1, C_2)$ is minimum. For example, let $C_1 = \{P, Q, R\}$ and $C_2 = \{X, Y, Z\}$. Then $Cut(C_1, C_2) = 1$, which is minimum among any set of vertices chosen between C_1 and C_2.

It is possible to formalize the minimum cut expression by considering the following:

- Let $I_i = 0$, if $X_i \in C_1$ and $I_i = 1$ if $X_i \in C_2$. Note that $(I_i - I_j)^2 = 1$ if X_i and X_j are in different clusters and it is 0 if they are in the same cluster. So, $Cut(C_1, C_2) = \sum_{x_i \in C_1, x_j \in C_2} A_{ij}(I_i - I_j)^2$.
- It is possible to show that $Cut(C_1, C_2) = I^t L I$, where I is the index vector of size n (where there are n vertices in the graph), Laplacian $L = D - S$, where D is the degree matrix and S is the adjacency matrix. So, minimizing the Cut amounts to finding the index vector I such that $I^t D I$ is minimized; once I is known, it is possible to obtain the clusters based on the polarity of the entries in I.
- L exhibits symmetry, similar to S and D. The eigenvalue with the lowest magnitude in L is 0, with its corresponding eigenvector being denoted as $\mathbf{1} = (1, 1, \dots, 1)^t$ since $L\mathbf{1} = 0 = 0\mathbf{1}$.
- By utilizing the eigenvector $\mathbf{1}$ as the value of I, it can be demonstrated that $I^t L I$ equals 0 because $LI = L\mathbf{1} = 0$. However, this particular choice of I does not yield a two-partition outcome as it results in a single (positive) cluster.
- Instead of selecting the smallest eigenvalue, opt for the subsequent smallest eigenvalue to maintain a small value for $I^t L I$. In this case, I represents the eigenvector corresponding to the second smallest eigenvalue. Moreover, because L is symmetric, its eigenvectors can be chosen to be orthonormal, and the eigenvalues are real. Consequently, by choosing I as the eigenvector linked to the second smallest eigenvalue, a new I is obtained that is orthogonal to $\mathbf{1}$.
- This implies that I comprises both negative and positive entries capturing two clusters. Hence, I is determined to be the eigenvector associated with the second smallest eigenvalue.
- For Fig. 7.14, the Laplacian matrix L is given by

$$L = \begin{bmatrix} 2 & -1 & -1 & 0 & 0 & 0 \\ -1 & 3 & -1 & -1 & 0 & 0 \\ -1 & -1 & 2 & 0 & 0 & 0 \\ 0 & -1 & 0 & 3 & -1 & -1 \\ 0 & 0 & 0 & -1 & 2 & -1 \\ 0 & 0 & 0 & -1 & -1 & 2 \end{bmatrix}$$

- The eigenvalues of L are 0, $\frac{5-\sqrt{17}}{2}$, 3, 3, 3, $\frac{5+\sqrt{17}}{2}$. The first two eigenvectors are $\mathbf{1}$ and $\left(1, \frac{-3+\sqrt{17}}{2}, 1, \frac{3-\sqrt{17}}{2}, \frac{-7+\sqrt{17}}{4}, \frac{3-\sqrt{17}}{2}\right)^t$. Note that the first three entries are positive and the remaining three are negative in the second eigenvector. So, the clusters are $C_2 = \{P, Q, R\}$ and $C_1 = \{X, Y, Z\}$, where C_1 is the negative cluster and C_2 is the positive cluster.
- Figure 7.14 is an example of a two-partition scenario. However, the number of clusters (k) can be more than two in general. In such cases, we have to consider k eigenvectors corresponding to the k smallest eigenvalues. A point to be noted is that each eigenvector has n dimensions (that is, number of data points), so the k eigenvectors can provide a k-dimensional representation of the n patterns by arranging them as k columns in a matrix of size $n \times k$. Moreover, these k eigenvectors are orthogonal to each other, allowing us to cluster the n rows (data points) into k clusters.

7.10 CLUSTERING LARGE DATA SETS

The development of technology has led to advancements in data acquisition, collection and processing capabilities. With these advancements, the size of the data sets has also increased, to the point where they cannot fit entirely into the main memory of the system being used.

Instead, they reside in secondary storage devices and need to be transferred to the main memory as required. However, accessing data from secondary storage is significantly slower compared to accessing data from the main memory. When comparing the time complexity of clustering algorithms, it is possible for them to have the same time complexity. However, the feasibility of the algorithms can differ based on the specific conditions. For example, the k-means algorithm has a time complexity of $\mathcal{O}(n)$ for fixed values of k (number of clusters) and the number of iterations, p.

If the value of p is, say 100, the algorithm needs to scan the entire data set 100 times. In the case of large data sets, where the data needs to be fetched from secondary storage, the k-means algorithm may not generate the desired partition within an acceptable time frame due to the slower access of secondary storage. Therefore, the number of data set scans also affects the performance of the clustering algorithm.

Another important consideration is handling incremental changes in the data set. In some cases, the data in the data set may change incrementally, and the clustering algorithm should be capable of accommodating these changes without referring to the previously seen data.

These requirements are addressed by the following two types of algorithms:

Incremental clustering algorithms: These algorithms cluster new data points based on existing cluster representatives and specific threshold parameters. They do not reconsider data points that were seen earlier.

Leader clustering algorithm is an example of an incremental clustering algorithm. It can be used in both offline scenarios, where the entire data set is available from the beginning, and online scenarios, where data is made available incrementally. An important characteristic of this algorithm is that the clusters are formulated using a *single scan of the data set*.

The core concept of this algorithm revolves around grouping data points that are close to each other into the same cluster, utilizing a predefined distance threshold. In essence, if a data point is located within the threshold distance of a cluster's representative (leader), it is assigned to that cluster. However, if there are no existing clusters within the proximity (threshold distance) of the data point, a new cluster is created with the point designated as the leader of this newly formed cluster.

EXAMPLE 19: Consider Table 7.15. If the data is processed in the sequence of $x_1, x_2, x_3, x_4, x_5,$ x_6, x_7 and x_8, with a threshold of 3, x_1 becomes the leader of cluster $C1$ and is assigned to it. x_2 and x_3 are also assigned to $C1$ since their distance from x_1 is less than the threshold. x_4 is assigned to a new cluster $C2$ and becomes the leader of it, while x_5 and x_6 are assigned to cluster $C2$. x_7 is put into the new cluster $C3$, as its distance from the leaders of $C1$ and $C2$ exceeds the threshold, and x_7 becomes the leader of $C3$. x_8 is assigned to $C3$.

The cluster leaders and clusters may differ if the data is processed in a different order. For instance, x_3 would have been the leader of $C1$ if it had appeared before x_1 and x_2. Also, if x_4 had appeared before x_3, and its distance from x_3 was less than the threshold, it would have been in $C1$, which might not happen if x_1 is the leader. Thus, the leader algorithm is order-dependent.

We discussed the Leader algorithm in Chapter 2.

Abstraction-based clustering: These algorithms generate abstractions from the data set, which are compact representations used for clustering. These abstractions can be generated by examining the data incrementally or non-incrementally. Figure 7.15 shows the abstraction-based incremental clustering activity.

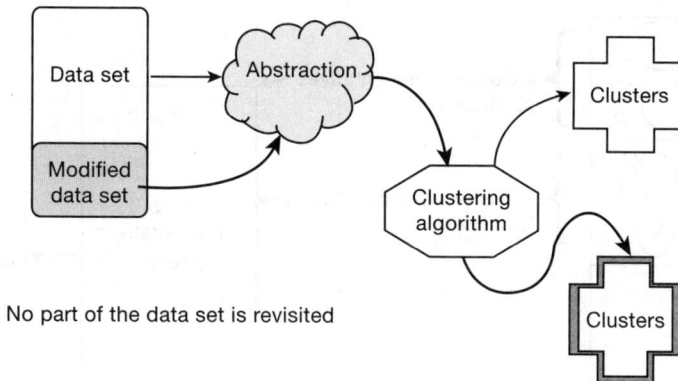

FIG. 7.15 Abstraction-based incremental clustering

Here, the abstraction generator updates the abstraction in an incremental fashion without referring to the data which is already considered. The pattern count tree (PC-tree)-based clustering, discussed in Chapter 5, is an example of an incremental approach to handling abstractions.

On the other hand, the divide-and-conquer method is commonly used for generating intermediate abstractions in a non-incremental manner, which can then be used for clustering. This method is discussed in the next section.

The approaches discussed above aim to address the challenges posed by large data sets and incremental changes in the data, allowing for efficient and effective clustering in such scenarios.

7.10.1 Divide-and-Conquer Method

The divide-and-conquer approach is an effective strategy for addressing the challenge of clustering large data sets that cannot be stored entirely in main memory. To overcome this limitation, a common solution is to process a portion of the data set at a time and store the relevant cluster representatives in memory.

The steps involved in the divide-and-conquer approach are as follows:

1. *Partitioning the data:* The data set, which has a size of $n \times l$, is divided into p blocks. The optimal value of p can be determined based on the chosen clustering algorithm and other factors.
2. *Processing each block:* Each block is transferred from secondary storage to main memory for clustering. A standard clustering algorithm is applied to each block, resulting in k clusters.
3. *Representative selection:* A representative pattern for each of the k clusters is identified, resulting in pk representatives over all the p blocks.
4. *Clustering the representatives:* The pk representatives are clustered into k clusters, potentially using the same or a different clustering algorithm. This step helps to organize and group the representative patterns.
5. *Re-labelling the original data:* The cluster labels of the representative patterns obtained in the previous step are used to re-label the original data set. This ensures that each original data point is assigned to the appropriate cluster based on the clustering of the representatives.

Figure 7.16 shows the divide-and-conquer process.

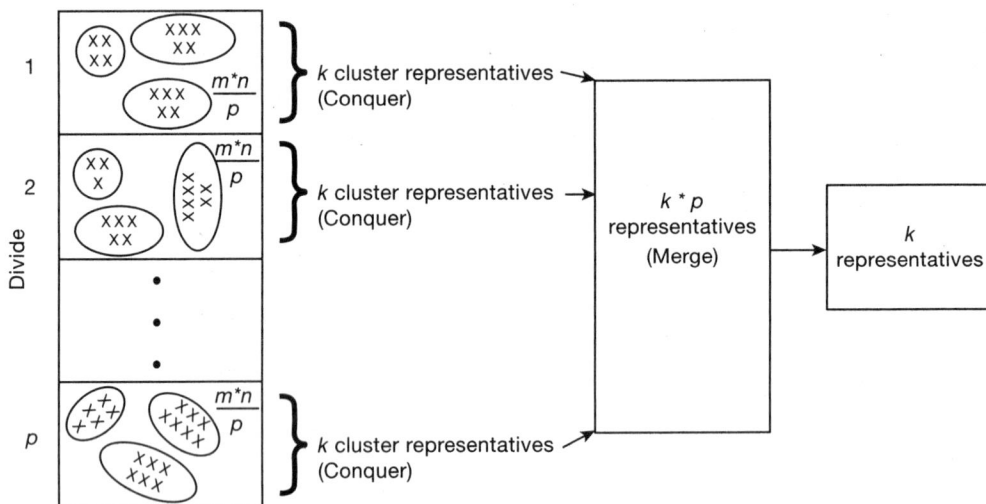

FIG. 7.16 Divide-and-conquer clustering process

By dividing the data set into manageable blocks and processing them incrementally, the divide-and-conquer approach allows clustering algorithms to handle large data sets that exceed the capacity of main memory.

The divide-and-conquer method is illustrated using the following example.

EXAMPLE 20: Consider a data set having 18 data points (patterns) with two features (f1 and f2) as shown in Table 7.25. Further, assume that the number of clusters to be formulated, $k = 2$. Visual representation of the data set is given in Fig. 7.17.

If the data points from x_1 to x_6 are considered as one block (b1), data points from x_7 to x_{12} as the second block (b2) and the remaining data points, x_{13} to x_{18} as the third block (b3), then each block has to be clustered separately.

TABLE 7.25 Example of data set with two features

Data points	f1	f2
x_1	5.5	5.5
x_2	7	6
x_3	6	6.5
x_4	5	7
x_5	5	8
x_6	7	8
x_7	9.5	7
x_8	9	6
x_9	10	6
x_{10}	10	7
x_{11}	11	6
x_{12}	11	7
x_{13}	11	8
x_{14}	10	8
x_{15}	10	8.5
x_{16}	9	9
x_{17}	10	9
x_{18}	10.5	9.5

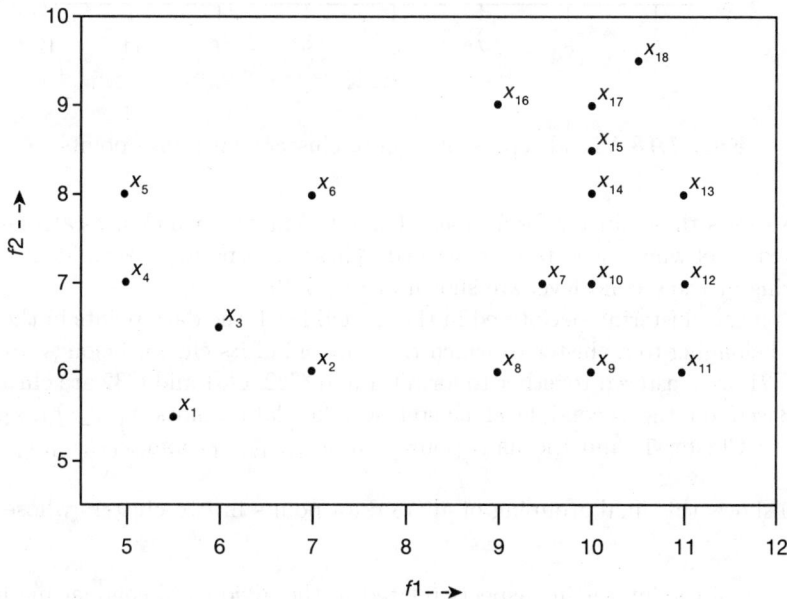

FIG. 7.17 Visual representation of data points listed in Table 7.25

In the first block, two clusters are formed, namely, one consisting of data points x_1, x_2 and x_3 with the centroid at $C11 = (6.17, 5.67)$ and the second consisting of data points x_4, x_5 and x_6 with the centroid at $C12 = (5.67, 7.67)$.

In the second block, two clusters are formed, namely, one consisting of data points x_7, x_8 and x_9 with the centroid at $C21 = (9.5, 6.33)$ and the second consisting of data points x_{10}, x_{11} and x_{12} with the centroid at $C22 = (10.67, 6.67)$.

In the third block, two clusters are formed, namely, one consisting of data points x_{13}, x_{14} and x_{15} with the centroid at $C31 = (10.33, 8.17)$ and the second consisting of data points x_{16}, x_{17} and x_{18} with the centroid at $C32 = (9.83, 9.17)$.

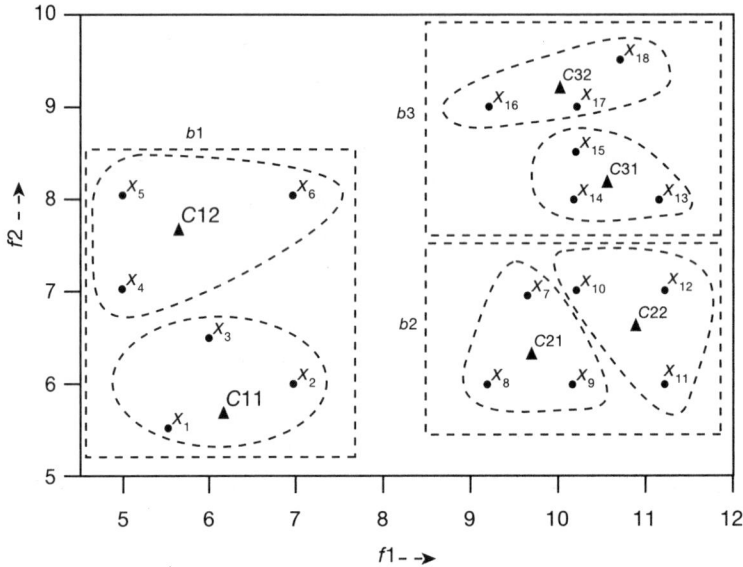

FIG. 7.18 Visual representation of clusters with data points

Figure 7.18 shows these cluster formations. These 6 centroids which are calculated form data points in the next level which need to be clustered. These data points (centroids in the first level) used for clustering at the second level are shown in Fig. 7.19.

Depending on the clustering performed in the second level, the data points in the first level are also labelled as belonging to a cluster to which the centroid of its cluster belongs. In our example, $C11$, $C12$ and $C21$ are clustered together to form $C1$ and $C22$, $C31$ and $C32$ are clustered together to form $C2$. Based on the second level clustering, the data points $x_1, x_2, x_3, x_4, x_5, x_6, x_7, x_8$ and x_9 belong to Cluster 1, and the data points $x_{10}, x_{11}, x_{12}, x_{13}$ and x_{14}, x_{16}, x_{17} and x_{18} will be in $C2$.

Figure 7.20 shows this final grouping of all 18 data points in two clusters whose centroids are $C1$ and $C2$.

Following are some of the interesting aspects related to the divide-and-conquer method:

- The divide-and-conquer method is not limited to only two levels of clustering. When dealing with large data sizes compared to the available memory, a multi-level approach can be employed.
- The divide-and-conquer method is well-suited for parallel processing, thereby enhancing the performance of the clustering process. By dividing the data set into p blocks, the clustering algorithm can be applied to each block independently in parallel, resulting in the generation of cluster centres.

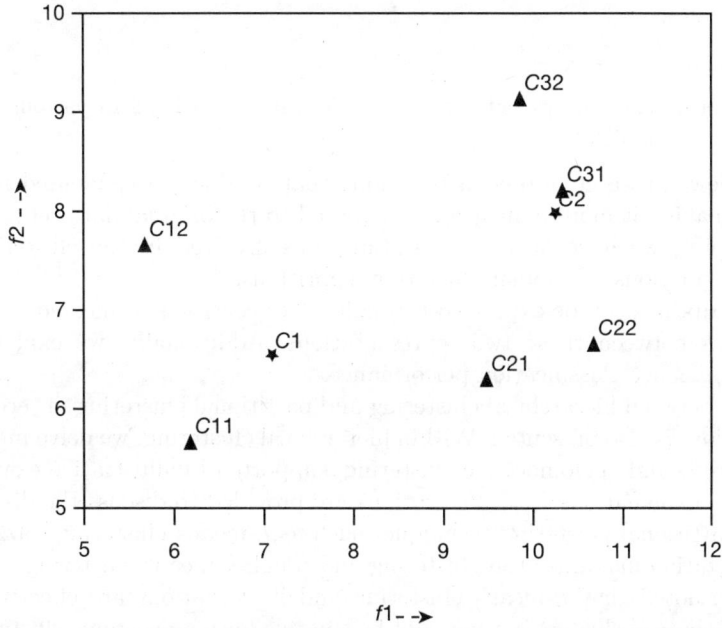

FIG. 7.19 Visual representation of cluster formation at second level

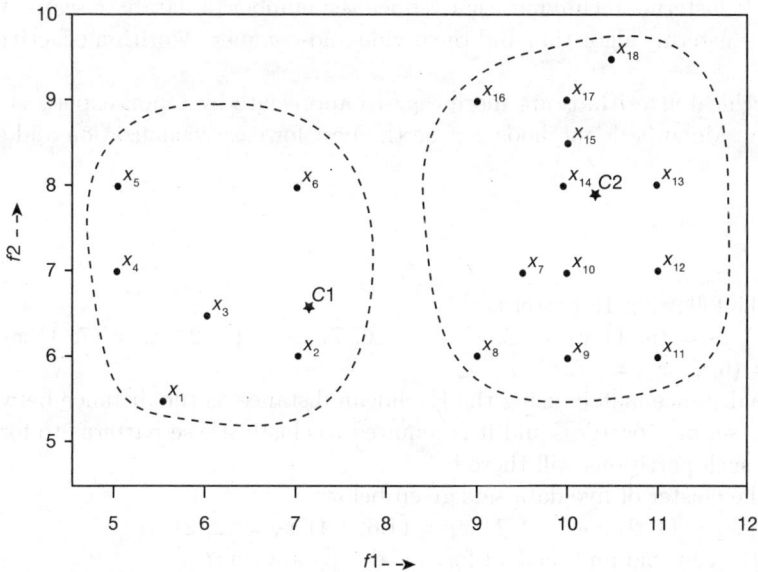

FIG. 7.20 Visual representation of final cluster formation using the divide-and-conquer method

- It is possible to utilize different clustering algorithms at different levels, giving rise to **hybrid clustering approaches**. For example, one can use the k-prototypes algorithm at the first level to obtain representatives and then apply a hierarchical algorithm on these representatives.

SUMMARY

Clustering plays a crucial role in pattern recognition and is utilized in various decision-making scenarios, including classification.

- It can be viewed as a compression technique that produces a condensed representation of data, often making it more manageable compared to the original data set.
- In this chapter, we have outlined essential steps involved in the clustering process and provided explanations of popular clustering algorithms.
- Cluster descriptions can be expressed through either centroids or medoids, and we discussed the variations between these two representations. Additionally, we explored how cluster descriptions enhance classification performance.
- Distinctions between hierarchical clustering and partitional clustering in terms of procedures and applications is also presented. Within hierarchical clustering, we delve into the differences between divisive and agglomerative clustering, supported by illustrative examples.
- Furthermore, explanations with illustrations are provided to discuss the disparities between important partitional clustering techniques such as k-means clustering, fuzzy c-means clustering, expectation maximization clustering and rough k-means clustering.
- It should be noted that k-means clustering and its variations are effective for data with isotropic or spherical clusters but may not be suitable for non-isotropic clustering. To address this, we discussed spectral clustering, which considers the data as a graph.
- As data volume and velocity increase, greater emphasis is placed on incremental clustering and parallel clustering techniques that reduce the number of database scans. We discuss how the Leader clustering algorithm and the divide-and-conquer algorithm effectively handle such scenarios.
- The hierarchical algorithms are more easy to appreciate by a non-expert as they provide a dendrogram. Monothetic methods can be the best for easy visualization and explanation.

EXERCISES

1. Consider the following 10 patterns:
   $x_1 = (1, 1)$ $x_2 = (6, 1)$ $x_3 = (2, 1)$ $x_4 = (6, 7)$ $x_5 = (1, 2)$ $x_6 = (7, 1)$ $x_7 = (7, 7)$ $x_8 = (2, 2)$ $x_9 = (6, 2)$ $x_{10} = (7, 6)$.
   Obtain the distance matrix using the Euclidean distance as the distance between two points.
2. If there is a set of n patterns and it is required to cluster these patterns to form two clusters, how many such partitions will there be?
3. Consider the cluster of five data sets given below:
   $x_1 = (1, 1)$ $x_2 = (1, 2)$ $x_3 = (2, 1)$ $x_4 = (1.6, 1.4)$ $x_5 = (2, 2)$
   What are the centroid and medoid for the above data sets?
4. If we add the data point $x_6 = (1.6, 7.4)$ or $x_6' = (1.6, 17.4)$ to the data sets given in Q3, what are the centroids and medoid? What is your observation?
5. Consider the collection of two-dimensional patterns: (1, 1, 1), (1, 2, 1), (2, 1, 1), (2, 1.5, 1), (3, 2, 1), (4, 1.5, 2), (4, 2, 2),(5, 1.5, 2), (4.5, 2, 2), (4, 4, 3), (4.5, 4, 3), (4.5, 5, 3), (4, 5, 3), (5, 5, 3), where each pattern is represented by feature 1, feature 2 and the class. Find the centroid and medoid of each class.

6. Consider the data set of six data points, each with two features - {(7,4), (5,9), (3,3), (1,3), (8,3), (10,1)}. Use k-means++ clustering to find the initial centroids for $k = 3$.

7. Consider the following data points:
   A(10, 5), B(1, 4), C(5, 8), D(9, 2), E(12, 10), F(15, 8), G(7,7)
   Obtain the dendrogram using monothetic clustering algorithm.

8. For the one-dimensional data set {40, 14, 20, 56, 70}, perform hierarchical clustering and plot the dendogram to visualize it. For this use (a) single linkage method and (b) complete linkage method.[Note: In complete linkage method, the distance between any pair of clusters Cp and Cq is $max_{x_i \in Cp \ and \ x_j \in Cq}(d(x_i, x_j))$.]

9. Give a two-dimensional example having eight 2-dimensional data points to be grouped into three clusters. Use a scheme to select initial 3 centroids ($k = 3$) such that the k-means algorithm leads to local optimum.

10. Consider a data set of eight data points, each with two features, as given in the table below. With the initial centroids x_1, x_4 and x_7, show the clusters generated using k-means clustering algorithm.

Data points	f1	f2
x_1	2.5	10
x_2	2.5	4.5
x_3	7.5	4
x_4	4.5	8
x_5	7	4.5
x_6	6.5	4
x_7	1.5	2.5
x_8	4	9

11. Consider the one-dimensional data set $\mathfrak{X} = \{2.1, 5.1, 1.9, 4.9\}$ with two clusters, c_1 and c_2. Assume that these data sets are drawn from two normals, each with the same variance ($\sigma^2 = 1$). Let $\theta_0 = (2, 4)^t$ be the initial set of mean values. Use the EM algorithm compute the probabilities (Z_{ij}) and mean values (μ) at each iteration. At what μ values of c_1 and c_2 will the algorithm terminate?

PRACTICAL EXERCISES

1. Use the Wisconsin Breast Cancer data set available under sklearn. There are 569 samples corresponding to two classes. Each is a 30-dimensional vector. For more details, visit https://scikit-learn.org/stable/modules/generated/sklearn.datasets.load\ _breast_cancer.html. There are two tasks. For both the tasks, split the data set into train and test parts using train_size = 0.8. Perform this splitting randomly 10 times and report the average accuracy. The tasks are:

   a. **Task 1:** Here, you are supposed to reduce the dimensionality of the data set by clustering the 30 features into 12, 20 and 30 clusters obtained using the k-means algorithm. Note that the resulting feature values are obtained by the centroids of the K clusters in each case. Compute the **percentage accuracy** using Gaussian naïve Bayes classifier on the test data. So, the resulting training data set is of size $455 \times K$ and the test data set is of size $114 \times K$.

b. **Task 2:** Repeat the task in (a) using k-means++ in place of the k-means algorithm.

Bibliography

- Murty, MN and Susheela Devi, V. 2015. *Introduction to Pattern Recognition and Machine Learning*, World Scientific Publishing Co. Pte. Ltd.: Singapore.

- Duda, RO, Hart, PE and Stork, DG. 2007. *Pattern Classification*, New York: John Wiley & Sons.

- Han, J, Kamber, M, and Pei, J. 2012. *Data Mining Concepts and Techniques*, Morgan Kaufmann Publishers, Waltham.

- Witten, IH, Frank, E and Hall, MA. 2011. *Data Mining: Practical Machine Learning Tools and Techniques, Third Edition*, Morgan Kaufmann Publishers, Burlington.

- Jain, AK, Murty, MN and Flynn, P. 1999. Data Clustering: A Review, *ACM Computing Surveys*, 31(3):264–323.

- Aggarwal, CA and Reddy, CK. 2013. *Data Clustering*, CRC Press.

CHAPTER 8
Linear Discriminants for Machine Learning

Learning Objectives

At the end of this chapter, you will be able to:

- Explain the concept of linear discriminants
- Describe important models such as perceptron, support vector machine, logistic regression and multi-layer perceptron network

8.1 INTRODUCTION TO LINEAR DISCRIMINANTS

In the previous chapters, we discussed classifiers that work on classes which are separated by linear, non-linear and piece-wise linear boundaries. Specifically, k-nearest neighbor (KNN) and its variants are easy to understand and implement. They can deal with **non-linear decision boundaries between classes**. They typically work on numerical features, even though distances like Gower distance can be used to classify pattern vectors that are of mixed type, that is, they have both numerical and categorical features. They are ideally suited for applications that have low-dimensional small-size training data sets.

Decision tree classifier (DTC) can capture **piece-wise linear boundaries.** It can deal with pattern vectors that have mixed type features. However, it is ideally suited for low-dimensional applications only because of its greedy nature in building the decision tree. Combinational classifiers, based on DTC, like random forest and gradient boost are more versatile and are suited for dealing with large-scale data sets. So, their implementations are popularly used in several practical applications of current interest. DTC and its variants are embedded-feature-selection schemes.

Bayes' classifier and its variants can deal with mixed type data and can handle classes separated by **non-linear boundaries.** Bayes' classifier is optimal and can be used when the underlying probability structure is known. Its variant, the naïve Bayes classifier (NBC), is a simplified version and can work on mixed type data. It deals with classes having **linear decision boundaries**. It can also be used as an embedded method for feature selection. All these models can also be used for function prediction or regression.

In this chapter, we deal with a collection of ML models that are ideally suited for dealing with large-scale data sets; they are chosen when the classes are **linearly separable**. They are designed to deal with classification of patterns that have only numerical features as these models depend on dot product computation between pattern vectors. Learning in these cases involves starting with some weight structure and updating the weights based on training data. The specific models are:

- **Perceptron:** It is the earliest ML model and also forms the building block for deep neural networks that are popular state-of-the-art ML models. It provides the appropriate decision boundary in a reasonable time when the classes are linearly separable. It can be used when the classes are not linearly separable, provided the functional form of the non-linearity is known.

- **Support Vector Machine (SVM):** This can be viewed as an artificial neural network but a major strength of SVM is that it offers global optimal solutions to a well-formulated optimization problem that maximizes the margin of separation between two classes. The simple model can be extended to deal with even non-linear decision boundaries between classes with the help of the well-known *kernel trick*.

  It captures the non-linear boundaries in the given input data by viewing it as though the data is projected into a higher-dimensional, theoretically even infinite-dimensional, space called the feature space, where the classes are linearly separable. Under some acceptable conditions, it permits us to make all the computations in the low-dimensional input space instead of in the high-dimensional feature space.

- **Logistic Regression:** A simple interpretation is that logistic regression is a variant of perceptron. Perceptron typically employs a threshold function, the linear threshold function, at the output that gives a binary output of 0 or 1, whereas logistic regression employs a function that can provide values in the range [0,1]; these values can be interpreted as probabilities. A more general view is that logistic regression is a linear model; this generalized linear model enables us to deal with non-linear boundaries.

- **Artificial Neural Networks (ANNs):** An ANN is made up of multiple levels of perceptrons that learn the required input–output mapping. It employs a simple gradient-descent scheme for learning or updating the weights. An edge between a pair of neurons is associated with a weight; learning in ANNs pertains to learning the appropriate weight matrix/structure. It is said to have the capacity to learn the features automatically, unlike the other ML models.

We will study all these ML models in this chapter. One unifying thread behind all these models is the notion of a **linear discriminant function**, which we consider next.

8.2 LINEAR DISCRIMINANTS FOR CLASSIFICATION

Let $w = (w(1), w(2), \ldots, w(l))$ be an l-dimensional weight vector and let $x = (x(1), x(2), \ldots, x(l))$ be a pattern represented as a feature vector. Then

$$g(w, x) = \sum_{i=1}^{l} w(i)x(i) + b$$

is called a **linear discriminant function**, where b is a scalar that is called the **bias or threshold**. Sometimes, $w(0)$ is used instead of b in the literature. However, b is more popular, so we use b in this chapter.

EXAMPLE 1: Let $x(1) + x(2) > 0$ or $x(1) + x(2) - 0.5 > 0$ be a linear discriminant function; here $w = (1, 1)$ and $b = 0$ for the former and $w = (1, 1)$ and $b = -0.5$ for the latter.

Both of them represent the Boolean OR function. Consider Fig. 8.1, which shows the Boolean OR on the left and the Boolean AND on the right. We can appreciate them better by looking at their truth table, given in Table 8.1.

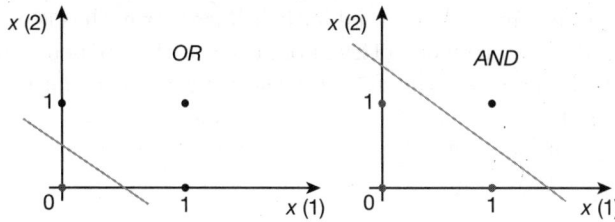

FIG. 8.1 Boolean functions: *OR* and *AND*

TABLE 8.1 Truth table for Boolean *OR* and *AND*

$x(1)$	$x(2)$	$OR(x(1), x(2))$	$x(1) + x(2) - 0.5 > 0$	$AND(x(1), x(2))$	$x(1) + x(2) > 1.5$
0	0	0	0	0	0
0	1	1	1	0	0
1	0	1	1	0	0
1	1	1	1	1	1

Here, $x(1)$ and $x(2)$ are binary features taking values of either 1 (*True*) or 0 (*False*). Note that $OR(x(1), x(2))$ is 0 only when both $x(1)$ and $x(2)$ are 0, else it is 1.

Consider $x(1) + x(2) - 0.5 > 0$. It is 0 (*False*) only when both $x(1)$ and $x(2)$ are 0, else it is 1 (*True*). So, Boolean *OR* is captured by $x(1) + x(2) - 0.5 > 0$. Similarly, $AND(x(1), x(2))$ is 1 (*True*) only when both $x(1)$ and $x(2)$ are 1 (*True*), else it is 0. Now consider $x(1) + x(2) > 1.5$. It is 1 (*True*) only when both $x(1)$ and $x(2)$ are 1, else 0. So, Boolean *AND* is captured by $x(1) + x(2) > 1.5$.

Note that in Fig. 8.1, output value 1 is shown by a black dot and output value 0 is indicated by a grey dot for both *OR* and *AND* functions. In each case, the line that separate 0s from 1s is the decision boundary. It separates the two classes of points; points falling above the line are 1s and those falling below are 0s.

In the case of *OR*, the decision boundary (line) characterizes points that satisfy $x(1) + x(2) = 0.5$. Points that give output 1 satisfy the property that $x(1) + x(2) > 0.5$ and a point for which output is 0 satisfies $x(1) + x(2) < 0.5$. In the case of *AND*, a point, $x = (x(1), x(2))$, on the line satisfies $x(1) + x(2) = 1.5$. A point that gives 1 as output satisfies $x(1) + x(2) > 1.5$ and points that give 0 as output satisfy $x(1) + x(2) < 1.5$.

Note that there could be infinite ways of representing the *OR* and *AND* functions. $OR(x(1), x(2))$ can be represented using $x(1) + x(2) - \alpha > 0$, where $\alpha \in [0, 1)$; α can take any value in the interval [0,1] excluding the value 1. Similarly, $AND(x(1), x(2))$ can be represented using $x(1) + x(2) > \beta$, where $\beta \in [1, 2)$; β can take any value in the interval [1,2] excluding the value 2.

8.2.1 Parameters Involved in the Linear Discriminant Function

The generic form of any linear discriminant function is

$$g(w, x) = w^t x + b = \sum_{i=1}^{l} w(i)x(i) + b$$

It is linear because the components $x(i)$ are used in their linear form, that is, with exponent 1. There are no non-linear terms like the product $x(1)x(2)$ or terms with exponents like $x(1)^2$. $g(x) = 0$ or equivalently $w^t x + b = 0$ characterizes the *decision boundary* between the *positive class* $(g(x) > 0)$ and the *negative class* $(g(x) < 0)$. So, it can be used as a linear classifier; once we know the weight vector w and the threshold or bias b, we decide a given x to be from the positive class if $w^t x + b > 0$, else x is from the negative class.

EXAMPLE 2: Let $w = (-1, 2)$ and $b = 1$ for a linear discriminant. Now we can classify patterns as follows:

Let $x_1 = (1, 1)$. $g(w, x_1) = w^t x_1 + b = -1 \times 1 + 2 \times 1 + 1 = 2$. So, $g(w, x_1) > 0 \Rightarrow$ Assign x_1 to the positive class.

Let $x_2 = (4, 1)$. $g(w, x_2) = w^t x_2 + b = -1 \times 4 + 2 \times 1 + 1 = -1$. So, $g(w, x_2) < 0 \Rightarrow$ Assign x_2 to the negative class.

We can visualize the boundary and other regions associated with the linear discriminant function as shown in Fig. 8.2.

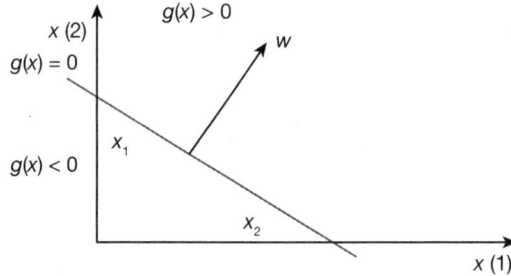

FIG. 8.2 Geometric visualization of the linear discriminant function

Some observations:

- It corresponds to two-dimensional data classification.
- The grey line indicates the decision boundary between two classes. Any point x on this line satisfies the property $g(x) = w^t x + b = 0$. In a higher dimensional space, that is, when the number of features is more than 2, the decision boundary will be a hyperplane.
- The decision boundary splits the whole space into two halves that are characterized by $g(x) > 0$ and $g(x) < 0$. Any point x in the *positive half space* satisfies the property that $g(x) > 0$. Similarly, any point x in the *negative half space* satisfies the property that $g(x) < 0$.
- Consider the two points x_1 and x_2 located on the decision boundary.

$$g(x_1) = g(x_2) = 0 = w^t x_1 + b = w^t x_2 + b \Rightarrow w^t (x_1 - x_2) = 0$$

This means w is orthogonal to $(x_1 - x_2)$ and $x_1 - x_2$ is the line joining the two points x_1 and x_2. So, w is orthogonal to the decision boundary.

- Consider the origin, that is, $x = 0$. So, $g(x) = w^t x + b = b$. So, $g(x) = b$.
  If $b > 0$, then origin is in the positive half space as $g(0) = b > 0$.
  If $b < 0$, then origin is in the negative half space as $g(0) = b < 0$.
  If $b = 0$, then origin is on the decision boundary as $g(x) = b = 0$.
- Consider the case where $b = 0$, as shown in Fig. 8.3. Consider the points x_3 and x_4. Let α be the angle between w and x_3 and β be the angle between w and x_4, as shown in the figure.

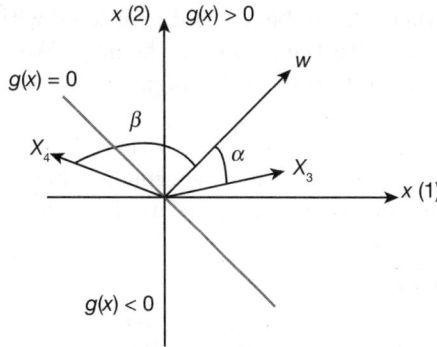

FIG. 8.3 Decision boundary passing through the origin ($b = 0$)

Note that x_3 is in the positive half space, so $g(x_3) = w^t x_3 > 0$ (as $b = 0$). We know that

$$cos\alpha = \frac{w^t x_3}{||w|| ||x_3||} \Rightarrow cos\alpha > 0 \Rightarrow |\alpha| < 90,$$

because the numerator is positive and both terms in the denominator are norms and so are positive. This holds good for any x in the positive half space. The angle between w and such an x is in the range $[-90, 90]$.

However, x_4 is in the negative half space. So, $g(x_4) = w^t x_4 < 0$.

$$cos\beta = \frac{w^t x_4}{||w|| ||x_4||} \Rightarrow cos\beta < 0 \Rightarrow |\beta| > 90$$

So, w is orthogonal to the decision boundary and points towards the positive half space, as shown in Fig. 8.3.

- On the other hand, b indicates the location of the decision boundary. If $b = 0$, then the decision boundary passes through the origin. If $b > 0$, then $g(0) = w^t 0 + b = b > 0$. So, the origin will be in the positive half space. Similarly, if $b < 0$, then the origin will be in the negative half space.

EXAMPLE 3: Consider a linear discriminant function $g_1(x) = x(1) - x(2) = w^t x + b$, where $x = \begin{pmatrix} x(1) \\ x(2) \end{pmatrix}$, $w = \begin{pmatrix} 1 \\ -1 \end{pmatrix}$ and $b = 0$. Here, a two-dimensional vector x such that

- $x(1) = x(2)$ will be on the decision boundary as $g_1(x) = 0$.
- $x(1) < x(2)$ will be in the negative half space as $g_1(x) < 0$.
- $x(1) > x(2)$ will be in the positive half space as $g_1(x) > 0$.

8.2.2 Learning w and b

We have seen that the weight vector w and b can be used to classify a test pattern x. However, such a classification scheme needs w and b to be obtained; a learning scheme is used to get these entities. This learning is achieved with the help of training data.

Instead of learning w and b separately, one can learn them together in one shot by augmenting. Note that $w^t x + b$ can be equivalently written as $w_a^t x_a$, where $w_a = [b, w]$ and $x_a = [1, x]$. That is, if w is an l-dimensional vector, then w_a will be $(l+1)$-dimensional with the first component being b; similarly, x_a is $(l+1)$-dimensional with the first entry being 1. We use x for x_a and w for w_a for the sake of brevity; whether we are referring to the augmented vectors or the original vectors will be clear from the context.

So, $\{(x_i, y_i), i = 1, 2, \ldots, n\}$ is linearly separable if there is w_* such that $w_* x_i > 0$ if $y_i = 1$, else $y_i = 0$. Such a w_* characterizes a separating hyperplane; there could be infinitely many such w_* vectors and the corresponding decision boundaries as shown in Fig. 8.4.

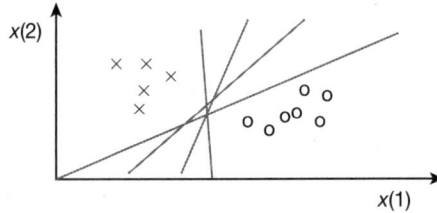

FIG. 8.4 Multiple decision boundaries (for colour figure, please see Colour Plate 2)

EXAMPLE 4: Consider $\{((1, 1), l_1 = 0), ((2, 2), l_2 = 0), ((4, 4), l_3 = 1)\}$, where the first two vectors are from class 0 and the third one is from class 1.

The augmented vectors in the resulting three-dimensional space are $x_1 = (1, 1, 1), x_2 = (1, 2, 2), x_3 = (1, 4, 4)$. A possible augmented w is $(-6, 1, 1)$ that can separate the three patterns. Note that $w^t x_1 = -4(< 0), w^t x_2 = -2(< 0)$ and $w^t x_3 = 2(> 0)$.

All the three are correctly classified; the first two patterns are from the negative half space (class 0) and the third from the positive half space (class 1). It is possible to check if $(-12, 2, 3)$ also correctly classifies all the three patterns. We leave it as an exercise.

We can compute the normal distance of a data point x from the decision boundary, as shown in Fig. 8.5. Let r be the magnitude of the distance between x and the decision boundary. We can view the vector x as a sum of two vectors x_p and $\frac{rw}{||w||}$, where x_p is the orthogonal projection of x onto the decision boundary and r is the unit of distance of point x from the decision boundary.

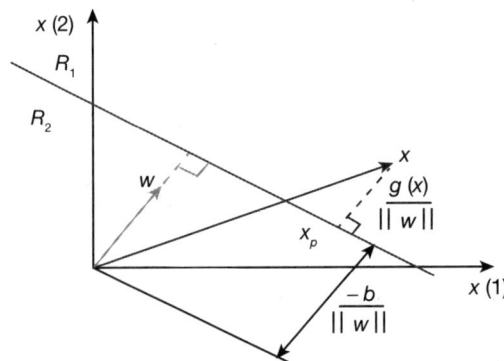

FIG. 8.5 Distance of a point from the decision boundary (for colour figure, please see Colour Plate 2)

The quantity $\frac{w}{||w||}$ gives us the unit vector in the direction of w. So, $r \times \frac{w}{||w||} = \frac{rw}{||w||}$ indicates the vector which when added to x_p gives us x.

We know that $g(x) = w^t x + b = w^t x_p + b + w^t \frac{rw}{||w||} = 0 + r||w||$ as x_p is on the decision boundary and so $w^t x_p + b = 0$. So, $r||w|| = g(x) \Rightarrow r = \frac{g(x)}{||w||}$.

8.3 PERCEPTRON CLASSIFIER

Perceptron is the earliest ML model that was used in classifying linearly separable data. Some related observations are as follows:

- We learn the augmented weight vector w with the help of some two-class training data.
- If the classes are linearly separable, then the perceptron learning algorithm will learn the augmented weight vector in a finite number of steps.
- The obtained weight vector w will classify any augmented pattern x to the negative class if $w^t x < 0$ and assign x to the positive class if $w^t x > 0$. We assume that the final w will classify every training pattern to one of the two classes and there is no x that gives $w^t x = 0$.
- It is likely that there are infinite possible solutions to the learning problem; that is, there could be possibly infinite w vectors. The perceptron learning algorithm will find one of them. It is based on starting with an initial weight vector and updating it until all the training patterns are correctly classified.
- If the weight vector at step k, w_k, misclassifies an x from the negative class, that is, $w_k^t x > 0$, then it is updated as $w_{k+1} = w_k - x$. Consider $w_{k+1}^t x = w_k^t x - x^t x$. So, $w_{k+1}^t x$ can be negative even if $w_k^t x$ is positive because $x^t x \geq 0$ and $w_{k+1}^t x \leq w_k^t x$.
- If the weight vector at step k, w_k, misclassifies an x from the positive class, that is, $w_k^t x < 0$, then it is updated as $w_{k+1} = w_k + x$. Consider $w_{k+1}^t x = w_k^t x + x^t x$. So, $w_{k+1}^t x$ can be positive even if $w_k^t x$ is negative because $x^t x \geq 0$ and $w_{k+1}^t x \geq w_k^t x$.
- So, this update scheme to get w_{k+1} from w_k is justified because w_{k+1} is in a better position, compared to w_k, to classify x correctly.
- There could be different ways of selecting the initial weight vector w_0. But a simple choice for theoretical analysis is to let $w_0 = 0$; so w_0 is chosen to be the zero vector initially.

8.3.1 Perceptron Learning Algorithm

The algorithm is very simple and it can be shown that the algorithm stops after a finite number of steps to give us a correct w if the classes are linearly separable.

Let the training data set be $\{(x_1, l_1), (x_2, l_2), \ldots, (x_n, l_n)\}$, where x_j is the jth pattern, l_j is its class label and $l_j \in \{-1, +1\}$ based on whether x_j is from the negative class or positive class. The algorithm follows the steps given below:

1. Let $i = 0$. Let the initial augmented weight vector w_i be the zero vector. Let $j = 1$. (Note that j is the pattern index and i is the weight vector index).
2. If x_j is incorrectly classified by w_i, then $i = i + 1$ and $w_i = w_{i-1} + l_j x_j$, set $j = (j + 1) mod\ n$ and go to step 2. Else set $j = (j + 1) mod\ n$ and go to Step 2.
3. Terminate the algorithm if there is no change in weight vector w_i for n successive steps.

Some observations:

- x_j is incorrectly classified by w_i (Step 2) if $l_j(w_i^t x_j) \leq 0$.
- Updating j in steps 2 and 3 involves using addition modulo n, that is, after considering the nth pattern, x_n, it considers the first pattern, x_1. So, it considers the data points cyclically till the termination condition, specified in Step 3, is satisfied.
- If point x_j from the negative class is misclassified, then $w_i = w_{i-1} - x_j$ as $l_j = -1$ and if x_j from the positive class is misclassified, then $w_i = w_{i-1} + x_j$ as $l_j = +1$.
- If there is no update for n successive steps, it means that the current w_i has classified all the patterns correctly and so the algorithm did not visit Step 2 in the recent n steps.
- If a pattern is correctly classified, then we do not update the weight vector and consider the next pattern in the sequence cyclically.

We illustrate the working of the algorithm using an example.

EXAMPLE 5: Let the training set be $\{((-1,1), l_1 = -1), ((-2,-2), l_2 = -1), ((4,4), l_3 = 1)\}$. So, the augmented patterns are $(1,-1,1)$, $(1,-2,-2)$ and $(1,4,4)$ with their respective labels.

1. $w_0 = (0,0,0)$. It misclassifies $(1,-1,1)$ as the dot product between w_0 and $(1,-1,1)$ is 0 (we expect it to be less than 0). So, $w_1 = w_0 - (1,-1,1) = (-1,1,-1)$.
2. w_1 correctly classifies the next pattern $(1,-2,-2)$ as the dot product is -1 (< 0).
3. w_1 misclassifies $(1,4,4)$ as the dot product with $(1,4,4)$ is -1 (we expect it to be greater than 0). So, $w_2 = w_1 + (1,4,4) = (-1,1,-1) + (1,4,4) = (0,5,3)$.
4. Note that $w_2 = (0,5,3)$ correctly classifies all the three augmented patterns as

   - The dot product of w_2 with $(1,-1,1)$ is -2 (< 0).
   - The dot product of w_2 with $(1,-2,-2)$ is -16 (< 0).
   - The dot product of w_2 with $(1,4,4)$ is 32 (> 0).

5. So, the algorithm terminates and gives us the weight vector $(0,5,3)$.

We consider one more example. It deals with learning of Boolean OR.

EXAMPLE 6: Consider the truth table of Boolean OR shown in the first three columns of Table 8.1. The inputs are $x(1)$ and $x(2)$ and the output is either a 0 or 1; we have a two-class problem based on these two output values. So, there is one pattern from Class 0 and three patterns from Class 1. The augmented patterns are shown in Table 8.2.

TABLE 8.2 Augmented patterns for Boolean OR

1	$x(1)$	$x(2)$	$OR(x(1), x(2))$
1	0	0	0
1	0	1	1
1	1	0	1
1	1	1	1

We show the details of the updating of the weight vector starting with $w_0 = (0,0,0)$ in Table 8.3. It takes 9 steps of weight updates to give the weight vector $w_9 = (-1,2,2)$ which classifies all the four augmented patterns correctly.

TABLE 8.3 Updating the weight vector

Weight vector	Pattern misclassified	$w^t x$	Weight vector	Pattern misclassified	$w^t x$
$w_0 = (0,0,0)$	(1,0,0)	0	$w_1 = (-1,0,0)$	(1,0,1)	-1
$w_2 = (0,0,1)$	(1,1,0)	0	$w_3 = (1,1,1)$	(1,0,0)	-1
$w_4 = (0,1,1)$	(1,0,0)	0	$w_5 = (-1,1,1)$	(1,0,1)	0
$w_6 = (0,1,2)$	(1,0,0)	0	$w_7 = (-1,1,2)$	(1,1,0)	0
$w_8 = (0,2,2)$	(1,0,0)	0	$w_9 = (-1,2,2)$	none	correct $\forall x$

Note that in the process, the algorithm subtracted the pattern $(1,0,0)$ five times, added the pattern $(1,1,0)$ twice and the pattern $(1,0,1)$ twice to the initial weight vector w_0. So, $w = w_9 = (0,0,0) - 5(1,0,0) + 2(1,1,0) + 2(1,0,1) = (-1,2,2)$. So, it is convenient to view the learnt vector as a linear combination of the training data points. Generically, $w = \sum_{i=1}^{n} \alpha_i l_i x_i$, where $\alpha_i \geq 0$. Note that in Example 1,

- $x_1 = (1,0,0), \alpha_1 = 5,$ *and* $l_1 = -1$
- $x_2 = (1,0,1), \alpha_2 = 2,$ *and* $l_2 = 1$
- $x_3 = (1,1,0), \alpha_3 = 2,$ *and* $l_3 = 1$
- $x_4 = (1,1,1), \alpha_4 = 0,$ *and* $l_4 = 1$

8.3.2 Convergence of the Learning Algorithm

It is possible to show that the learning algorithm finds the correct weight vector, a weight vector that classifies all the patterns correctly, if the classes are linearly separable.

Let the augmented patterns be $x_1, x_2, \ldots, x_n$ with their respective class labels being $l_1, l_2, \ldots, l_n$ and $l_i \in \{-1, +1\}$ for all i. If they are linearly separable, then there exists a vector w that correctly classifies all the patterns. That is, if x is in the positive class, then $w^t x > 0$, and if x is in the negative class, then $w^t x < 0$. In either case, $w^t l x > 0$, where l is the class label of x because $l = +1$ if x is in the positive class and $l = -1$ if x is in the negative class.

We have $w_0 = 0$. Let the first pattern misclassified by w_0 be x^0 and let $l(0)$ be its class label. So, $w_1 = w_0 + l(0)x^0$. In general, if augmented pattern x^{k-1} is misclassified by w_{k-1}, then $w_k = w_{k-1} + l(k-1)x^{k-1}$. By expanding recursively, we have $w_k = w_0 + l(0)x^0 + l(1)x^1 + \cdots + l(k-1)x^{k-1}$, where $w_0 = 0$.

Considering the dot product between the correct w and w_k, we have

$$w^t w_k = w^t (l(0)x^0 + l(1)x^1 + \cdots + l(k-1)x^{k-1})$$

Let $\alpha_i = w^t l(i)x^i$. So, $w^t w_k = \alpha_0 + \alpha_1 + \cdots + \alpha_{k-1}$. We know that $\alpha_i > 0$ for $i = 0, 1, \ldots, k-1$ because of w. So, $w^t w_k > 0$.

Let α_p be the $minimum(\alpha_0, \alpha_1, \ldots, \alpha_{k-1})$. So, $w^t w_k > k\alpha_p$. Consider $w_k^t w_k = l(0)^2 \|x^0\|^2 + l(1)^2 \|x^1\|^2 + \cdots + l(k-1)^2 \|x^{k-1}\|^2 = \|x^0\|^2 + \|x^1\|^2 + \cdots + \|x^{k-1}\|^2$ as $l(i)^2 = 1$ for $i = 0, 1, \ldots, k-1$.

Let $\beta_i = \|x^i\|^2$ for $i = 0, 1, 2, \ldots, k-1$. So, clearly $\beta_i > 0$ for all i. So, $w_k^t w_k = \beta_0 + \beta_1 + \cdots + \beta_{k-1}$ is positive. Let $\beta_q = maximum(\beta_0, \beta_1, \ldots, \beta_{k-1})$. Then $w_k^t w_k < k\beta_q$.

Let θ be the angle between w and w_k. We know that $cos\theta = \frac{w^t w_k}{\|w\|\|w_k\|}$. Note that both the numerator and each of the two terms in the denominator is positive as $w^t w_k > 0$; this is because

w_k is the sum of the given patterns. So, $cos\theta > 0$. So, $1 > cos\theta = \frac{w^t w_k}{||w|| ||w_k||} > \frac{k\alpha_p}{||w||\sqrt{k\beta_q}}$. So, $||w||\sqrt{k\beta_q} > k\alpha_p$. Simplifying this, we get $k < \frac{||w||^2\beta_q}{\alpha_p^2}$. All the terms in the RHS of this inequality are positive.

We can make it simpler by choosing the w to be an unit vector. Then $||w||^2 = 1$. So, the inequality simplifies to $k < \frac{\beta_q}{\alpha_p}^2$. We can simplify even further by normalizing all the augmented patterns so that $||x_i||^2 = 1$, then $\beta_q = 1$. So, the inequality is $k < \frac{1}{\alpha_p}$. So, the value of k can be larger if α_p is smaller. This can occur if $min(w^t l(0)x^0, \ldots, w^t l(k-1)x^{k-1})$ is small. If an x^i for $i \in \{0, 1, \ldots, k-1\}$ is very close to being orthogonal to w, then $w^t l(i)x^i \geq 0$. However, it is not possible for $w^t l(i)x^i = 0$ as w correctly classifies x^i.

8.3.3 Linearly Non-Separable Classes

We have seen that the perceptron learning algorithm converges to a correct w if the classes are linearly separable. However, when the classes are not linearly separable, the learning algorithm may not stop. We have two practical solutions:

- Keep updating the weight vector till most of the training data points are correctly classified. In this case, some of the training patterns are misclassified and the resulting weight vector can still be used to classify test patterns.
- If we know the form of the non-linearity, then we can use perceptron as shown in the example below.

EXAMPLE 7: Consider the data shown in Table 8.4.

TABLE 8.4 Non-linearly separable data

Pattern	x	Class
1	1	-1
2	-1	-1
3	2	-1
4	-2	-1
5	3	1
6	4	1
7	-3	1
8	-4	1

Here, the patterns are one-dimensional and specified by the value of x. Suppose we know that the function $g(x) = a + bx + cx^2$. Then we can view it as a linear function in a higher dimensional space.

Let $w = (a, b, c)$ and $x_a = (1, x, x^2)$. Then $w^t x_a = a + bx + cx^2 = g(x)$. So, $g(x)$ is non-linear in x. However, it is linear in x_a. We show, in Table 8.5, the augmented vectors corresponding to each value of x in Table 8.4. The algorithm terminates after 21 steps with the weight vector $w_{21} = (-11, 0, 2)$. So, $g(x) = 2x^2 - 11$. We leave working out the details as an exercise.

TABLE 8.5 Augmented non-linearly separable data

Pattern	1	x	x^2	Class
1	1	1	1	-1
2	1	-1	1	-1
3	1	2	4	-1
4	1	-2	4	-1
5	1	3	9	1
6	1	4	16	1
7	1	-3	9	1
8	1	-4	16	1

8.3.4   Multi-Class Problems

Classifiers based on linear discriminant functions are ideally suited for two-class problems or **binary classification** problems. It is possible to extend a binary classifier to deal with C-class ($C > 2$) problems. Two popular schemes are:

- **One-vs-One:** Here, we consider a *pair of classes at a time*; there are $\frac{C(C-1)}{2}$ such pairs if the total number of classes is C. Learn a linear discriminant function for each pair of classes. Combine these decisions to arrive at the class label.

   EXAMPLE 8:   Let us consider a 4-class problem where the classes are C_1, C_2, C_3 and C_4. So, there are 6 pairs of classes. Let a test pattern x be assigned for each of these pairs as follows:

   C_1 vs C_2 classification: Let x be assigned to class C_1.
   C_1 vs C_3 classification: Let x be assigned to class C_1.
   C_1 vs C_4 classification: Let x be assigned to class C_4.
   C_2 vs C_3 classification: Let x be assigned to class C_3.
   C_2 vs C_4 classification: Let x be assigned to class C_4.
   C_3 vs C_4 classification: Let x be assigned to class C_4.

   So, by using majority voting, we assign x to class C_4 as 3 out of 6 pairs have voted for C_4.

   However, there could be problems associated with this scheme:

   - The number of classifier pairs that need to be trained is $\mathcal{O}(C^2)$ for a C-class problem.
   - There could be ties. For example, in step f of Example 8, if the decision had gone in favour of C_3, then there will be a tie in taking majority voting between C_1, C_3 and C_4.

- **One-vs-Rest**: For each class C_i, let

$$\bar{C}_i = \cup_{j=1; j \neq i}^{C} C_j$$

be the complementary class. Learn a linear discriminant function to classify to C_i or $\bar{C}_i$ for each i. Combine these linear discriminants to classify a pattern.

   EXAMPLE 9:   Consider again a 4-class problem where the classes are C_1, C_2, C_3 and C_4. There are 4 classifiers to be trained here. Let the decisions be:

   C_1 vs $\bar{C}_1$: Let x be assigned to C_1.
   C_2 vs $\bar{C}_2$: Let x be assigned to $\bar{C}_2$.

C_3 vs $\bar{C}_3$: Let x be assigned to $\bar{C}_3$.
C_4 vs $\bar{C}_4$: Let x be assigned to $\bar{C}_4$.

We decide to assign x to class C_1.

This scheme can also have problems:

- There could be ties. For example, in step d of Example 9, if the decision had gone in favour of C_4, there would have been a tie in taking majority voting between C_1 and C_4.
- If the number of classes C is large, there can class imbalance between each C_i and $\bar{C}_i$. Specifically, a class C_i could be a minority class having less than $\frac{1}{C}$ fraction of training data points and $\bar{C}_i$, the majority class, could have more than $\frac{C-1}{C}$ fraction. This leads to problems while using several classifiers.

Even though these techniques have been well understood for more than four decades, it is support vector machines (SVMs) (binary classifiers) that have popularized them.

8.4 SUPPORT VECTOR MACHINES

We have seen in Fig. 8.4 that there could be theoretically infinite decision boundaries if the classes are linearly separable. Perceptron selects one of them. However, if we want to select a decision boundary that is evenly centred between the two class boundaries, then support vector machine (SVM) is the best choice to obtain the optimal solution. Consider Fig. 8.6.

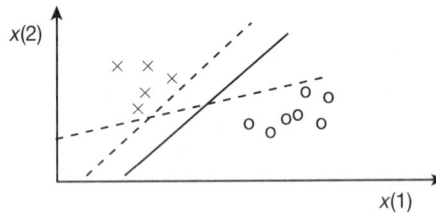

FIG. 8.6 Bad and good decision boundaries

There are two classes in a two-dimensional space, specified by $x(1)$ and $x(2)$. There are 5 Xs from the negative class and 7 circles from the positive class. These two classes are linearly separable.

There are two decision boundaries indicated by broken lines. Both of them are bad; one of them is very close to the negative class and the other one is unable to exploit the margin of separation between the two classes. The third decision boundary shown by a solid line appears to be good as it evenly splits the margin of separation between the two classes.

The SVM classifier formulates an optimization problem to maximize the margin between the two classes; the decision boundary will be evenly positioned between the two classes. Some of its important features are as follows:

- It considers a two-class problem with class label $y_i \in \{-1, +1\}$ for x_i in the training set for $i = 1, 2, \ldots, n$; if $y_i = -1$, then x_i is from the negative class and if $y_i = +1$, then x_i is from the positive class.

- Any point x on the decision boundary will satisfy $w^t x + b = 0$, where w is the weight vector and b is the bias.
- For any point x from the negative class, $w^t x + b < 0$ and any point x from the positive class, $w^t x + b > 0$.
- There is a finite number, n, of training patterns. So, one can identify an $\epsilon > 0$ such that:

$w^t x_i + b \geq \epsilon$ for all i with $y_i = 1$
$w^t x_i + b \leq -\epsilon$ for all i with $y_i = -1$

- By dividing both sides of the two inequalities by ϵ, we get

$w'^t x + b' \geq 1$ for all i with $y_i = 1$
$w'^t x + b' \leq -1$ for every x for all i with $y_i = -1$, where $w' = \frac{1}{\epsilon} w$ and $b' = \frac{b}{\epsilon}$

- We can combine the two inequalities into one by exploiting the value of $y_i \in \{-1, 1\}$. The resulting inequality is

$$y_i(w^t x_i + b) \geq 1, \forall i$$

Note that we have dropped the superscripts for w and b for the sake of brevity. So, we have $w^t x + b = 1$ for a boundary pattern x from the positive class and $w^t x + b = -1$ for a boundary pattern x from the negative class.

These two equalities characterize two parallel lines when w and x are two-dimensional vectors and they correspond to two parallel planes if the vectors are higher-dimensional (dimensionality > 2). The data points closest to the hyperplanes are the *support vectors*.

- The distance from any point x on a plane $w^t x + b = 1$ to the decision boundary given by $w^t x + b = 0$ is $\frac{1}{||w||}$, as discussed in Section 8.2.2. Similarly, the distance from any point x on plane $w^t x + b = -1$ to the decision boundary is $\frac{1}{||w||}$. So, the total distance between the two parallel planes is $\frac{2}{||w||}$. This is called the **margin**.
- The margin is maximized by SVM. This optimization problem has constraints in the form

$$y_i(w^t x_i + b) \geq 1, \forall i$$

Instead of maximizing $\frac{2}{||w||}$, we can minimize $\frac{||w||^2}{2}$. So, the optimization problem with constraints is *minimize* $\frac{1}{2} || w ||^2$ *subject to* $1 - y_i(w^t x_i + b) \leq 0$ *for* $i = 1, \ldots, n$.
The Lagrangian is $\mathcal{L} = \frac{1}{2} w^t w + \sum_{i=1}^{n} \alpha_i (1 - y_i(w^t x_i + b))$.

- The necessary and sufficient conditions for a (w, b) pair to be the optimal solution are (by setting the gradient of $\mathcal{L}$ with respect to w to zero, we get)

$$w + \sum_{i=1}^{n} \alpha_i(-y_i)x_i = 0 \Rightarrow w = \sum_{i=1}^{n} \alpha_i y_i x_i$$

$$\frac{\delta \mathcal{L}}{\delta b} = 0 \Rightarrow \sum_{i=1}^{n} \alpha_i y_i = 0$$

$$1 - y_i(w^t x_i + b) \leq 0, \forall i$$

$$\alpha_i \geq 0 \text{ and } \alpha_i(1 - y_i(w^t x_i + b)) = 0, \forall i$$

- Consider the condition $\alpha_i(1 - y_i(w^t x_i + b)) = 0$, $\forall i$. If $\alpha_i > 0$, then $1 - y_i(w^t x_i + b) = 0 \Rightarrow w^t x_i + b = 1$ if $y_i = 1$ and $w^t x_i + b = -1$ *if* $y_i = -1$. If $\alpha_i = 0$ and $1 - y_i(w^t x_i + b) \neq 0$, then $w^t x_i + b > 1$ if $y_i = 1$ and $w^t x_i + b < -1$ if $y_i = -1$. Such points are called well-classified points; these points are not on the boundary of the class and are within the proper region of the respective class. If $\alpha_i = 0$ and $1 - y_i(w^t x_i + b) = 0$, then such x_i lie on the support planes, but $\alpha_i = 0$.

- So, if a data point x_i lies away from the support plane, that is, $y_i(w^t x_i + b) > 1$, then $\alpha_i = 0$. Now consider $w = \sum_{i=1}^n \alpha_i y_i x_i$. If $\alpha_i = 0$ for x_i, then it will not contribute to w. So, well-classified points do not contribute to w. Only training data points on the support planes with $\alpha > 0$ can contribute to w. Such vectors are called **support vectors**. So, $w = \sum_{i=1}^n \alpha_i y_i x_i = \sum_{i \in S} \alpha_i y_i x_i$, where S is the set of support vectors and S is a subset of the training set.

EXAMPLE 10: Consider the training points from the positive and negative classes, as shown in Fig. 8.7. There are two points from the negative class and one point from the positive class. These are:

- Negative Class: $(2, 1)$ and $(1, 3)$
- Positive Class: $(6, 3)$

So, they need to satisfy:

$$2w(1) + w(2) + b = -1$$

$$w(1) + 3w(2) + b = -1$$

$$6w(1) + 3w(2) + b = 1$$

These are three equations in three unknowns. By solving them, we get

$$w(1) = \frac{2}{5}, w(2) = \frac{1}{5} \text{ and } b = -2$$

Note that the point $(4, 2)$ is on the boundary because $4 \times \frac{2}{5} + 2 \times \frac{1}{5} - 2 = 0$. Further, $\sum_{i=1}^3 \alpha_i y_i = 0 \Rightarrow -\alpha_1 - \alpha_2 + \alpha_3 = 0 \Rightarrow \alpha_3 = \alpha_1 + \alpha_2$.

We know that $w = (\frac{2}{5}, \frac{1}{5}) = -\alpha_1 \times (2, 1) - \alpha_2 \times (1, 3) + \alpha_3 \times (6, 3) \Rightarrow \alpha_1 = \alpha_3 = \frac{1}{10}; \alpha_2 = 0$.

So, the set of support vectors is $S = \{(2, 1), (6, 3)\}$. The point $(1, 3)$ will not contribute to w as $\alpha_2 = 0$. So, a boundary point or a point on the hyperplane need not contribute to w and so need not be a support vector.

Consider the origin, that is, $x = (0, 0)$. Here, $w^t x + b = -2 < -1$. So, it cannot be a support vector even if it is added to the training data. Similarly, a point $(7, 6)$ will be a well-classified point.

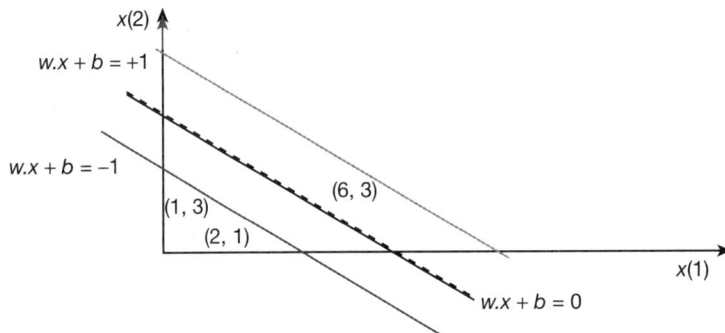

FIG. 8.7 An example to illustrate SVM (for colour figure, please see Colour Plate 2)

The SVM classifier can also be used as a feature selector. Let the w vector obtained at the end of training be $(0.001, -20, 10.9, 0)$ in a four-dimensional space. So, the second and the third features are important. The remaining two features are unimportant. The second feature is the most important and it votes in favour of the negative class. The next important feature is the third feature and it is in favour of the positive class.

8.4.1   Linearly Non-Separable Case

Consider the data shown in Fig. 8.8.

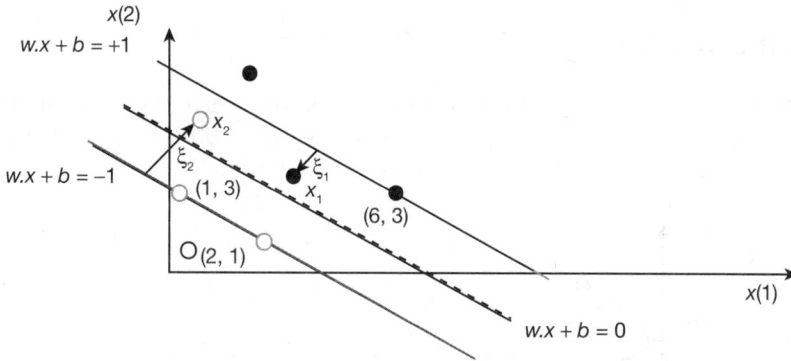

FIG. 8.8 An example to illustrate the margin violators (for colour figure, please see Colour Plate 2)

There are three points from the positive class; each of these is indicated by a blackened circle. There are four data points from the negative class; each of these is depicted using a whitened circle. There are two violators; they are labelled x_1 and x_2.

Data point x_1 is from the positive class and it violates the constraint given by $w^t x_1 + b \geq 1$ as it falls to the left of the positive support line. However, it is still on the correct side of the decision boundary. Similarly, data point x_2 is from the negative class and it violates the constraint $w^t x_2 + b \leq 1$. It clearly violates both the support line and the decision boundary.

The extent of violation is captured by values ξ_1 and ξ_2, respectively, for the two violators x_1 and x_2. Even though x_1 is a violator, it is still correctly classified as belonging to the positive class. However, x_2 is classified incorrectly as a member of the positive class. So, the violators need to be handled properly. Violations need to be tolerated to some extent.

This forms the basis for the so-called *soft-margin formulation* of the SVM given by

$$minimize \ \frac{1}{2} w^t w + C \sum_{i=1}^{n} \xi_i$$

$$subject \ to \ 1 - \xi_i - y_i(w^t x_i + b) \leq 0, \ i = 1, \dots, n$$

$$- \xi_i \leq 0, \ i = 1, \dots, n.$$

The Lagrangian is $\mathcal{L} = \frac{1}{2} w^t w + C \sum_{i=1}^{n} \xi_i + \sum_{i=1}^{n} \alpha_i (1 - \xi_i - y_i(w^t x_i + b)) - \sum_{i=1}^{n} \lambda_i \xi_i$. So, the necessary and sufficient conditions are

- $\nabla_w \mathcal{L} = 0 \Rightarrow w = \sum_{i=1}^{n} \alpha_i y_i x_i$
- $\frac{\delta \mathcal{L}}{\delta b} = 0 \Rightarrow \sum \alpha_i y_i = 0$

- $\frac{\delta \mathcal{L}}{\delta \xi_i} = 0 \Rightarrow \alpha_i + \lambda_i = C$
- $1 - \xi_i - y_i(w^t x_i + b) \le 0$; $\xi_i \ge 0$, $\forall i$
- $\alpha_i \ge 0$; $\lambda_i \ge 0$, $\forall i$
- $\alpha_i(1 - \xi_i - y_i(w^t x_i + b)) = 0$; $\lambda_i \xi_i = 0$, $\forall i$

So, the expressions for w and b have not changed. In the hard-margin (earlier) case, the Lagrange multipliers satisfied $\alpha_i \ge 0$. Here, it is possible to show that they will satisfy $\alpha_i \in [0, C]$; they are bounded by C; as $C \to \infty$, the soft-margin solution collapses to the earlier hard-margin one. If $C \to 0$, we tolerate any amount of violation (or any value of ξ). So, we need to find the appropriate value of C to obtain a good performance.

8.4.2 Non-linear SVM

It is possible that a non-linear decision boundary is better when there are margin violators. Consider, for example, the non-linear decision boundary (grey broken line) shown in Fig. 8.9. We will consider a simple example to illustrate a possible solution.

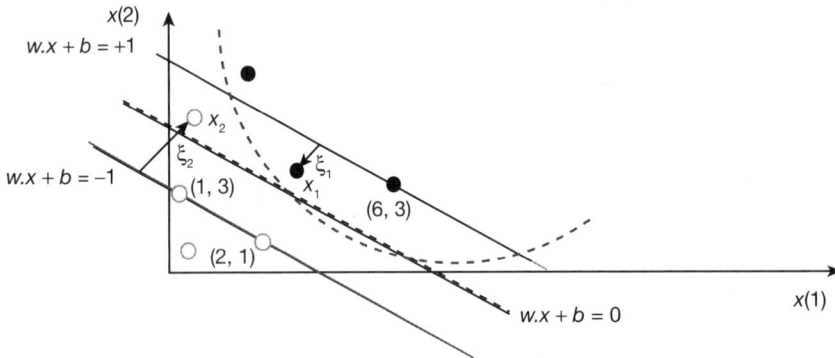

FIG. 8.9 An example to illustrate a non-linear SVM (for colour figure, please see Colour Plate 2)

EXAMPLE 11: Consider the truth table of the Boolean exclusive OR (XOR) shown in Table 8.6. Let us see whether we can have a linear decision boundary to separate the two classes based on outputs 0 and 1.

TABLE 8.6 Truth table of XOR

$x(1)$	$x(2)$	$XOR(x(1), x(2))$
0	0	0
0	1	1
1	0	1
1	1	0

Let the 0 output value correspond to the negative class and output 1 correspond to the positive class. Let the linear function be of the form $ax(1) + bx(2) + c$.

Input: $(0, 0)$ - Output: 0. So, $a \times 0 + b \times 0 + c < 0 \Rightarrow c < 0$.

Input: $(0, 1)$ - Output: 1. So, $a \times 0 + b \times 1 + c > 0 \Rightarrow b + c > 0 \Rightarrow b = -c + \delta_1$, where $\delta_1 > 0$.
Input: $(1, 0)$ - Output: 0. So, $a \times 1 + b \times 0 + c > 0 \Rightarrow a + c > 0 \Rightarrow a = -c + \delta_2$, where $\delta_2 > 0$.
Input: $(1, 1)$ - Output: 0. So, $a \times 1 + b \times 1 + c < 0 \Rightarrow a + b + c < 0$. Plugging in for b and a, we get $-c + \delta_1 - c + \delta_2 + c = -c + \delta_1 + \delta_2 < 0$. This is not possible because $c < 0 \Rightarrow -c > 0$ and $\delta_1 > 0$ and $\delta_2 > 0$.

So, it is not possible to capture XOR using a linear function in $x(1)$ and $x(2)$.

Let us include the third input in the form of $x(1)x(2)$, the conjunction of $x(1)$ and $x(2)$. The resulting truth table along with a linear form in the three variables is shown in Table 8.7. A linear form $x(1) + x(2) - 2x(1)x(2)$ using the variables $x(1)$, $x(2)$ and $x(1)x(2)$ is shown in the fifth column of the table. Note that the output of XOR and $x(1) + x(2) - 2x(1)x(2)$ are identical in all the four cases. So, XOR can be represented as a linear function in a three-dimensional space based on variables $x(1)$, $x(2)$ and $x(1)x(2)$.

TABLE 8.7 Truth table of XOR in three variables

$x(1)$	$x(2)$	$x(1)x(2)$	$XOR(x(1), x(2))$	$x(1) + x(2) - 2x(1)x(2)$
0	0	0	0	0
0	1	0	1	1
1	0	0	1	1
1	1	1	0	0

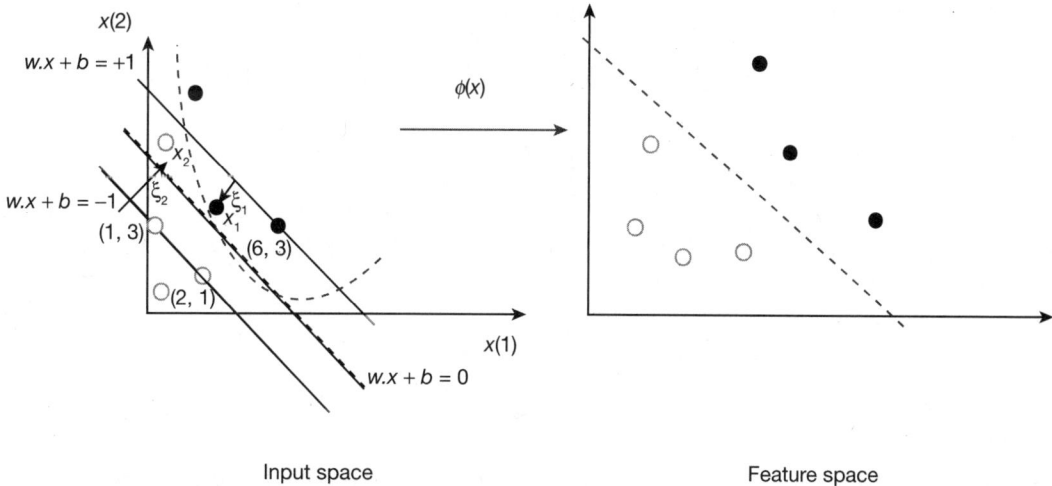

FIG. 8.10 Mapping the data to a high-dimensional feature space (for colour figure, please see Colour Plate 2)

So, we map the data points in the **input space** to a high-dimensional **feature space** using a mapping ϕ, as depicted in Fig. 8.10, with the hope that the data points are linearly separable in the higher-dimensional (feature) space. In the case of XOR, we have mapped the two-dimensional vector $(x(1), x(2))$ to a three-dimensional space given by $(x(1), x(2), x(1)x(2))$. Here, data points are linearly separable in the three-dimensional space.

We demonstrated the linear separability of *XOR* using the three-dimensional representation explicitly. In general, there are two problems associated with this approach:

- The data points may not be linearly separable in the feature space. We handle this issue by using the soft-margin formulation in the feature space.
- We may not know the explicit representation of the vectors in the feature space. Further, computation in the feature space may be unwieldy even if we know the explicit representation. This problem is handled using the **kernel trick**.

8.4.3   Kernel Trick

We are transforming a vector x in the input space to a higher dimensional space ($\phi(x)$) and look for a possible linear separation in the new space that is called the feature space. The most important observation is that we do not need to map ϕ explicitly. In theory, ϕ could map points into an infinite dimensional space. What we need is the inner product in the feature space. If x_i and x_j are two data points in the input space, then our computations will need terms of the form $\phi(x_i)^t \phi(x_j)$.

Many similarity functions can be viewed as inner products. If vectors $\phi(x_i)$ and $\phi(x_j)$ are unit norm vectors, then $cos\theta = \phi(x_i)^t \phi(x_j)$, where θ is the angle between the two vectors $\phi(x_i)$ and $\phi(x_j)$. We use the **kernel function**, $K(x_i, x_j)$, that maps a pair of vectors $x_i \in \Re^l$ and $x_j \in \Re^l$ to a real number. Specifically,

$$K(x_i, x_j) = \phi(x_i)^t \phi(x_j)$$

We know that $w = \sum \alpha_i y_i \phi(x_i)$ if we want to use the vectors of the form $\phi(x_i)$ explicitly. However, in order to classify a pattern x, we need to compute

$$w^t \phi(x) + b = \sum_i \alpha_i y_i \phi(x_i)^t \phi(x) + b = \sum_i \alpha_i y_i K(x_i, x) + b$$

Even b can be computed as follows. Let (x_p, y_p) be a training data pair. So, we have $w^t \phi(x_p) + b = y_p$. So, $b = y_p - w^t \phi(x_p) = y_p - \sum_i \alpha_i y_i K(x_i, x_p)$. So, $w^t \phi(x) + b = \sum_i \alpha_i y_i K(x_i, x) + y_p - \sum_i \alpha_i y_i K(x_i, x_p)$. So, all the computations can be made using the kernel function and the α_is. However, we need to have a suitable kernel function that satisfies $K(x_i, x_j) = \phi(x_i)^t \phi(x_j)$. Some popularly used kernel functions are:

- Polynomial of degree p: $K(x_i, x) = (x_i^t x)^p$
- Polynomial kernel up to degree p: $K(x_i, x) = (1 + x_i^t x)^p$.
- Gaussian kernel: $K(x_i, x) = e^{-\frac{||x_i - x||^2}{\sigma^2}}$.
- Sigmoidal kernel: $K(x_i, x) = tanh(a x_i^t x + b)$

We will see the underlying feature mapping for kernel that represents a polynomial of degree 2. Consider the kernel on two-dimensional patterns given by

$$K(x_i, x) = (x_i^t x)^2 = (x_i(1)x(1) + x_i(2)x(2))^2 = x_i(1)^2 x(1)^2 + x_i(2)^2 x(2)^2 + 2x_i(1)x_i(2)x(1)x(2).$$

This kernel computation can be achieved by using a feature mapping from the two-dimensional input space to a three-dimensional feature space given by:

$$\phi(\begin{pmatrix} x_i(1) \\ x_i(2) \end{pmatrix}) = \begin{pmatrix} x_i(1)^2 \\ x_i(2)^2 \\ \sqrt{2}x_i(1)x_i(2) \end{pmatrix}$$

Now we can verify that $K(x_i, x) = \phi(x_i)^t \phi(x)$ because the RHS is the dot product of $\phi(x_i)$ and $\phi(x)$ and

$$\phi(x_i)^t \phi(x) = x_i(1)^2 x(1)^2 + x_i(2)^2 x(2)^2 + 2x_i(1)x_i(2)x(1)x(2)$$

So, using the polynomial kernel of degree 2 is the same as applying the mapping ϕ that maps a two-dimensional vector to a three-dimensional vector as specified and computing the dot product in the three-dimensional space.

Note that the Gaussian kernel employs an exponential function and so its expansion can have infinite terms. So, the corresponding feature map ϕ can map a finite dimensional vector to an infinite dimensional vector. Because of the kernel trick, our computations are all performed in the finite dimensional input space but not in the feature space. So, the kernel function is a similarity function and the kernel trick can be employed even in the case of other ML models where similarity needs to be computed. We explore its use along with nearest neighbor classifier (NNC) in an exercise.

8.5 LOGISTIC REGRESSION

We have seen how linear discriminants are used by both perceptron and SVM in classification.

Perceptron employs a linear model $g(x) = w^t x + b$ and learns w and b using the training data. It can employ a generalized linear model to handle non-linear decision boundaries by augmenting w and x vectors suitably; it then uses a linear model on the augmented data.

SVM inherently deals with a linear model of the form $w^t x + b$, where w corresponds to the maximum margin given by $maximize \frac{2}{||w||}$. It employs a kernel to deal with non-linear decision boundaries by looking for a linear separation in the feature or kernel space. It can also combine a soft-margin formulation to permit some margin violators; this helps us in using a simpler model.

We will see that in logistic regression, another generalized linear model is employed. Before we get into the details of logistic regression, we will consider linear regression.

8.5.1 Linear Regression

Here, we assume that the model obeys a linear relationship. We would like to obtain the weight vector w from the data given. Specifically, let $A_{n \times l}$ be the data matrix having n patterns, each pattern forming a row, and l columns, one column per feature. Let $y_{n \times 1}$ be the vector of n target or the given output values. Let w be the vector of unknowns that linearly links x_i with $\hat{y}_i$, that is, an estimate of y_i. Note that the pattern x_i forms the ith row of A and the value $\hat{y}_i$ is $w^t x_i$.

In this simple linear model, we are assuming that there is no noise and the bias (b) is not explicitly shown. We assume that x_i and w are suitably augmented to handle the bias term if required, like in the case of perceptron. We assume that the weight vector w is such that $w^t x_i = x_i^t w = y_i$; that is, x_i and y_i are linearly related. In practice, the estimated $\hat{y}_i$ can differ from its target value y_i. So, error vector e is given by $e = Aw - y$.

So, we consider the squared norm of e which is

$$||e||^2 = ||Aw - y||^2 = (Aw - y)^t (Aw - y) = w^t A^t Aw - y^t Aw - (Aw)^t y + y^t y$$

$= w^t A^t Aw - 2y^t Aw + y^t y$. We find w that minimizes this squared norm.

We need to find w. So, by taking the gradient of the squared norm of e with w and equating to 0, we get

$$2A^t Aw - 2A^t y = 0 \Rightarrow w = (A^t A)^{-1} A^t y$$

So, w gives us the estimates of the coefficients involved.

EXAMPLE 12: Let us consider a simple example of three vectors $x_1 = (1,1)$, $x_2 = (1,2)$ and $x_3 = (2,2)$. Let $y_1 = 3$, $y_2 = 5$ and $y_3 = 6$. So, the data matrix A consisting of the augmented vectors is: $A = \begin{bmatrix} 1 & 1 & 1 \\ 1 & 2 & 1 \\ 2 & 2 & 1 \end{bmatrix}$ and $w = (w(1), w(2), w(3))$.

We have seen that $w = (A^t A)^{-1} A^t y$; so we have

$$A^t A = \begin{bmatrix} 1 & 1 & 2 \\ 1 & 2 & 2 \\ 1 & 1 & 1 \end{bmatrix} \times \begin{bmatrix} 1 & 1 & 1 \\ 1 & 2 & 1 \\ 2 & 2 & 1 \end{bmatrix} = \begin{bmatrix} 6 & 7 & 4 \\ 7 & 9 & 5 \\ 4 & 5 & 3 \end{bmatrix}.$$

and

$$(A^t A)^{-1} = \begin{bmatrix} 2 & -1 & -1 \\ -1 & 2 & -2 \\ -1 & -2 & 5 \end{bmatrix}$$

$$A^t y = \begin{bmatrix} 1 & 1 & 2 \\ 1 & 2 & 2 \\ 1 & 1 & 1 \end{bmatrix} \begin{pmatrix} 3 \\ 5 \\ 6 \end{pmatrix} = \begin{pmatrix} 20 \\ 25 \\ 14 \end{pmatrix}. \text{ So, } w = (A^t A)^{-1} A^t y =$$

$$\begin{bmatrix} 2 & -1 & -1 \\ -1 & 2 & -2 \\ -1 & -2 & 5 \end{bmatrix} \begin{pmatrix} 20 \\ 25 \\ 14 \end{pmatrix} = \begin{pmatrix} 1 \\ 2 \\ 0 \end{pmatrix}$$

So, the coefficient corresponding to the augmented position is 0 ($b = 0$). Here, the linear form is $\hat{y} = 1x(1) + 2x(2)$. We can verify that for pattern $x_1 = (1,1)$, $\hat{y}_1 = 1 \times 1 + 2 \times 1 = 3 = y_i$. Similarly it is possible to verify for the other x_i also.

8.5.2 Sigmoid Function

We use a non-linear function of the linear form given by $f(w^t x + b)$, where f is the sigmoid function. The sigmoid or logistic function is given by $f(a) = \frac{1}{1 + e^{-a}}$, where $a = w^t x + b$. Some of its properties are given below:

- $f : \Re \to [0, 1]$. It maps a real number to a value in the interval $[0,1]$.
- The function assumes its maximum value of 1 as $a \to \infty$ and its minimum value is reached as $a \to -\infty$.
- When $a = 0$, $f(a) = \frac{1}{2}$.
- If $y = f(a) = \frac{1}{1 + e^{-a}}$, then $a = f^{-1}(y) = ln(\frac{y}{1-y})$. So, the function f is invertible.
- The derivative of $f(a)$ is

$$f'(a) = \frac{(-1)(-e^{-a})}{(1 + e^{-a})^2} = f(a)(1 - f(a))$$

- We can interpret $f(a)$ as probability by viewing $a = w^t x + b$.

$$f(a) = P(y = 1|x) = \frac{1}{1 + e^{-w^t x - b}}$$

- If we let $p_1 = P(y = 1|x)$, then

$$p_1 = \frac{1}{1 + e^{-w^t x - b}} \Rightarrow p_1(1 + e^{-w^t x - b}) = 1 \Rightarrow w^t x + b = ln\left(\frac{p_1}{1 - p_1}\right)$$

- The ratio of p_1 to $1 - p_1$ is called the **odds ratio** because $1 - p_1$ is the probability of $y = 0$ given x, denoted by $P(y = 0|x)$.
- The logarithm of the odds ratio is called the **logit function**.

8.5.3 Learning w and b in Logistic Regression

We can train the classifier, given a set of training patterns $\{(x_1, y_1), (x_2, y_2), \cdots, (x_n, y_n)\}$, in different ways:

- We can compute the inverse $y_i^{lor} = f^{-1}(y_i) = ln(\frac{y_i}{1-y_i})$, $i = 1, 2, \ldots, n$ and use the data $\{(x_1, y_1^{lor}), (x_2, y_2^{lor}), \ldots, (x_n, y_n^{lor})\}$ to learn w and b using least squares fit linear regression.

EXAMPLE 13: Let us consider the following data. $x_1 = (1, 2)$, $x_2 = (2, 1)$, $x_3 = (6, 7)$, $x_4 = (7, 6)$. Let $y_1 = 0$, $y_2 = 0$, $y_3 = 1$, $y_4 = 1$. The corresponding logarithms of the odds ratios cannot be computed as we end up with $ln(0)$ and $ln(\frac{1}{0})$. So, we approximate 1 by 0.99 and 0 by 0.01.
Using the approximate values, we compute

$$y_1^{lor} = ln\left(\frac{0.01}{1 - 0.01}\right) = -4.6 = y_2^{lor}, \text{ and } y_3^{lor} = y_4^{lor} = 4.6$$

We have $A = \begin{bmatrix} 1 & 2 & 1 \\ 2 & 1 & 1 \\ 6 & 7 & 1 \\ 7 & 6 & 1 \end{bmatrix}$.

$$(A^t A)^{-1} = \frac{1}{400} \begin{bmatrix} 104 & -96 & -512 \\ -96 & 104 & -32 \\ -32 & -32 & 356 \end{bmatrix}$$

$$A^t y^{lor} = \begin{bmatrix} 1 & 2 & 6 & 7 \\ 2 & 1 & 7 & 6 \\ 1 & 1 & 1 & 1 \end{bmatrix} \times \begin{pmatrix} -4.6 \\ -4.6 \\ 4.6 \\ 4.6 \end{pmatrix} = \begin{pmatrix} 46 \\ 46 \\ 0 \end{pmatrix}$$

So, $w = \frac{1}{400} \begin{bmatrix} 104 & -96 & -512 \\ -96 & 104 & -32 \\ -32 & -32 & 356 \end{bmatrix} \times \begin{pmatrix} 46 \\ 46 \\ 0 \end{pmatrix} = \begin{pmatrix} \frac{23}{25} \\ \frac{23}{25} \\ -\frac{184}{25} \end{pmatrix}$. So, the discriminant function is given by

$$\hat{y} = \frac{23}{25} \times x(1) + \frac{23}{25} \times x(2) - \frac{184}{25}$$

We can verify that $\hat{y}_1 = \frac{23}{25} \times 1 + \frac{23}{25} \times 2 - \frac{184}{25} \times 1 = -4.6 = y_1^{lor}$. Similarly, we can verify for the other three augmented patterns.
If we want the original value of $p \in [0, 1]$, we need to compute p from $ln(\frac{p}{1-p}) = \hat{y}$. We can use the value of p to obtain the class label. If $p < 0.5$, then the class is 0, else it is 1.

- The other option is to train it like a perceptron neural network with the sigmoid function as the output and a specific loss function which permits us to minimize the loss.
In perceptron with a linear threshold function, we get a binary output. When the sigmoid function is used, the output is in the range $[0, 1]$. It permits us to model non-linear functions. Further, it is amenable to differentiation.

In the terminology of neural functions, it is called **sigmoid activation function**. It is also called logistic function and is denoted by $\sigma(a)$, where a is the activation input. Here, $a = w^t x$ assuming that w and x are augmented vectors.

We have seen that $\sigma'(a) = \sigma(a)(1 - \sigma(a))$. We are given training data in the form $\{(x_i, y_i), i = 1, 2, \ldots, n\}$. These y_is could be viewed as target outputs or required outputs; so, we can denote them as y_i^{tar}.

Based on the value of w, we obtain $y_i^{obt} = \sigma(w^t x)$, as shown in Fig. 8.11.

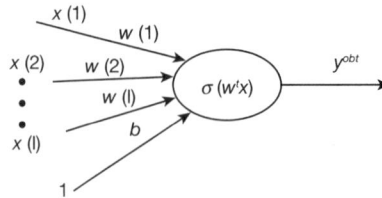

FIG. 8.11 Logistic regression using perceptron

The error is $E = \sum_{i=1}^{n}(y_i^{tar} - y_i^{obt})^2$ based on squared error loss; it is the square of the difference between the target output and the obtained output based on the current w. The partial derivative of E with respect to $w(j)$ is

$$\frac{\partial E}{\partial w(j)} = \sum_{i=1}^{n}(y_i^{tar} - y_i^{obt})\sigma'(w^t x)x_i(j) = \sum_{i=1}^{n}(y_i^{tar} - y_i^{obt})\sigma(w^t x)(1 - \sigma(w^t x))x_i(j)$$

Note that

$$\frac{\partial E}{\partial b} = \sum_{i=1}^{n}(y_i^{tar} - y_i^{obt})\sigma(w^t x)(1 - \sigma(w^t x)) \times 1$$

because $x_i(j) = 1$ in the augmented vector x for the position corresponding to b.

EXAMPLE 14: Let us consider the data used in Example 10.

$x_1 = (1, 2)$, $x_2 = (2, 1)$, $x_3 = (6, 7)$, $x_4 = (7, 6)$
$y_1 = 0$, $y_2 = 0$, $y_3 = 1$, $y_4 = 1$

Let $w_0 = (0, 0, 0)$ be the initial augmented vector. Using the update equation $w_1(j) = w_0(j) + \eta(\sum_{i=1}^{n}(y_i^{tar} - y_i^{obt})\sigma(w^t x)(1 - \sigma(w^t x))x_i(j)$ for $j = 1, 2$, we get $w_1(1) = 0.0125 = w_1(2)$ using $\eta = 0.01$, where η is the learning rate.

For $w_1(3) = b_1$, we get $w_1(3) = 0.01 \times (-0.5) \times 0.25 \times 2 + 0.01 \times (0.5) \times 0.25 \times 2 = 0$. So, $w_1 = (0.0125, 0.0125, 0)$.

The corresponding $\sigma(w_1^t x_i)$ values for the patterns are 0.509, 0.509, 0.541 and 0.541, assigning all the patterns to Class 1 as the output values are larger than 0.5 for all the four patterns.

After some iterations, we get $w_9 = (0.084, 0.084, -0.01)$. The corresponding $\sigma(w_9^t x_i)$ values are 0.2531, 0.2531, 0.747, 0.747 for the four patterns, where the first two patterns are assigned to Class 0 and the remaining two patterns to Class 1 based on thresholding with 0.5; that is, if the value is less than 0.5, it is treated as 0, else as 1.

8.6 MULTI-LAYER PERCEPTRONS (MLPS)

An artificial neuron forms the building block of any MLP and is shown in Fig. 8.11. A neuron can have a finite number of inputs. The inputs are represented as $x(1), x(2), \ldots, x(l)$. Each input is associated with a weight; input $x(i)$ uses weight $w(i)$ for $i = 1, 2, \ldots, l$. Bias threshold can be considered as an input with a fixed input value of 1. The corresponding weight is b or $w(0)$ or $w(l+1)$ based on how the augmentation is performed.

Output or input to the activation unit is $a = w^t x = w(1)x(1) + w(2)x(2) + \cdots + w(l)x(l) + w(l+1)1$. If $\sigma(a)$ is the activation output, then using sigmoid activation we have the output as $\frac{1}{1+e^{-a}} = \frac{e^a}{1+e^a}$. The problem is to learn the weights $w(1), w(2), \ldots, w(l+1)$ using a training data set.

In an MLP network, we typically have multiple layers of connected neurons, as shown in Fig. 8.12. It has an input layer where the augmented vector x is input. The number of input units is $l+1$ as there will be $l+1$ entries in the augmented vector. There are 5 input units in this example.

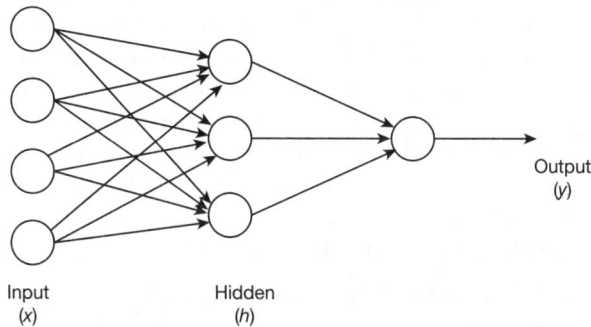

FIG. 8.12 An example of an MLP network

There is an output layer that outputs y. In general, there can be N_o output units. In this example, there is only one output node.

There is a layer of neurons in between the input and output layers in the example. It is called the **hidden layer** and is denoted by h; it is hidden from the input–output activity. However, learning representations from the input data takes place here. In the example, there are three nodes in the hidden layer. We also call it a **feed-forward neural network** because outputs of the neurons in each layer form the inputs for the neurons in the next layer. This forward propagation continues till we reach the output layer for which there will be no more layers.

In general, there could be more than one hidden layer and each of them can have a different number of processing elements or neurons. If there are p hidden layers, then we call them h_1, h_2, $\cdots$, h_p. There is no specific relationship between the number of neurons in the input, output and hidden layers. The number of output neurons can be different from the number of input neurons.

In Fig. 8.12, we have two layers having weights. So, we say that it has two layers and one of them is a hidden layer. It is possible to show that one hidden layer or a network with two layers is adequate to approximate any continuous function for inputs in a specified range.
The architecture of MLP has:

- No connections among the neurons within a layer.
- Typically no direct connections between neurons in the input layer and neurons in the output layer.
- Neurons between layers that are fully connected. In the MLP network shown in Fig. 8.12, each of the five input layer neurons is connected to all the three neurons in the hidden layer. So, there are 15 weights between layers. Similarly, each of the three hidden neurons is connected to the output neuron; here we have 3 weights between the hidden and output layers.

- Each hidden neuron is a perceptron with output $y_i = f(w^t x_i)$, where x_i is its input. Even each of the output neurons is typically a perceptron.

Recall that we had a linear threshold function or a step function as the activation function in the case of perceptron. However, this function is not differentiable and we require a differentiable function to deal with the learning process. So, a simple activation function is the linear activation function. However, if we use the linear activation function throughout the MLP network, we have, with respect to Fig. 8.12,

$$h = A_1 x; y = A_2 h \Rightarrow y = A_2 A_1 X = Bx,$$

where A_1 is the matrix of weights in the layer between the input and hidden layers and A_2 is the matrix of weights between the hidden and output layers.

The size of A_1 is 5×3 and of A_2 it is 3×1 in the example. In general, A_1 can be of size $l + 1 \times n_h$ and A_2 can be of size $n_h \times n_o$, where n_h is the number of neurons in the hidden layer and n_o is the number of neurons in the output layer. So, if we let the product of the matrices $A_2 \times A_1 = B$, then effectively the input–output mapping will be $y = Bx$, that is, a simple linear mapping. So, non-linear mappings cannot be realized if we use linear activation throughout. Further, $y = Bx$ in effect corresponds to a single-layer computation. This is applicable not only to the example in Fig. 8.12, but for any general MLP.

We will illustrate the working of an MLP with an example.

EXAMPLE 15: Consider the character 7 shown in Fig. 8.13. It is in a grid of size 6×6 cells or 36 pixels. The grey portion indicates the background and the whitened portion represents character 7. We will look at an MLP network that can be used to detect whether the given image of 6×6 pixels is a 7 or not.

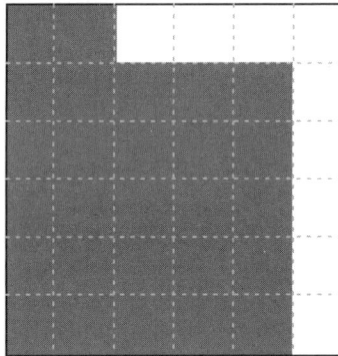

FIG. 8.13 The character 7 in a 6×6 grid

We have 36 pixels in the input image. The number of input neurons is 36, as shown in Fig. 8.14. We typically initialize the MLP with small random weights. The number of hidden nodes is m, where m is a tunable parameter. The training and test data sets consist of a number of patterns, each of size 6×6.

For a given initialization of weights in the MLP, we get an output y_i^{obt} when we input x_i. Based on the difference between y_i^{tar} and y_i^{obt}, the weights in the network are adjusted so that the MLP gives an output of 1 when the input character is 7 and 0 otherwise. This updating of weights is performed with the help of a training algorithm called backpropagation which we will consider next.

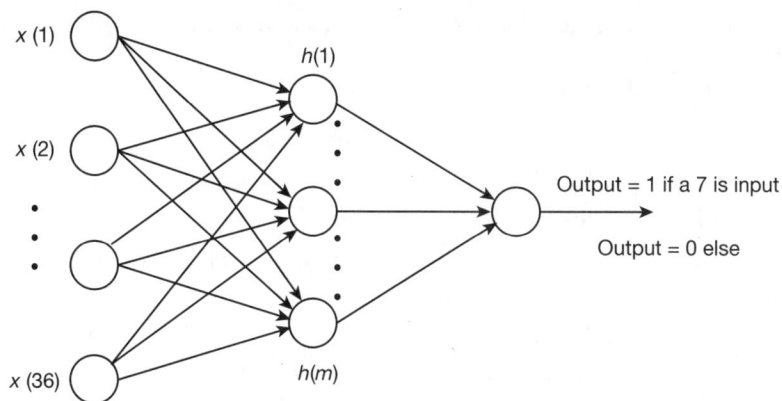

FIG. 8.14 MLP to classify 7

8.6.1   Backpropagation for Training an MLP

Backpropagation is a well-known and simple optimization algorithm that helps us in learning the augmented weight vectors in MLP.
It has two passes:

- **Forward Pass:** The network is initialized with some random weights. Input x is applied at the input layer. The output of a layer of neurons is computed and is forwarded to the next layer. This process is continued till the output of the entire MLP is computed. Let the obtained output be y^{obt}, where the required or target output is y^{tar}.
- **Backward Pass:** The error at the output is a function of the difference between y^{tar} and y^{obt}. Here, y^{obt} depends on the current weights of the network. An error is formulated as a function of current weights and these weights are updated to move in the direction of minimizing the error using gradient descent. This is achieved by backpropagating the error.

Backpropagation employs a **gradient descent** scheme. Gradient descent may be explained using Fig. 8.15. There are two local minima, $w*$ and w_3. If we start at w_2 and try to move in the negative

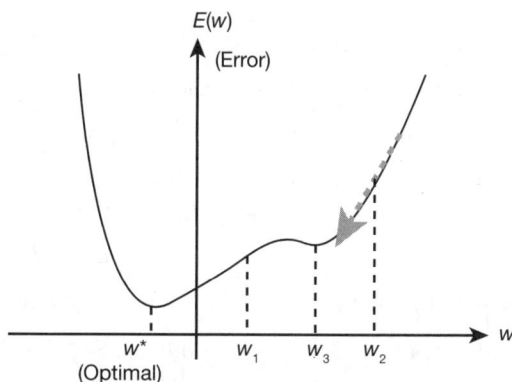

FIG. 8.15 Gradient descent for minimizing error

direction of the gradient at w_2, as shown using the broken grey line in Fig. 8.15, we keep moving towards the local minimum w_3.

The update equation will typically be

$$w_{updated} = w_{current} - \eta \nabla E(w_{current}),$$

where $\nabla E(w_{current})$ is the gradient of E with respect to entries in w and the result is evaluated at the current value of w. We illustrate it with an example.

EXAMPLE 16: Let $w_{current} = (0.5, 0)$ and the pattern $x = (1, 2)$. Let $E = (w(1)x(1) + w(2) x(2) - 1)^2$. Here, $\nabla E(w_{current})$ is obtained as follows:

$$\frac{\delta E}{\delta w(1)} = 2(w(1)x(1) + w(2)x(2) - 1)x(1)$$

$$\frac{\delta E}{\delta w(2)} = 2(w(1)x(1) + w(2)x(2) - 1)x(2)$$

Plugging in the values of $w(1), w(2), x(1)$ and $x(2)$, we have, at the value of $w_{current} = (0.5, 0)$,

$$\frac{\delta E}{\delta w(1)} = -1$$

Similarly,

$$\frac{\delta E}{\delta w(2)} = -2$$

The gradient $\nabla E(w_{current})$ is a vector of size 2 as $w_{current}$ is a two-dimensional vector, where the first component is the partial derivative with respect to $w(1)$ evaluated at $w_{current}$ and the second component is the partial derivative with respect to $w(2)$ evaluated at $w_{current}$. In this example, $\nabla E(w_{current}) = (-1, -2)$.

If we start at a point like w_1 and move in the direction of the negative of the gradient, we may reach the local optimum $w*$ that is also the global optimum. η is the *learning rate*. The value of η plays an important role. If it is too small, then it takes a large number of steps and updates to reach the local minima. If the value of η is large, then it may oscillate around the local minimum point without reaching it.

Forward Pass

Let us consider a simple MLP network shown in Fig. 8.16.

EXAMPLE 17: We will illustrate the forward pass of the backpropagation algorithm. The forward pass will give us the following outputs by considering the input $\begin{pmatrix} 0 \\ 0 \end{pmatrix}$. The activation inputs to both C and D are the same and are equal to zero. So, their outputs are also equal and they are $\frac{1}{1+e^{-0}} = 0.5$. So, the input to node O is $0.5 \times (-30) + 0.5 \times 25 = -2.5$. So, the output of node O is $\frac{1}{1+e^{2.5}} = 0.07586$.

For the input $\begin{pmatrix} 0 \\ 1 \end{pmatrix}$, the activation input to C is 1 and output of C is $\frac{1}{1+e^{-1}} = 0.731$. The activation input to D is 8 and output of D is $\frac{1}{1+e^{-8}} = 0.9996$. So, the input to node E is $0.731 \times (-30) + 0.9996 \times 25 = 3.06$. So, the output of node O is $\frac{1}{1+e^{-3.06}} = 0.9552$. Computation of the outputs in the forward pass for inputs $\begin{pmatrix} 1 \\ 0 \end{pmatrix}$ and $\begin{pmatrix} 1 \\ 1 \end{pmatrix}$ are left as part of an exercise.

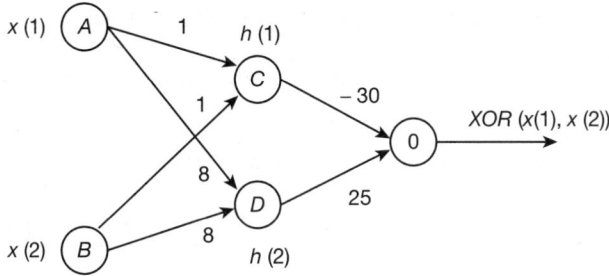

FIG. 8.16 Backpropagation for learning weights

Backward Pass

Here, we compute the errors at the output node/nodes.

EXAMPLE 18: Recall that the value of y^{obt} for the input $\binom{0}{0}$ in Example 17 is 0.07586. We know that the target value y^{tar} is 0. So, the error is $(0.07586 - 0) = 0.07586$.

These errors are backpropagated to the previous layer and play a role in updating the weights for the edges CO and DO. We will see the general framework for updating the weights using the cross-entropy loss function given by

$$-(y^{tar}log(y^{obt}) + (1 - y^{tar})log(1 - y^{obt}))$$

Note that it is 0 when $y^{tar} = y^{obt}$.

Let us consider the partial derivative of the error E with respect to the weight w_{CO}. It is

$$\frac{\partial E}{\partial w_{CO}} = \frac{\partial E}{\partial y^{obt}} \times \frac{\partial y^{obt}}{\partial a_O} \times \frac{\partial a_O}{\partial w_{CO}}$$

using the chain rule of differentiation. Here, a_O is the association input to the output node O. Let us consider each component of the product.

$$\frac{\partial E}{\partial y^{obt}} = \frac{d}{dy^{obt}} - (y^{tar}log(y^{obt}) + (1 - y^{tar})log(1 - y^{obt}))$$

$$= -\left(\frac{y^{tar}}{y^{obt}} - \frac{1 - y^{tar}}{1 - y^{obt}}\right) = \frac{y^{obt} - y^{tar}}{y^{obt}(1 - y^{obt})}$$

$$\frac{\partial y^{obt}}{\partial a_O} = \frac{\partial \sigma(a_O)}{\partial a_O} = \sigma(a_O)(1 - \sigma(a_O)) = y^{obt}(1 - y^{obt}), \text{ where } \sigma(a_O) = \frac{1}{1 + e^{-a_O}}.$$

Finally

$$\frac{\partial a_O}{\partial w_{CO}} = \frac{\partial}{\partial w_{CO}} \sum_k w(k)h(k) = \frac{\partial}{\partial w_{CO}}(w_{CO}h_1 + w_{DO}h_2) = h(1)$$

So, the partial derivative of error with respect to one weight w_{CO} is $\frac{\partial E}{\partial w_{CO}} = \frac{\partial E}{\partial y^{obt}} \times \frac{\partial y^{obt}}{\partial a_O} \times \frac{\partial a_O}{\partial w_{CO}} = \frac{(y^{obt} - y^{tar})}{y^{obt}(1 - y^{obt})} \times y^{obt}(1 - y^{obt})h(1) = (y^{obt} - y^{tar})h(1)$. If η is the learning rate, then the update equation is

$$w_{CO}^{k+1} = w_{CO}^k + \eta(y^{obt} - y^{tar})h(1)$$

TABLE 8.8 Weight update computations

x	y^{tar}	y^{obt}	$y^{obt} - y^{tar}$	$h(C)$	Δw_{CO}
(0,0)	0	0.075858	0.075858	0.5	0.037929
(0,1)	1	0.9552	−0.0448	0.731	−0.0327488
(1,0)	1	0.9552	−0.0448	0.731	−0.0327488
(1,1)	0	0.194	0.194	0.88	0.17072

The various components that are required for updating the weight w_{CO} are shown in Table 8.8. In the last column, we have the incremental value of weight w_{CO} for the four inputs. So, $w_{CO}^{new} = w_{CO}^{old} - \eta \times \frac{1}{4}[0.037929 - 0.0327488 - 0.0327488 + 0.17072] = 0.0035788$ for $\eta = 0.1$. In general, we need to add m terms and divide the sum by m if there are m input–output training pairs. In the current example, we have $m = 4$. The weight w_{DO} can be updated similarly and it is left as an exercise.

Backpropagation Through Hidden Nodes

Here also, we use a similar kind of chain rule of differentiation.

Consider $\frac{\partial E}{\partial w_{AC}}$. It may be expressed using the chain rule as

$$\frac{\partial E}{\partial y^{obt}} \times \frac{\partial y^{obt}}{\partial a} \times \frac{\partial a_O}{\partial y_C} \times \frac{\partial y_C}{\partial a_C} \times \frac{\partial a_C}{\partial w_{AC}}$$

We have seen earlier that

$$\frac{\partial E}{\partial y^{obt}} = \frac{d}{dy^{obt}} - (y^{tar}log(y^{obt}) + (1 - y^{tar})log(1 - y^{obt}))$$

$$= -\left(\frac{y^{tar}}{y^{obt}} - \frac{1 - y^{tar}}{1 - y^{obt}}\right) = \frac{y^{obt} - y^{tar}}{y^{obt}(1 - y^{obt})}$$

$$\frac{\partial y^{obt}}{\partial a_O} = \frac{\partial \sigma(a_O)}{\partial a_O} = \sigma(a_O)(1 - \sigma(a_O)) = y^{obt}(1 - y^{obt}), \text{ where } \sigma(a_O) = \frac{1}{1 + e^{-a_O}}.$$

So, $\frac{\partial E}{\partial y^{obt}} \times \frac{\partial y^{obt}}{\partial a_O} = y^{obt} - y^{tar}$

$$\frac{\partial a_O}{\partial y_C} = -30$$

$$\frac{\partial y_C}{\partial a_C} = y_C \times (1 - y_C)$$

$$\frac{\partial a_C}{\partial w_{AC}} = x_A, \text{ where } x_A \text{ is the input to node } A.$$

So, $\frac{\partial E}{\partial w_{AC}} = (y^{obt} - y^{tar}) \times (-30) \times y_C \times (1 - y_C)x_A.$

Similarly, $\frac{\partial E}{\partial w_{BC}} = (y^{obt} - y^{tar}) \times (-30) \times y_C \times (1 - y_C)x_B.$

$$\frac{\partial E}{\partial w_{AD}} = (y^{obt} - y^{tar}) \times 25 \times y_D \times (1 - y_D)x_A$$

$$\frac{\partial E}{\partial w_{BD}} = (y^{obt} - y^{tar}) \times 25 \times y_D \times (1 - y_D)x_B$$

EXAMPLE 19: Consider the data shown in Table 8.8 and Example 18. We can compute the partial derivatives as:

$\frac{\partial E}{\partial w_{AC}} = \frac{1}{4}[-0.0448 \times -30 \times 0.731 \times 0.269 + 0.194 \times -30 \times 0.88 \times 0.12] = -0.087577296$ as x_A is 0 for 2 of 4 inputs.

Similarly,

$\frac{\partial E}{\partial w_{AD}} = \frac{1}{4}[-0.0448 \times 25 \times 0.9996 \times 0.0004 + 0.194 \times 25 \times 0.99999 \times 0.00001] = -0.00009983$ as x_A is 0 for 2 of 4 inputs.

So, $w_{AC} = 1 - (0.1) \times (-0.087577296) = 1.0087577296$ and $w_{AD} = 8 - (0.1) \times (0.00009983) = 8.000009983$ assuming $\eta = 0.1$. One can compute the updates for the weights w_{BC} and w_{BD} in the same manner, which is left as an exercise.

8.7 RESULTS ON THE DIGITS DATA SET

We have conducted experiments on the Digits data set available with sklearn. There are approximately 180 patterns from each of the handwritten digits 0 to 9. Each is a 8×8-pixel image. We conducted experiments using perceptron, SVM, logistic regression and MLP and present the results in Fig. 8.17. Here, the Y-axis shows the test accuracy and the X-axis shows the test data size. Note that the performance of the perceptron classifier is inferior to the other three ML models. There is no significant difference in the performance of the other three models.

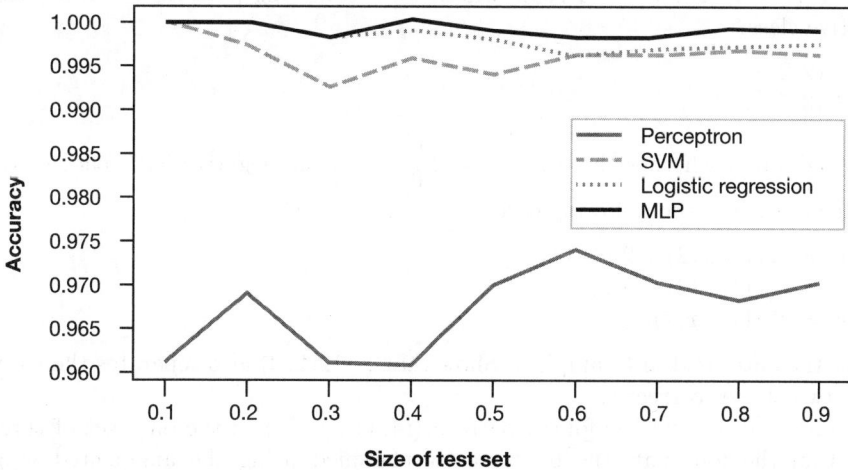

FIG. 8.17 Comparison of linear classifiers on the Digits data set (for colour figure, please see Colour Plate 2)

SUMMARY

We have considered various ML models based on linear discriminant functions. These are perceptron, SVM, logistic regression and MLP. Some important observations about these models are as follows:

- These are more or less data dependent. Logistic regression may be viewed as a probabilistic model dealing with logarithm of odds ratio, the values being estimated using statistics.
- Perceptron converges to a correct w if the classes are linearly separable. It can work on non-linearly separable problems provided we know the form of the non-linearity.
- SVM gives us a w that corresponds to the maximum margin in the linearly separable case.
- We can employ the kernel trick and use the SVM even when the classes are not linearly separable.
- Logistic regression can capture the linearity of the exponent of the sigmoid or logistic function.
- MLP with one hidden layer can theoretically approximate any input–output mapping.
- The ML models based on linear discriminant functions are well-known to be *opaque*. However, perceptrons and linear SVMs give the importance of the features based on the w vector; so, some explanation can be offered through the weight vectors.

EXERCISES

1. Let $w = (1, -3, 2)$ and $b = 2$. Decide whether the following patterns are from positive class or negative class:

   a. $x_1 = (2, 2, 2)$
   b. $x_2 = (3, 5, 1)$

2. Find w and b and whether the origin $x = \begin{pmatrix} 0 \\ 0 \end{pmatrix}$ is in the negative half space or the positive half space in each of the following cases:

   a. $g_1(x) = x(1) - x(2) + 2$
   b. $g_2(x) = x(1) - x(2) - 4.2$
   c. $g_3(x) = 2x(1) - x(2)$

3. Consider the data used in Example 4. Show that $(-12, 2, 3)$ also separates the three patterns into the two classes correctly.

4. What happens if we add the fourth pattern $((4, 1), l_4 = 1)$ to the data set of three patterns in Q3? Can the four patterns be correctly classified using the augmented weight vector $(-12, 2, 3)$? How about the augmented weight vector $(-12, 4, 1)$?

5. Consider the data given in Example 7, in Table 8.5. Find the w vector that classifies all the augmented vectors correctly. Find the α_i value for pattern x_i, where $i = 1, 2, \ldots, 8$ by representing $w = \sum_{i=1}^{8} \alpha_i l_i x_{a_i}$.

6. Consider a point x on the line $w^t x + b = c$. Find the normal distance of x from the decision boundary given by $w^t x + b = 0$.

7. Consider Example 10. Note that $w = \left(\frac{2}{5}, \frac{1}{5} \right)$ and $b = -2$. Classify the patterns $(7, 1)$ and $(2, 0)$ using the w and b used in Example 10.

8. Consider Example 10. Add two points $(0, 1)$ from the negative class and $(8, 4)$ from the positive class to the training data set having three points. Find the w, b and the set, S, of support vectors.

9. Consider the following training data points:

   - Class 1: $(1,2)$, $(2,1)$
   - Class 2: $(6,7)$, $(7,6)$

   Let the test pattern $x = (1, 3)$. Use the Gaussian kernel similarity with $\sigma = 1$ to classify x using the nearest neighbor classifier.

10. Consider the data given in Example 14. Verify that w learnt works correctly in predicting the values of ys for x_2 and x_3 also. Obtain w if the y_is are changed as $y_1 = 5$, $y_2 = 7$ and $y_3 = 8$.

11. Consider Example 13.

    a. Verify that $\hat{y}_i = y_i^{lor}$ for i = 2,3,4 using

    $$\hat{y}_i = \frac{23}{25} \times x_i(1) + \frac{23}{25} \times x_i(2) - \frac{184}{25}$$

    b. Predict the value of y, that is, obtain $\hat{y}$, if augmented $x = (1, 1, 1)$ using the w estimated in Example 13. Obtain the value of p and get the class label.

    c. Classify the augmented pattern $(8, 8, 1)$.

12. Show the forward pass computations for the inputs $\begin{pmatrix} 1 \\ 0 \end{pmatrix}$ and $\begin{pmatrix} 1 \\ 1 \end{pmatrix}$ using the network shown in Example 18.

13. Consider the weight update for w_{CO} in Example 18. Compute the update required for weight w_{DO}.

14. Consider the data in Example 19. Calculate the updates for weights w_{BC} and w_{BD} using backpropagation.

PRACTICAL EXERCISES

1. Download the Olivetti Face data set. There are 40 classes (corresponding to 40 people), each class having 10 faces of the individual; so there are 400 images in total. Here each face is viewed as an imgae of size 64 × 64 (= 4096) pixels; each pixel having values 0 to 255 which are ultimateley converted into floating numbers in the range [0,1]. Visit `https://scikit-learn.org/0.19/datasets/olivetti\_faces.html` for more details.

   Your Tasks: There are two subtasks. For both the subtasks, split the data set into the train and test parts. Vary the test size. The tasks are:

   a. **Task 1:** Reduce the dimensionality of the data set from 4096 to 400 using PCA and classify the P-dimensional data set using perceptron, SVM, logistic regression and MLP.

   b. **Task 2:** Repeat Task 1 using SVs instead of PCs.

Bibliography

- Minsky, M and Papert, S. 2017. *Perceptrons: An Introduction to Computational Geometry*, MIT Press.

- Burges, CJC. 1998. A tutorial on support vector machines for pattern recognition, *Data Mining and Knowledge Discovery*, 2(2): 121–167.

- Schölkopf, B and Smola, AJ. 2001. *Learning with Kernels*, MIT Press.

- Sastry PS: Pattern Recognition, *NPTEL*, IISc. `https://nptel.ac.in/courses/117/108/117108048`

- Murty, MN and Susheela Devi, V. 2015. *Introduction to Pattern Recognition and Machine Learning*, World Scientific Publishing Co. Pte. Ltd.: Singapore.

CHAPTER 9

Deep Learning

Learning Objectives

At the end of this chapter, you will be able to:

- Define deep learning
- Explain the autoencoder and how it is used in non-linear dimensionality reduction
- Describe convolutional neural networks, recurrent neural networks, long short-term memory and generative adversarial networks

9.1 INTRODUCTION TO DEEP LEARNING

Representation of data affects the performance of any machine learning system. For example, *weight of objects* alone might be adequate to classify objects as *chairs* and *adult humans*. Such a choice is simple in real life and is often based on common sense. However, in most practical ML applications, it is not so easy to come out with a representation of the data. The options we have are:

- Select or extract features using a dimensionality reduction tool and use the resulting representation in learning the ML model. This process is popular in conventional ML.
- Learn the representation automatically. Deep learning systems based on artificial neural networks (ANNs) claim to learn representations with the help of hidden nodes/layers.

In deep learning, the underlying learning system is realized using a cascade of systems that successively process data and pass on the information to subsequent levels. The size of the cascade is an indication of the *depth* of the learning system. Some architectures have more than 152 hidden layers. Successive layers are expected to learn more and more refined representations of the input data.

Learning in neural networks is achieved by changing the weights suitably to capture the required input–output behaviour. Some important properties of deep learners are as follows:

- They require more training data as the depth of the network grows.
- They typically employ dot product and matrix multiplications. So, they are ideally suited for numerical data.
- Selection of initial weights needs to be performed properly.
- We can train a neural network for a task and then use it for other related tasks with a simple fine-tuning of the parameters. This is called **transfer learning**.

- It is possible to choose an appropriate activation function to realize the associated/pre-specified non-linearity.
- Even though some of the ANNs including support vector machines (SVMs) normalize the data as a processing step, in theory, *normalization* is not required for using ANNs. For example, if a component $\phi_i(X)$ is more important than another component $\phi_j(X)$, the associated weight w_i can be chosen to be larger than w_j. However, in practice some kind of normalization is used.

9.2 NON-LINEAR FEATURE EXTRACTION USING AUTOENCODERS

An autoencoder is an unsupervised model for dimensionality reduction. It is a special type of neural network that uses a non-linear transformation on the input to encode/compress it to h, as shown in Fig. 9.1.

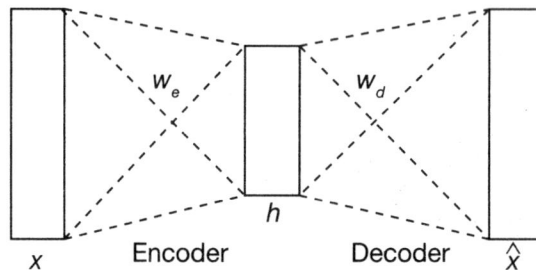

FIG. 9.1 Autoencoder architecture

Note that the input is x and the output required is ideally x or at least an approximate version of x, say $\hat{x}$, as shown in the figure.

An autoencoder uses an encoder. The encoder compresses the input x that is a vector in an l-dimensional space to a vector h in a k-dimensional space. A popular choice is $k < l$; this is called an **undercomplete autoencoder**. This is done so that the original data x can be reconstructed as $\hat{x}$ using the low-dimensional representation h.

If we can obtain $\hat{x} = x$ by using h, then h is a loss-less encoding of x; principal component analysis (PCA) displays a similar behaviour. In fact, it is possible to show that under some conditions, the autoencoder can work like representation using principal components (PCs). In general, PCA is a linear feature extraction scheme and autoencoder is a non-linear encoder. The decoder decompresses the encoded input to get back the original input x; note that ideally we want x and $\hat{x}$ to be the same.

Another important component is the code, also known as the bottleneck, h, which is the compressed representation of the input. This code or latent representation h is the required low-dimensional output from the autoencoder for non-linear dimensionality reduction. It is a non-linear scheme because it employs a non-linear activation function, like the sigmoid function.

The code h is the output of the encoder and is the input to the decoder. The encoder comprises the input and the hidden layer, while the decoder is made up of the hidden and output layers. Input, output and hidden layers can have any number of units (neurons). It is possible to have deep autoencoders where one can have many hidden layers. $w_{e_{k \times l}}$ and $w_{d_{l \times k}}$ indicate the weights in the encoding and decoding layers, respectively; biases are augmented appropriately. In some cases, these weights can be tied together, such that $w_e = w_d^t$, which is sometimes used to avoid overfitting as the number of trainable parameters is less in this setting.

An autoencoder computes the compressed representation of the input as follows:

- The encoder receives the input (x) (input layer) and computes the latent representation or (code) at the hidden layer as $h = \sigma(w_e x)$. This is fed to the decoder which outputs the reconstructed input $w_d h$, where σ is an activation function.
- When the input and output are real vectors, we use $h = sigmoid(w_e x)$ and the output activation is linear.

Autoencoders are trained using gradient descent, and parameters are learned using the back-propagation rule as used by multi-layer perceptrons (MLPs). The loss function of the autoencoder depends on the input x and the output z as the loss should reflect the deviation of the reconstructed input (output of the autoencoder) from the input.

One possible loss function is the L_2 norm. Specifically, it is useful when the data vectors are real. Let a data set contain $x_1, x_2, \ldots, x_n$, where n is the number of samples in the data set. The L_2 norm loss is given by:

$$Loss = \frac{1}{2} \sum_{i=1}^{n} ||x_i - w_d h||^2 \qquad (9.1)$$

For backpropagation, we need to compute the gradients:

- Let x be an input for which the decoded output is $w_d h = \hat{x}$. We have squared error loss to be $\mathcal{L}(w) = ||x - \hat{x}||^2$, where w is the set of weights and biases to be learnt.
- If we want the gradient with respect to w_d, we have, using the chain rule,

$$\nabla_{w_d}(\mathcal{L}(w)) = \frac{\partial \mathcal{L}(w)}{\partial \hat{x}} \times \frac{\partial \hat{x}}{\partial a} \times \frac{\partial a}{\delta w_d},$$

where $a = w_d h$ and the partial derivatives need to be carefully understood as follows:

- $\frac{\partial \mathcal{L}(w)}{\partial \hat{x}}$ is the gradient of a scalar with respect to a vector. In this case, it is $2(\hat{x} - x)$ as we are using the squared error loss.
- $\frac{\partial \hat{x}}{\partial a} = I$ when we use a linear activation function.
- $\frac{\partial a}{\delta w_d} = h$
- Similarly, we can show using the chain rule that

$$\nabla_{w_e}(\mathcal{L}(w)) = \frac{\partial \mathcal{L}(w)}{\partial \hat{x}} \times \frac{\partial \hat{x}}{\partial a} \times \frac{\partial a}{\delta h} \times \frac{\partial h}{\partial a_1} \times \frac{\partial a_1}{\partial w_e},$$

where a_1 is the activation reaching the hidden nodes. We leave the details as an exercise.

EXAMPLE 1: Consider the linear autoencoder shown in Fig. 9.2.

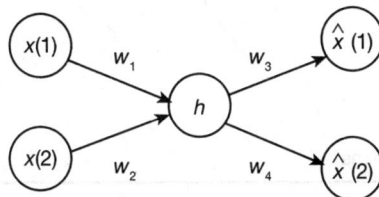

FIG. 9.2 Example of linear autoencoder

Let us consider that all the activations are linear.

So, if $x(1) = 1$ and $x(2) = 2$ is the input, then $h = a_1 = 1 \times w_1 + 2 \times w_2 = 5$, assuming that $w_1 = 1$ and $w_2 = 2$. So, $\hat{x}(1) = 5w_3$ and $\hat{x}(2) = 5w_4$. Let us consider one more pattern $x(1) = 2$ and $x(2) = 4$. The corresponding $h = a_1 = 10$ and $\hat{x}(1) = 10w_3$ and $\hat{x}(2) = 10w_4$.

We can calculate the optimal values of w_3 and w_4 that minimize the squared error loss as follows:

The squared error loss is

$$\mathcal{L}(w) = (5w_3 - 1)^2 + (5w_4 - 2)^2 + (10w_3 - 2)^2 + (10w_4 - 4)^2$$

If we equate the partial derivative of $\mathcal{L}(w)$ with respect to w_3 to zero, we get

$$5(5w_3 - 1) + 10(10w_3 - 2) = 0 \Rightarrow 125w_3 = 25 \Rightarrow w_3 = 0.2$$

Similarly, equating the partial derivative of $\mathcal{L}(w)$ with respect to w_4 to zero, we get

$$5(5w_4 - 2) + 10(10w_4 - 4) = 0 \Rightarrow 125w_4 = 50 \Rightarrow w_4 = 0.4$$

These weight values give a zero squared error on the training data of the two patterns because:

- When $x(1) = 1$ and $x(2) = 2$ is the input, then $h = 5$ and $\hat{x}(1) = 5 \times 0.2 = 1$ and $\hat{x}(2) = 5 \times w_4 = 2$. So, the input and output are identical, contributing a value of zero to the loss.
- Similarly, when $x(1) = 2$ and $x(2) = 4$ is the input, then $h = 10$ and $\hat{x}(1) = 10 \times 0.2 = 2$ and $\hat{x}(2) = 10 \times w_4 = 4$. So, the input and output are identical again, contributing a value of zero to the loss. So, the squared error loss is zero.

In general, when we use non-linear activation functions, we will have non-linear dimensionality reduction.

9.2.1   Comparison on the Digits Data Set

We compare the performance of dimensionality reduction using PCA, SVD and autoencoder (AE) on the sklearn Digits data set, as shown in Fig. 9.3.

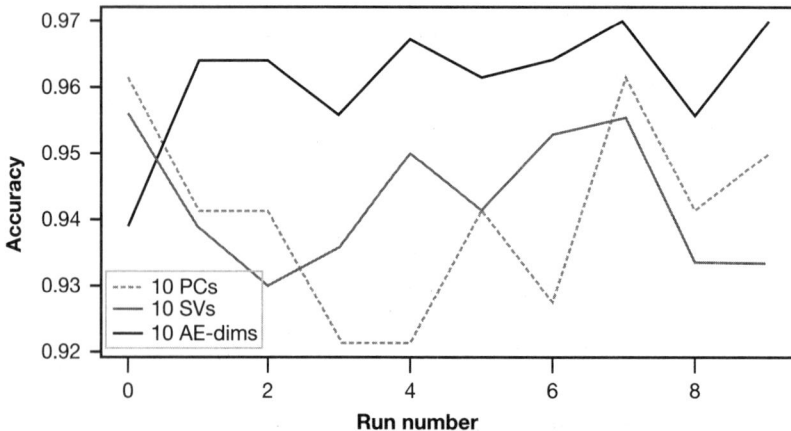

FIG. 9.3 Performance comparison in the Digits data set

There are a total of 1797 patterns from the handwritten digits 0 to 9. They are randomly split, 10 times, into 1437 training and 360 test patterns. Each training pattern is a binary image of size $8 \times 8 = 64$ pixels. The dimensionality is reduced from 64 to 10 using the linear feature extraction schemes PCA and SVD and a non-linear feature extraction scheme using autoencoder.

The results on each of these 10 random splits are shown in the figure. The X-axis depicts the random-split-based–run indexed from 0 to 9 for the 10 random splits. The Y-axis shows the accuracy in the range [0,1]. Note that in this case, the performance based on 10 AEs is superior to the performance based on 10 PC and 10 SV dimensions.

It is possible to obtain the importance/weights of the features selected using the autoenoder as shown in Fig. 9.4. This is performed as explained below.

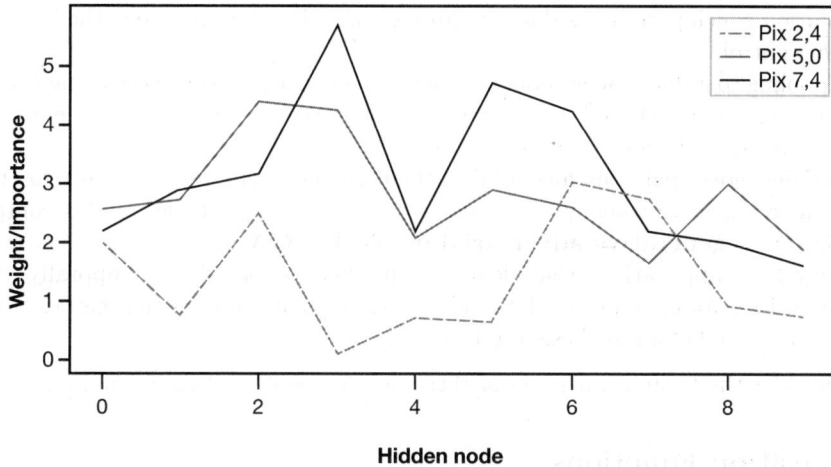

Fig. 9.4 Weights associated with input pixels

In the application, we have 64 input and output units. There are 10 hidden units. Each input node is connected to each hidden node with a weight. So, each pixel in the image has a weight associated with a hidden node. So, we can view each input node or pixel as a 10-dimensional weight vector based on the weights from the pixel to the 10 hidden nodes.

In Fig. 9.4, the X-axis denotes the number of hidden nodes from 0 to 9; note that there are 10 hidden nodes. The Y-axis denotes the magnitude of the weight between a pixel in the image and a hidden node. In the figure, the weight vectors of three pixels are shown. These are the 4th pixel on the second row, the 0th pixel on the fifth row and the 4th pixel on the seventh row. A similar analysis is possible even when we use PCA and SVD for dimensionality reduction. The autoencoder features perform better here because the data has some non-linearity.

9.3 DEEP NEURAL NETWORKS

Even though backpropagation was proposed and used in training MLPs in the 1980s, the area of deep learning using neural networks became practical more recently. There are several important contributions that have facilitated its growth. A major impetus is in terms of computing power. In addition, there was better appreciation of various components, including the following:

- How to exploit the non-linear activation functions in an optimal way
- Backpropagation employs a gradient descent scheme that is well-known for getting stuck in a locally optimal solution or not even reaching the local minimum of the error function. Several novel optimization tricks for better performance of gradient descent have been proposed.
- Learning in neural networks is achieved by selecting and changing the weights appropriately. Better schemes for weight initialization and normalization have been proposed.
- It is known that a single-hidden-layer–based neural network is adequate to model any input–output behaviour. However, a deeper structure that employs a large number of hidden layers offers greater flexibility; but, it requires large-sized training data to train the network. Some of the solutions are based on:
  - Convolution which permits us share weights so that the effective size of the learning problem is under control.
  - Using pooling to reduce the number of weights to be learnt so that overfitting can be avoided.
  - Optimization which is based on calculus and using calculus-friendly loss functions that both capture the objective and yield good gradients.
  - Synthesizing novel patterns and adding them to the small training data so that training can be more effective when small training data sets are available. In this context a major contribution is **generative adversarial networks (GANs)**.
  - One may have applications that deal with predictions based on temporally varying data or time-series data or sequence data. There are sequence models that can exploit recurrent behaviour in prediction in these applications.

We will consider the technical aspects of these developments in this chapter.

9.3.1 Activation Functions

A deep neural network is made up of several layers of artificial neurons or units. Each of these neurons employs an activation function f. The input to the function is a real number, a, that is input to the unit. The output of the unit is $f(a)$, where f is an activation.

Even though it is not mandatory, it is convenient for all the units in a layer to employ the same activation function. We have seen the linear threshold activation function that is used by a perceptron. We have also considered the linear activation function in Chapter 8.

Two activation functions, sigmoid and tanh, were proposed and used in the early days of backpropagation. There are some more activation functions that are popular today. We will discuss them in this section.

Linear Activation

A simple linear activation unit will take input a and return it as the output. So, $f(x) = x$ if f is a linear activation function. We show this in Fig. 9.5. Consider a simple network, as shown in Fig. 9.6.

The input to the network is x, we also call it h_0. It is a feed-forward network with three layers, one neuron per layer, with activation functions f_1, f_2, and f_3 as shown. In general, these three activation functions can be different. However, here we consider all of them to be linear. So, $h_1 = a_1 = w_1 h_0$; $h_2 = a_2 = w_2(w_1 h_0)$; finally $y = w_4 a_3 = w_4(w_3 h_2) = w_4 w_3 w_2 w_1 h_0$. So, the relation between y and x is of the form $y = wx$, where $w = w_4 w_3 w_2 w_1$.

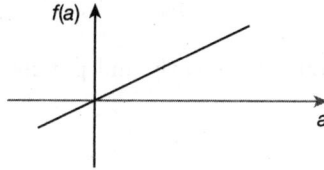

FIG. 9.5 Linear activation function

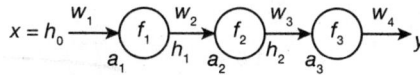

FIG. 9.6 Cascade of linear units

Therefore, cascading multiple (any number of) linear units will give linear input–output behaviour. We cannot capture non-linearity using a cascade of such linear units. The derivative of a linear activation is a constant.

$$\frac{df(x)}{dx} = 1$$

Linear Threshold Function

In perceptron, with augmented w and x, we compute the activation input as $w^t x$ and the output of the perceptron as

$$f(w^t x) = \begin{cases} 1 & if \quad w^t x > T \\ 0 & if \quad w^t x \leq T \end{cases}$$

It may be depicted as shown in Fig. 9.7.

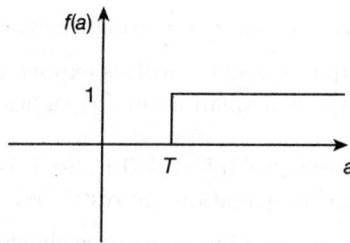

FIG. 9.7 Linear threshold activation

It has only two possible outputs based on whether a is greater than the threshold T or not. So, it is not a natural choice for multi-class classification. It is a hard limiter. It cannot offer smooth output transition from 0 and 1.

In a more formal sense, it is not calculus friendly. Its derivative is 0, making it difficult for backpropagation to be used.

Sigmoid Activation

The sigmoid activation function is one of the early and popular activation functions. It is characterized by

$$f(a) = \frac{1}{1 + e^{-a}},$$

where a is the activation input to the sigmoid unit.

It may be viewed as a softer version of the linear threshold function. Its derivative is $f'(a) = f(a)(1 - f(a))$. The sigmoid along with its derivative is shown in Fig. 9.8.

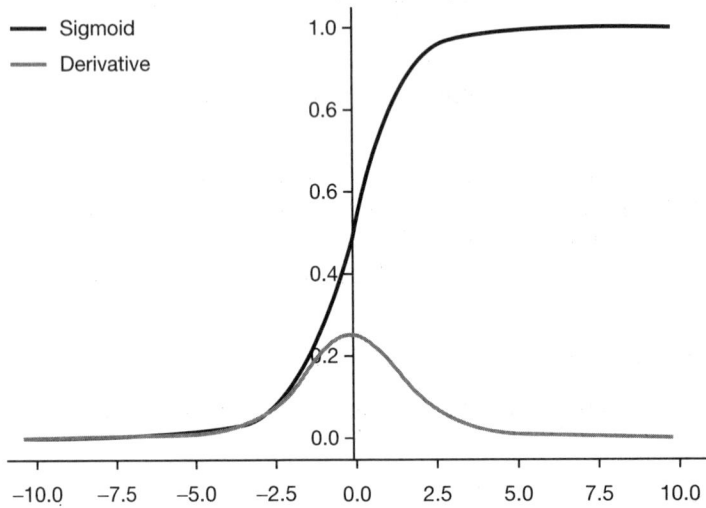

FIG. 9.8 Sigmoid activation function

Some of the problems associated with sigmoid activation function are as follows:

- Computing sigmoid values is expensive as it involves exponentiation.
- Its gradient tends to zero at large and small input (a) values as the function does not change much in these regions.
- Note that the sigmoid takes values in the interval [0,1]. So, it is not zero-centred. A consequence is that the weight updates in backpropagation are restricted.

Consider logistic regression. The corresponding network is shown in Fig. 9.9. Here, the activation function of the unit is a logistic or sigmoid function. If we want to update the weight w using gradient descent in backpropoagation, we have

$$w_{k+1} = w_k - \eta \nabla w, \text{ where } \nabla w = \frac{\partial \mathcal{L}(w)}{\partial w}$$
$$= (f(wx) - y) \times f(wx) \times (1 - f(wx)) \times x$$

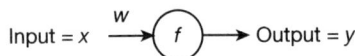

FIG. 9.9 Logistic regression

We can update the weight using this scheme. However, if wx is large, then $f(wx)$ tends to 1 and $(1 - f(wx))$ tends to 0. So, the gradient becomes very small. If wx is too small (negative), $f(wx)$ tends to 0 and again the gradient becomes too small.

This leads to the notion of a saturated neuron where the gradient tends to zero.

Let us consider the network shown in Fig. 9.6. If we consider $\nabla w_1 = \frac{\partial \mathcal{L}(w)}{\partial w_1}$, then we need to compute product of terms including

$$\frac{\partial h_3}{\partial a_3}; \frac{\partial h_2}{\partial a_2}; \frac{\partial h_1}{\partial a_1}$$

If the activation functions f_1, f_2 and f_3 are all sigmoids, then we will have the product, $(f(a_3)(1 - f(a_3))) \times (f(a_2)(1 - f(a_2))) \times (f(a_1)(1 - f(a - 1)))$. So, the gradient updates will not be effective as we are multiplying three derivatives of sigmoids here. Note that as shown in Fig. 9.8, the derivative has a maximum value of 0.25.

So, in a deeper neural network, we will have a larger number of these small derivative values in the computation of the product corresponding to the weight update, making the gradient tend to zero. This is called the **vanishing gradient problem** because of saturated neurons in a chain. The problem with not being zero-centred may be illustrated using the network shown in Fig. 9.10.

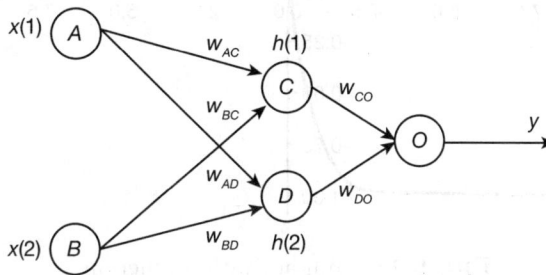

FIG. 9.10 An example to illustrate the problem with sigmoid outputs

Let a_O be the activation input to O and $h_O = y$ be the output of O. The equations for updating the weights w_{CO} and w_{DO} are given by

$$\nabla w_{CO} = \frac{\partial \mathcal{L}(w)}{\partial h_O} \times \frac{\partial h_O}{\partial a_O} \times \frac{\partial a_O}{\partial w_{CO}} = \frac{\partial \mathcal{L}(w)}{\partial h_O} \times \frac{\partial h_O}{\partial a_O} \times h(1)$$

Similarly, for w_{DO}, we have

$$\nabla w_{DO} = \frac{\partial \mathcal{L}(w)}{\partial h_O} \times \frac{\partial h_O}{\partial a_O} \times \frac{\partial a_O}{\partial w_{DO}} = \frac{\partial \mathcal{L}(w)}{\partial h_O} \times \frac{\partial h_O}{\partial a_O} \times h(2)$$

If C and D are sigmoid units, then $h(1)$ and $h(2)$ are in the range $[0,1]$; they are non-negative. The other two terms, $\frac{\partial \mathcal{L}(w)}{\partial h_O}$ and $\frac{\partial h_O}{\partial a_O}$, are common to both ∇w_{CO} and ∇w_{DO}. So, ∇w_{CO} and ∇w_{DO} are both positive or both negative based on whether the value of $\frac{\partial \mathcal{L}(w)}{\partial h_O} \times \frac{\partial h_O}{\partial a_O}$ is positive or negative.

It is possible to show that if we employ sigmoid units everywhere, then because they output non-negative values, all the gradients of the weights, at a layer, of edges reaching a sigmoid unit are positive or all the gradients are negative. We leave showing that the updates to w_{AC}, w_{BC} are either positive or negative if $x(1) \in [0, 1]$ and $x(2) \in [0, 1]$ as an exercise.

Tanh Activation

Tanh or hyperbolic tangent activation function is also one of the early activation functions which were proposed and used in neural networks. It may be viewed as a refined version of the sigmoid function. It is characterized by

$$f(a) = \frac{e^a - e^{-a}}{e^a + e^{-a}},$$

where a is the activation input to the tanh unit.

Its derivative is $f'(a) = 1 - f^2(a)$. Tanh along with its derivative is shown in Fig. 9.11.

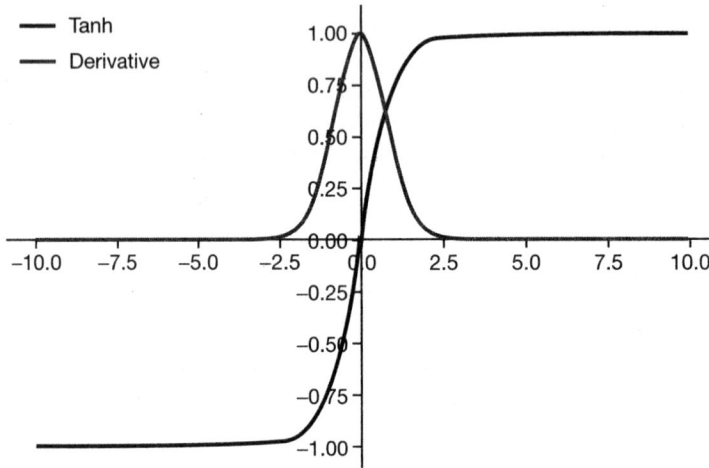

FIG. 9.11 Tanh activation function

Tanh and sigmoid are related. It is possible to show that $tanh(a) = 2sigmoid(2a) - 1$. We will leave the proof as an exercise.

Some of the problems associated with the tanh activation function are as follows:

- Computing tanh values is expensive as it involves exponentiation.
- Its gradient tends to zero at large and small input (a) values, where tanh $= 1$ or -1, as the function does not change much in these regions.
- Note that tanh takes values in the interval $[-1, 1]$. So, it is zero-centred. So, it is an improvement over sigmoid.

Rectified Linear Unit (ReLU)

It is characterized by

$$f(a) = max(0, x),$$

where a is the activation input to the ReLU unit.

ReLu has no derivative at $a = 0$. Its derivative is $f'(a) = \begin{cases} 0 & if \quad a < 0 \\ 1 & if \quad a > 0 \\ NotThere & if \quad a = 0 \end{cases}$

ReLU along with its derivative, at points other than $a = 0$, is shown in Fig. 9.12. It does not saturate in the positive region. Computationally efficient, it converges faster than sigmoid/tanh. It is a non-linear activation function.

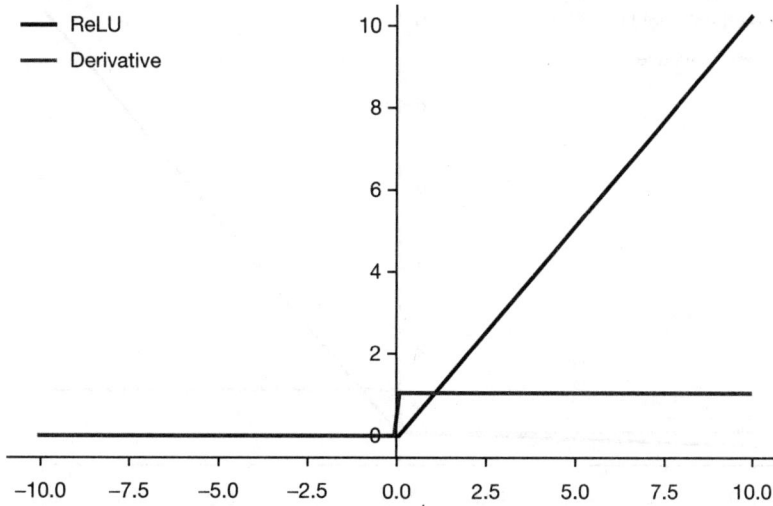

FIG. 9.12 ReLU activation function

A major problem associated with ReLU is that it can have difficulty in updating weights in backpropagation because of a dead neuron.

Consider the network shown in Fig. 9.10. Let C and D be ReLU units. If at some time during training, the activation goes below 0, then neuron C will output 0 ($h(1) = 0$) because it is a ReLU unit. So, $\frac{\partial h(1)}{\partial a(1)} = 0$.

Now neither w_{AC} nor w_{BC} will get updated. For w_{AC}, we have

$$\nabla w_{AC} = \frac{\partial \mathcal{L}(w)}{\partial h_O} \times \frac{\partial h_O}{\partial a_O} \times \frac{\partial a_O}{\partial h(1)} \times \frac{\partial h(1)}{\partial a(1)} \times x(1),$$

where the fourth term is 0. So, $\nabla w_{AC} = 0$.

The same thing takes place for the weight w_{BC} also. So, there will be no updates to the weights getting into neuron C and its activation will not change anytime in the future. So, it will become a dead neuron.

Leaky ReLU

It is characterized by

$$f(a) = max(0.01x, x),$$

where a is the activation input to the leaky ReLU unit.

Its derivative is $f'(a) = \begin{cases} 0.01 & if \quad a < 0 \\ 1 & if \quad a \geq 0 \end{cases}$

The leaky ReLU along with its derivative is shown in Fig. 9.13. It does not saturate. It avoids the dead neuron problem associated with ReLU and it is computationally efficient. It has the flavour of zero-centred units.

Softmax Activation Function

This is another important function that is useful in multi-class classification.

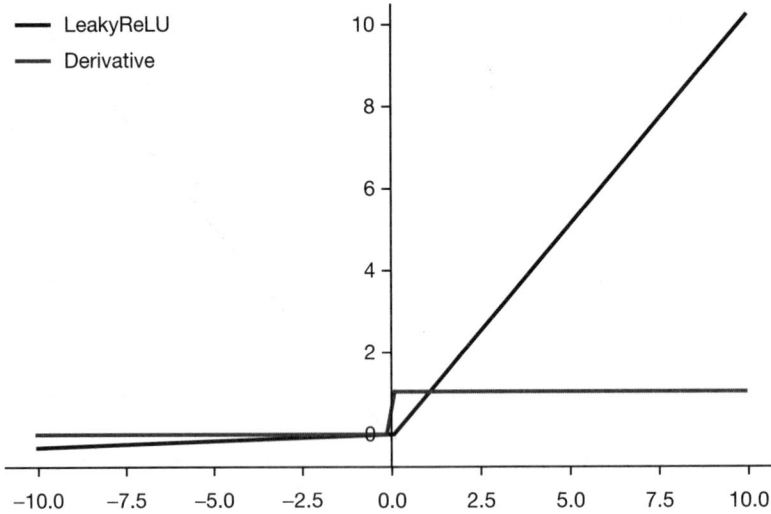

FIG. 9.13 Leaky ReLU activation function

It is typically used in the output layer to convert real-valued outputs $\hat{y}_1, \hat{y}_2, \ldots, \hat{y}_C$ into probabilities as follows:

$$p_i = \frac{e^{\hat{y}_i}}{\sum_{j=1}^{C} e^{\hat{y}_j}}, \; for \; i = 1, 2 \cdots, C$$

Consider the partial derivative $\frac{\partial p_i}{\partial \hat{y}_s} = \frac{-e^{\hat{y}_i} \times e^{\hat{y}_s}}{(\sum_j e^{\hat{y}_j})^2} = -p_i p_s, \; for \; i \neq s.$

If $i = s$, then

$$\frac{\partial p_i}{\partial \hat{y}_i} = \frac{\sum_j e^{\hat{y}_j} \times e^{\hat{y}_i}}{(\sum_j e^{\hat{y}_j})^2} - \frac{-e^{\hat{y}_i} \times e^{\hat{y}_i}}{(\sum_j e^{\hat{y}_j})^2}$$

$$= p_i - p_i^2 = p_i(1 - p_i)$$

So, $\frac{\partial p_i}{\partial \hat{y}_s} = \begin{cases} -p_i p_s & if \quad i \neq s \\ p_i(1 - p_i) & if \quad i = s \end{cases}$

9.3.2 Initializing Weights

We have seen that learning in neural networks amounts to starting with a good initial selection of weights for the edges in the network and an appropriate updation of these weights using error backpropagation. So, initialization of weights is an important topic.

Let us take the option of initializing all the weights in the network to 0. Let us illustrate it with the network shown in Fig. 9.10. So, activation input $a(1) = w_{AC} \times x(1) + w_{BC} \times x(2) = 0$; similarly, activation input $a(2) = 0$. So, h(1) = h(2) as $a(1) = a(2)$.

$$\nabla w_{AC} = \frac{\partial \mathcal{L}(w)}{\partial h_O} \times \frac{\partial h_O}{\partial a_O} \times \frac{\partial a_O}{\partial h(1)} \times \frac{\partial h(1)}{\partial a(1)} \times x(1)$$

Similarly, $\nabla w_{AD} = \frac{\partial \mathcal{L}(w)}{\partial h_O} \times \frac{\partial h_O}{\partial a_O} \times \frac{\partial a_O}{\partial h(2)} \times \frac{\partial h(2)}{\partial a(2)} \times x(1)$.

We know that $a(1) = a(2)$ and $h(1) = h(2)$. So, $\nabla w_{AC} = \nabla w_{AD}$. $w_{AC} + \nabla w_{AC} = w_{AD} + \nabla w_{AD}$. They remain equal during the entire training phase.

Similarly, $w_{BC} = w_{BD}$ till the end of training when the initial weights are chosen to be 0. This is called the **symmetry problem**. We encounter this problem when all the weights are chosen to have the same value initially. This constant need not be 0.

On the contrary, if we take larger weights, then the activation will be large, either positive or negative. This will mean that the gradients will vanish. So, we need to initialize weights so that the activation values are not too large.

This is accomplished by controlling the variance of the activations to be the same across layers to avoid saturation. This may be explained as follows:

Let the number of neurons in each layer be l. Let the input at the input layer be $x = (x(1), x(2), \ldots, x(l))$. So, the activation at the ith node in the hidden layer is

$$a_1(i) = \sum_{i=1}^{l} w_1(i) x(i)$$

Consider the variance of $a_1(i)$. It is

$$Var(a_1(i)) = Var\left(\sum_{i=1}^{l} w_1(i) x(i) \right) = \sum_{i=1}^{l} Var(w_1(i) x(i))$$

Here, we have assumed that $w_1(p) x(p)$ and $w_1(q) x(q)$ are uncorrelated for any $p, q \in \{1, 2, \ldots, l\}$, where $p \neq q$ for replacing the variance of the sum by the sum of variances.

We can show that the variance of a product of two random variables is the product of the variances under some conditions. Let X and Y be two random variables.

$$Var(XY) = E[X^2 Y^2] - E[XY]^2$$

If X and Y are independent, then

$$Var(XY) = E[X^2 Y^2] - E[XY]^2 = E[X^2] E[Y^2] - E[X]^2 F[Y]^2$$
$$= (Var(X) + E[X]^2)(Var(Y) + E[Y]^2) - E[X]^2 E[Y]^2$$
$$= Var(X) Var(Y) + Var(X) E[Y]^2 + Var(Y) E[X]^2 = Var(X) Var(Y)$$

if we assume that $E[X] = E[Y] = 0$.

So, $Var(XY) = Var(X) Var(Y)$ if $E[X] = E[Y] = 0$ and X and Y are independent. Using these assumptions, we have

$$Var(a_1(i)) = \sum_{i=1}^{l} Var(w_1(i)) Var(x(i)) = l Var(w) Var(x),$$

assuming that the weights and inputs are independent and their means are zero. This will simplify to

$$Var(a_1(i)) = l Var(w) Var(x)$$

If we want the variance to be the same across the input and the first hidden layer, that is, $Var(a_1(i)) = Var(x)$, it implies that we have $l Var(w) = 1$. It means that we need to choose weights so that $Var(w) = \frac{1}{l}$.

So, we need to choose weights that are independent and identically distributed with zero mean and $\frac{1}{l}$ variance. This is called the **LeCun initialization**. Some modifications need to be made to the LeCun initialization:

- We have assumed that the number of neurons is l in each layer. Instead, if there are l_{in} units in the input and l_{out} in the hidden layer, then $Var(w) = \frac{1}{l_{in}+l_{out}}$. This is called the **Xavier initialization**.
- For ReLU and leaky ReLU layers, it was noted that $var(w) = \frac{2}{l_{in}}$. This is called the **He initialization**.

We have compared these schemes using the Fashion MNIST image data set. It has a training set of 60,000 examples and a test set of 10,000 examples. Each example is a 28×28 greyscale image associated with a label from 10 classes corresponding to different types of garments.

Different initialization schemes were employed on an MLP classifier with one hidden layer employing tanh activation. The results are depicted in Fig. 9.14.

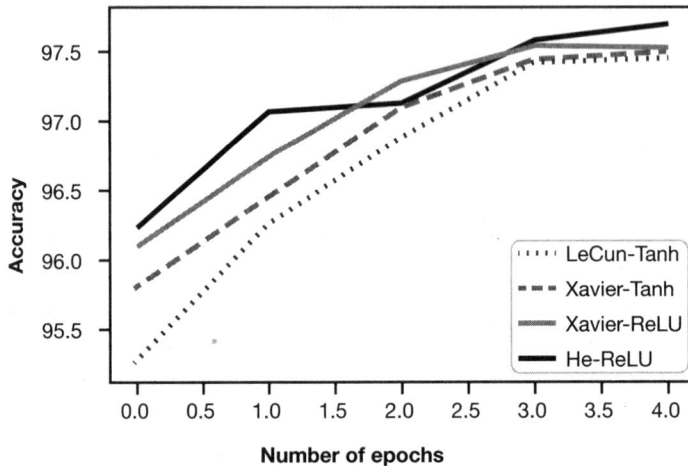

FIG. 9.14 Comparison of various weight initialization schemes (for colour figure, please see Colour Plate 2)

The code was run for 5 epochs. The X-axis in the figure depicts the number of epochs and the Y-axis shows the percentage accuracy. Note that He ReLU and Xavier ReLU are superior. The weight initialization based on LeCun tanh shows inferior performance.

9.3.3 Improved Optimization Methods

We have seen earlier that backpropagation employs gradient descent to update weights. We will examine some of the prominent types of gradient descent in this section.

Stochastic Gradient Descent (SGD)

This can be a gross approximation of the true gradient descent as we may be updating the weights based on the error of one pattern or a small number of patterns. The update equation for weight w is $w_{k+1} = w_k - \eta \nabla w_k$, where $\nabla w_k = \frac{\partial \mathcal{L}(w)}{\partial w}$; it is evaluated at w_k. It is possible to better understand the descent using the plot in Fig. 9.15.

If the curve is steep, the gradient $\frac{\nabla L2}{\nabla w2}$ is large and the magnitude of the update is also large. If the slope is gentle, the gradient $\frac{\nabla L1}{\nabla w1}$ is small and the magnitude of the update is also small.

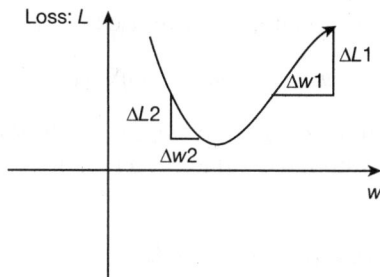

FIG. 9.15 Slope of the function and gradient

There are different schemes for updation frequency:

- In stochastic gradient descent (SGD), we update the weight after calculating the error for every pattern. It is faster. It has low memory requirements as the update is carried out immediately after viewing one pattern. It is ideally suited for streaming data. However, it may have a lot of fluctuations.
- In batch mode update, an update is made at the end of an epoch, that is, after viewing all the n patterns. Batch mode update can be slow, it requires memory to store all the gradient values corresponding to the n errors. It may have difficulty with streaming data. It converges to the global optimum for convex objective functions, that is, functions having a single minimum. It converges to a local optimum if the function is not convex.
- The weight is updated after viewing a mini-batch of patterns. If the size of the mini-batch is m, then it typically satisfies $m < n$, where n is the number of training patterns. It is popularly used in training neural networks as it strikes a balance between the two extremes of updating after a pattern or an epoch. It can reduce the variance of the weight updates leading to a more stable convergence. It is also called SGD because we do not consider all the n patterns for each update.

Momentum-Based Gradient Descent

In addition to the gradient at a point, we take into account the history of previous updates to decide on the update. We have seen that GD takes more time to reach the optimal point in regions where there is a gentler slope. Momentum-based update has the following basis: *if we have moved frequently in some direction, then take a bigger move in that direction.* It is implemented as follows: $ud_k = \alpha\, ud_{k-1} + \eta \nabla w_k$, where ud_k is the update at the kth step. The history update rate α is the weight given to history. Effectively it will be in the range $[0.8, 1.0]$. $w_{k+1} = w_k - ud_k$ is the update equation.

If $\alpha = 0$, then the momentum-based update boils down to a simple gradient descent–based update. It enhances updates for dimensions whose gradients are stable and point in the same direction and marginalizes updates for dimensions whose gradients are unstable and oscillate or change directions.

Nesterov Accelerated Gradient Descent

This can be thought of as a refinement of momentum-based GD. One problem with a momentum-based update is that it can cause the search to overshoot the minimum at the bottom of a basin. Nesterov momentum is a modification that overcomes this overshooting of the minima. Here, we first calculate the position of the variable using the change from the last update and then calculate the derivative at this position.

In the case of momentum-based update, we have used

$$ud_k = \alpha \; ud_{k-1} + \eta \nabla w_k$$

So, the update will be greater than or equal to $\alpha \; ud_{k-1}$. So, it is possible to improve the descent by calculating the gradient at this updated value rather than at w_k. So, the gradient update for Nesterov momentum is

$$w_{anticipated} = w_k - \alpha ud_{k-1}$$
$$ud_k = \alpha \; ud_{k-1} + \eta \nabla w_{anticipated}$$
$$w_{k+1} = w_k - ud_k$$

This update based on anticipated weight prevents us from accelerating too much; it pulls us back by a quantity proportional to ud_{k-1}. We show the results of an MLP output in Fig. 9.16.

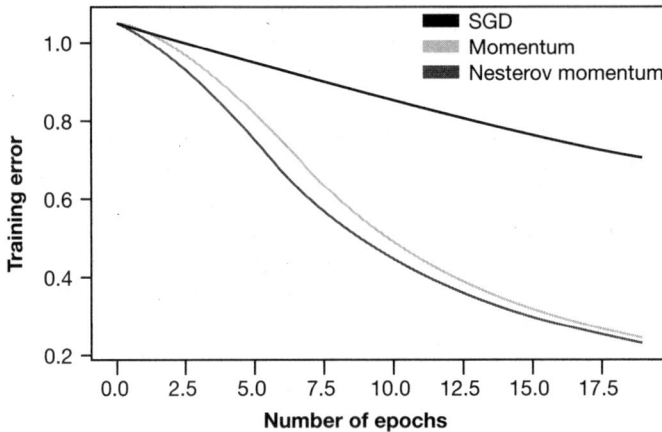

FIG. 9.16 Training error versus number of epochs on the Digits data set (for colour figure, please see Colour Plate 2)

This is based on SGD, momentum-based SGD and Nesterov momentum-based SGD. The Digits data set available under sklearn is used in the experiment. The plot shows how the training error decreases as the number of epochs increases up to 20. We can observe that momentum is better than SGD and the Nesterov momentum-based approach is the best in terms of reducing error.

9.3.4 Adaptive Optimization

Let us consider a sigmoid unit, as shown in Fig. 9.17. The activation input to the unit is $a = w^t x = w(1)x(1) + w(2)x(2) + w(3)x(3)$. We know that $y = \sigma(a) = \frac{1}{1+e^{-a}}$.

If there are n training pairs, $\{(x_1, y_1), (x_2, y_2), \ldots, (x_n, y_n)\}$, then we can use the gradients with respect to all the n patterns to update the weights $w(1), w(2)$ and $w(3)$. If $x(3) = 0$ in most of these n patterns, then $\nabla w(3) \approx 0$ and so $w(3)$ will not be updated.

However, if $x(3)$ is important for prediction, then learning $w(3)$ correctly is important. This will mean that we need to have different learning rates for different weight updates. Different schemes have been proposed and used to solve the adaptive learning rates. We will consider them next.

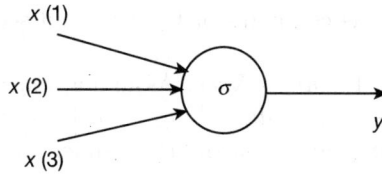

FIG. 9.17 A sigmoid unit

Root Mean Square Propagation (RMSProp)

This simple adaptive scheme is based on adaptive gradient descent or AdaGrad that works in favour of updates to weights that do not change frequently. The update rule is given as follows:

$$\gamma_k = \gamma_{k-1} + (\nabla w_k)^2$$

and

$$w_{k+1} = w_k - \frac{\eta}{\sqrt{\gamma_k + \epsilon}} \nabla w_k$$

If γ_k is large, the difference between successive weight values, that is, between w_k and w_{k+1}, will be small. This is because the effective learning rate is small. γ_k will be large when γ_{k-1} or ∇w_k is large. So, if a weight is frequently updated, then the quantum of its update will be smaller as γ_k will reduce such update magnitudes.

The value of ϵ is small; it is used to ensure that $\gamma_k = 0$ will not create a *zero division* problem. Adagrad decays the effective learning rate as $\frac{\eta}{\sqrt{\gamma_k + \epsilon}}$. So, the effective learning rate decreases as γ_k or $\sqrt{\gamma_k + \epsilon}$ increases. As a consequence, weights that are updated frequently will not receive proper attention.

RMSProp provides a solution to this problem by controlling the quantity $\sqrt{\gamma_k + \epsilon}$. This is based on the following updates to γ_k:

$$\gamma_k = dr(\gamma_{k-1}) + (1 - dr)(\nabla w_k)^2,$$

where dr is the decay rate parameter used to control γ_k; dr typically has a value of 0.9.

So, RMSprop improves, compared to AdaGrad, the scope for frequently updated weights to receive further updates.

Adaptive Moment Estimation (ADAM)

It is one of the most popular gradient descent variants. It updates γ_k exactly like RMSprop; $\gamma_k = dr_1 \times \gamma_{k-1} + (1 - dr_1) \times (\nabla w_k)^2$, where dr_1 is the decay rate that can be as large as 0.99. In addition, it estimates the gradient using $m_k = dr \times m_{k-1} + (1 - dr) \times \nabla w_k$, where dr can have a value of 0.9.

It employs m_k and γ_k in a normalized form as follows:

$$\hat{m}_k = \frac{m_k}{1 - dr^k}; \ \hat{\gamma}_k = \frac{\gamma_k}{1 - dr_1^k}$$

These normalized forms are used in weight updataion as

$$w_{k+1} = w_k - \frac{\eta}{\sqrt{\hat{\gamma}_k + \epsilon}} \times \hat{m}_k$$

It is possible to view m_k and γ_k as estimates of the first and second moments of the gradients, respectively.

We use $m_k = dr \times m_{k-1} + (1 - dr) \times \nabla w_k$. Assuming $m_0 = 0$, we have $m_1 = dr \times m_0 + (1 - dr) \times \nabla w_1$ and $m_2 = dr \times m_1 + (1 - dr) \times \nabla w_2 = dr(1 - dr)\nabla w_1 + (1 - dr) \times \nabla w_2$. We can show that, by expanding and rearranging in a similar manner, $m_3 = (1 - dr) \sum_{i=1}^{3} dr^{3-i} \nabla w_i$.

In general, we can show that $m_k = (1 - dr) \sum_{i=1}^{k} dr^{k-i} \nabla w_i$, which we leave as an exercise. Taking expectation on both sides of the equation, we get

$$E[m_k] = E[(1 - dr) \sum_{i=1}^{k} dr^{k-i} \nabla w_i]$$

Observing that dr is a deterministic constant, we have

$$E[m_k] = (1 - dr) \sum_{i=1}^{k} dr^{k-i} E[\nabla w_i]$$

Because all the weight updates come from the same distribution, we have

$$E[m_k] = E[\nabla w](1 - dr) \sum_{i=1}^{k} dr^{k-i} = E[\nabla w](1 - dr)(dr^{k-1} + dr^{k-2} + \cdots + dr^0),$$

where $E[\nabla w_i] = E[\nabla w]$.

So, $E[m_k] = E[\nabla w](1 - dr)\frac{1-dr^k}{1-dr} = E[\nabla w](1 - dr^k)$. So, $E[\frac{m_k}{1-dr^k}] = E[\nabla w] \Rightarrow E[\hat{m}_k] = E[\nabla w]$. This justifies why we are computing $\hat{m}_k$ in the update of the weight, instead of $E[\nabla w_k]$. So, our weight updates are based on expected value of gradients and not just on ∇w_k alone.

It may be viewed as a combination of RMSprop and momentum; the RMSprop equation is used in updating γ_k and updation of m_k is more like a momentum-based update. We show the results of using these optimizers on the Fashion MNIST data set, in Fig. 9.18.

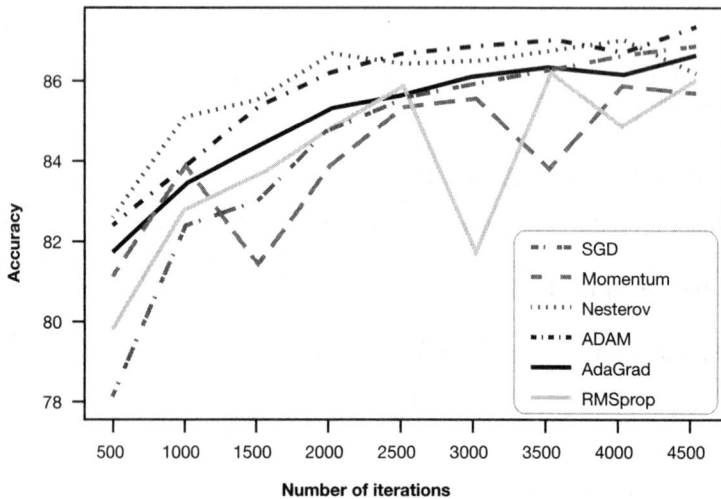

FIG. 9.18 Comparison of optimizers on the Fashion MNIST data set (for colour figure, please see Colour Plate 2)

We have used all the optimizers that we have considered in this section. The Nesterov momentum-based update and the ADAM update have given superior results in terms of accuracy. SGD, RMSprop and momentum-based updates are inferior in this case. The performance of the AdaGrad-based scheme is somewhere in the middle. We show the results on the MNIST data set in Fig. 9.19.

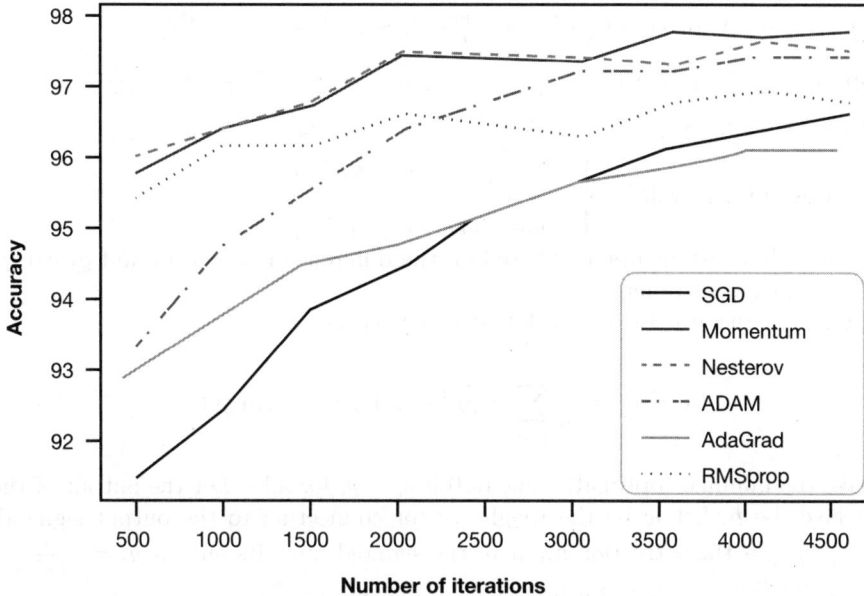

FIG. 9.19 Comparison of optimizers on the MNIST data set (for colour figure, please see Colour Plate 2)

We have used all the 6 optimizers in this experiment also. The Nesterov momentum-based update and the momentum-based update have given superior results in terms of accuracy on the MNIST data set. SGD and AdaGrad-based updates are inferior in this case. The performance of ADAM and RMSprop-based schemes are somewhere in the middle.

9.3.5 Loss Functions

Earlier, we had seen some loss functions. Loss functions help us in providing some score indicating how far the ML model's output deviates from the target values specified by the training data. Some of the loss functions are useful for regression while others are useful for classification.

- **Squared Error Loss:** If $\hat{y}_i$ is the target value and y_i is the predicted value for input x_i, then the squared error loss for n training patterns is

$$SE = \sum_{i=1}^{n}(y_i - \hat{y}_i)^2$$

We require the derivative of a loss function in backpropagation. Its partial derivative with respect to y_j is $2(y_j - \hat{y}_j)$. It is also called L_2 loss and dividing SE by n gives us mean squared error (MSE) loss.

It sums the squares of the deviations. So, minor deviations will become insignificant and major deviations will have a bigger impact. That is, the contribution of outliers can be huge.

- **Mean Absolute Error (MAE):** It is given by $MAE(y) = \frac{\sum_{i=1}^{n}|y_i - \hat{y}_i|}{n}$. It is considered to ensure that the contribution of outliers is not large. It is also called L_1 loss. Its partial derivative with respect to y_j is $\frac{y_j - \hat{y}_j}{|y_j - \hat{y}_j|}$. It is derived by using the following:

  Let $g(y_i) = (y_i - \hat{y}_i)$. So, $MAE(y) = \frac{\sum_{i=1}^{n}|g(y_i)|}{n}$. So, $\frac{\partial MAE(y)}{\partial y_j} = \frac{1}{n}\frac{\partial |g(y_j)|}{\partial y_j}$.

  We compute $\frac{\partial |g(y_j)|}{\partial y_j}$ by considering $|g(y_j)| = \sqrt{g(y_j)^2}$. So, $\frac{\partial |g(y_j)|}{\partial y_j} = \frac{\partial \sqrt{g^2(y_j)}}{\partial y_j} = \frac{1}{2}.(g^2(y_j))^{-\frac{1}{2}}$.

  $2.g(y_j).g'(y_j) = \frac{g(y_j)}{|g(y_j)|}$ as $g'(y_j) = 1$. So, $\frac{\partial MAE(y)}{\partial y_j} = \frac{1}{n}\frac{\partial |g(y_j)|}{\partial y_j} = \frac{1}{n}\frac{y_j - \hat{y}_j}{|y_j - \hat{y}_j|}$

  Note that this gradient will be $\begin{cases} \frac{1}{n} & if \quad y_j > \hat{y}_j \\ -\frac{1}{n} & if \quad y_j < \hat{y}_j \\ undefined & if \quad y_j = \hat{y}_j \end{cases}$

  So, it makes only constant updates based on the difference between y_j and $\hat{y}_j$ irrespective the quantity by which they differ.

- **Binary Cross Entropy Loss (BCE):** It is defined as

$$BCE(w) = \frac{1}{n}\sum_{i=1}^{n} -[(\hat{y}_i log(y_i) + (1 - \hat{y}_i)log(1 - y_i)]$$

It will have the minimal (optimal) value of 0 if $y_i = \hat{y}_i$ for all i. Let the output of the previous (hidden) layer be h. Let w be the weight vector connecting to the output sigmoid unit. So, $a_j = \sum_{p=1}^{l} w_p h_p$ is the activation input to the sigmoid unit. Its output $y_j = \frac{1}{1+e^{-a_j}}$. $\frac{\partial BCE(w)}{\partial w_p} = \frac{\partial BCE(w)}{\partial y_j}.\frac{\partial y_j}{\partial a_j}.\frac{\partial a_j}{\partial w_p}$. Each of the components is:

$$\frac{\partial BCE(w)}{\partial y_j} = -\frac{1}{n}(\frac{\hat{y}_j}{y_j}) + \frac{1 - \hat{y}_j}{1 - y_j} = \frac{1}{n}\frac{(y_j - \hat{y}_j)}{y_j(1 - y_j)}$$

$$\frac{\partial y_j}{\partial a_j} = y_j(1 - y_j)$$

$$\frac{\partial a_j}{\partial w_p} = x_p$$

So, $\frac{\partial BCE(w)}{\partial w_p} = \frac{1}{n}\frac{(y_j - \hat{y}_j)}{y_j(1 - y_j)} \times y_j(1 - y_j) \times x_p = \frac{1}{n}(y_j - \hat{y}_j) \times x_p$.

9.3.6 Regularization

Deep neural networks involve a reasonable number of hidden layers and each layer may have a good number of neurons. So, deep neural networks are highly complex models involving a lot of non-linear computations.

So, we need to learn many weights. It is easy for the complex networks to overfit the training data and make the training error vanish. The trained model may fail to work well on the test data.

In order to address this problem, several simplifications to the model and additional constraints on the optimization criterion are used. These constrained models are called *regularized* models. We examine some of them in this section.

- L_2 **Regularization:** Let $\mathcal{L}(w)$ be the loss function of the model. We add $||w||^2$ squared L_2 norm of the weight matrix to the loss function.

So,

$$\mathcal{L}_{reg}(w) = \mathcal{L}(w) + \frac{\lambda}{2}||w||^2,$$

where λ is the Lagrange variable. So, gradient of $\mathcal{L}_{reg}(w)$ with respect to w is

$$\nabla\mathcal{L}_{reg}(w) = \nabla\mathcal{L}(w) + \lambda w$$

$\Rightarrow$ Gradient descent is done using the update rule

$$w_{k+1} = w_k - \eta\nabla\mathcal{L}(w_k) - \eta\lambda w_k$$

It is also called *ridge regularization*. If $\lambda = 0$, it amounts to GD without regularization. If λ is chosen to be large, then the weights will become small. However, it will not force them to be 0 (zero) valued. It can learn complex class boundaries. It is not robust to outliers as it involves a squared term.

- **Adding Noise to Inputs:** This can be shown to be equivalent to L_2 regularization by imposing minor constraints on the distribution of noise. Let x be the input vector with components $x(1), x(2), \ldots, x(l)$. Let ϵ be the noise vector with components $\epsilon(1), \epsilon(2), \ldots, \epsilon(l)$. Let $\epsilon(i)$ for $i = 1, 2, \ldots, l$ be independently drawn from a zero-mean and λ variance Gaussian distribution. That is, $\epsilon(i) \sim \mathcal{N}(0, \lambda)$.

  Let the noisy version of x be x^n given by $x^n(i) = x(i) + \epsilon(i)$ for $i = 1, 2, \ldots, l$. Let $y^{obt} = a = \sum_{i=1}^{l} w(i)x(i)$ using a linear activation at the output. So, the noisy version, $y^n = \sum_{i=1}^{l} w(i)x^n(i) = \sum_{i=1}^{l} w(i)x(i) + \sum_{i=1}^{l} w(i)\epsilon(i) = y^{obt} + \sum_{i=1}^{l} w(i)\epsilon(i)$.

  So, $y^n = y^{obt} + \sum_{i=1}^{l} w(i)\epsilon(i)$. So, $E[(y^n - y^{tar})^2] = E[((y^{obt} - y^{tar}) + (\sum_{i=1}^{l} w(i)\epsilon(i)))^2]$, where y^{tar} is the target output for input x. So, $E[(y^n - y^{tar})^2] = E[(y^{obt} - y^{tar})^2] + E[(\sum_{i=1}^{l} w(i)\epsilon(i))^2] + 2E[(y^{obt} - y^{tar})\sum_{i=1}^{l} w(i)\epsilon(i)]$.

  So, $E[(y^n - y^{tar})^2] = E[(y^{obt} - y^{tar})^2] + E[(\sum_{i=1}^{l} w(i)^2\epsilon(i))^2] + 0$. This is because $w(i)$ and $\epsilon(i)$ and $\epsilon(i)$ and $\epsilon(j)$ are independent of each other. Further $E[\epsilon(i)] = 0$ as the mean of the distribution is 0. Note that $E[(\epsilon(i))^2] = $ Variance of noise $= \lambda$.

  So, $E[(y^n - y^{tar})^2] = E[(y^{obt} - y^{tar})^2] + \lambda E[\sum_{i=1}^{l} w(i)^2]$. This form is the same as the L_2 regularizer term on the weights.

- **Early Stopping**: Early stopping may be explained using Fig. 9.20.

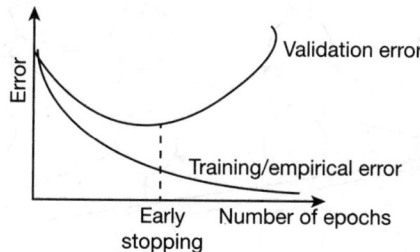

FIG. 9.20 Early stopping

Typically, the training error will decrease as the number of epochs to train the network increases. However, the validation error decreases up to a point and starts increasing beyond that point because of over training or overfitting. These are depicted in Fig. 9.20. In this context, early stopping helps us to stop training the network to avoid overfitting. Such a point at which training can be stopped is shown as *early stopping point*. Some popularly used schemes for early stopping are:

(a) Train the model for a fixed number of epochs. This scheme also depends on the learning rate. So, it needs some amount of trial and error to fix these values and so is not popular currently.

(b) Stop training when the update is less than a prefixed value. Such a stopping minimizes loss and saves further computing. The number of epochs is automatically decided. However, overfitting is still possible.

(c) Consider a point that has the following characteristics: increasing the number of epochs has no significant effect on the training error beyond this point and the validation error starts increasing. This is the most popular scheme for early stopping. A problem with this scheme is that it may take a large number of epochs to reach this point and might be computationally expensive.

(d) One can use a hybrid strategy by combining schemes (b) and (c) judiciously.

- **Data Set Augmentation**: This is a well-known and long-tested technique in ML. Synthesize additional patterns using some scheme and add them to the training data set to ensure that the model is trained on a bigger data set. Some possible ways are:

(a) Make minor and meaningful translations and rotations of the data to synthesize new patterns. This is easy to do in the case of images and speech patterns. In the case of text and other data sources, one has to use a legal operation to synthesize novel data. We have seen such a scheme based on bootstrapping in the context of nearest neighbors.

(b) Use a neural network model to synthesize additional patterns or learn the distribution with the help of generative adversarial networks. We will study these models later in this section.

- **Dropout**: Consider a simple MLP with two hidden layers, as shown in Fig. 9.21.

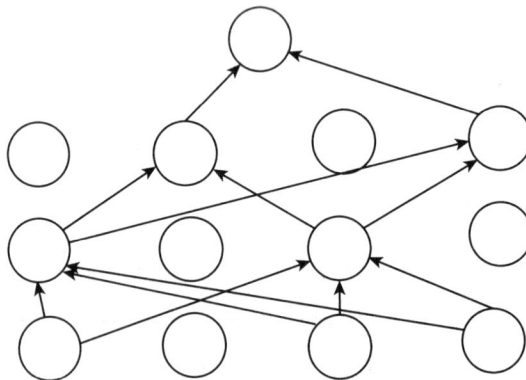

FIG. 9.21 Dropout

The MLP has 4 input nodes, 1 output node and 8 hidden nodes, four from each hidden layer. In the fully connected version, there will be 16 weights from the 4 input nodes to 4 first hidden layer nodes; there will be 16 weights from the 4 first hidden layer nodes to 4 second hidden layer nodes; and finally there will be 4 weights from the 4 hidden nodes to the output node. Initially all these weight values are chosen appropriately.

In dropout we randomly drop out some of the visible and hidden nodes. For example, in Fig. 9.21, one input node and four hidden nodes are dropped out. Typically hidden nodes are dropped out with a probability of 0.5 and input nodes are dropped out with a probability of 0.8. When a node is dropped, all its incoming and outgoing connections will be dropped or become inactive as shown in the figure.

It works as follows:

1. Present the first pattern and randomly drop out some of the nodes and their connections (weights). Use only the nodes and weights that are not dropped out in both forward computation and backpropagation to update only the active weights.
2. Now present the second pattern and again randomly drop out nodes and their connections and use forward and backpropagation to adjust only the active weights.
3. Continue this process till the training phase is completed.
4. So, if there are n training patterns and e epochs of training and out of ne dropouts, a node is active m times, then the probability of participation of the node is $p = \frac{m}{ne}$.
5. So, the node is active with probability p during training and during testing it will always be present, but the contributions are multiplied by p, as shown in Fig. 9.22.

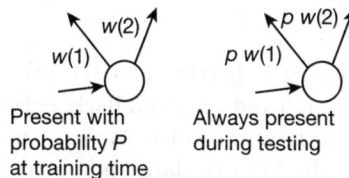

FIG. 9.22 Dropout regularizer

It may be viewed as an ensemble of classifiers without having to explicitly learn each classifier and then realize a combination.

9.3.7 Adding Noise to the Output or Label Smoothing

Label smoothing is a regularization technique that prevents overfitting of a neural network by softening the ground-truth labels. The idea behind label smoothing is to discourage the model to learn that a specific output is guaranteed for some inputs. In other words, it tries to reduce the confidence with which an ML model predicts.

Instead of dealing with hard labels, we can convert them to be soft to some extent. This will not have a significant impact on the performance of the model. The idea is to improve generalization by being soft on the output labels.

Consider for example, when the input is a handwritten digit 9, the training data will have the class label 9. However, it is possible that the digit is not a 9, but a handwritten 7. By softening the labels, we can provide some chance for the pattern to be classified as a 7. For a pattern from class 9, in the hard-label case, we expect 0, 0, 0, 0, 0, 0, 0, 0, 0, 1 at the output layer. Instead when we employ label smoothing, we can specify 0.01, 0.01, 0.01, 0.01, 0.01, 0.01, 0.01, 0.01, 0.01, 0.91 at the output layer. Such a softness permits some mislabelled 9s to be classified properly.

9.3.8 Experimental Results on the MNIST Data Set

We compared the performance of MLP classifier based on the following regularizers:

- L_2 regularizer
- Dropout regularizer
- Label smoothing

We tested performance across different regularization approaches by setting all other parameters as fixed. There are three layers. The input layer is of size 28×28 corresponding to the pixels in each of the MNIST digits.

The first hidden layer has 64 ReLU units and the second hidden layer has 32 ReLU units. Finally the output layer has 10 linear neurons.

We employed a Train : Val : Test split size of 0.63 : 0.07 : 0.3. Batch size is 512 and the network is trained for 20 epochs using cross-entropy loss for all the regularizers. We used the ADAM optimization algorithm with a learning rate of 0.001, the decay rates $dr = 0.9$ and $dr_1 = 0.99$ and ϵ as 1e-08 across all regularization techniques.

We used PyTorch; $nn.Dropout()$ was used for dropout, weight decay was used for L_2 regularization, and for noise in output, label_smoothing was used. The resulting test accuracies were:

- L_2 regularizer: 96.38%
- Dropout regularizer: 96.77%
- Label smoothing: 97%

Note that in this case, label smoothing is better than the other two because there are overlaps between classes of patterns: among 1, 4, 9 and 7 and similarly between 3, 5, 6 and 8. Lable smoothing permits all the overlapping classes to have a chance. We ran the experiment for different number of epochs using label smoothing on the MNIST data and the results are shown in Fig. 9.23. Note that when the number of epochs is 50, we have the maximum accuracy of 97.43%. This indicates that it is good to stop early, that is, at 50 rather than at 55 or 60 epochs.

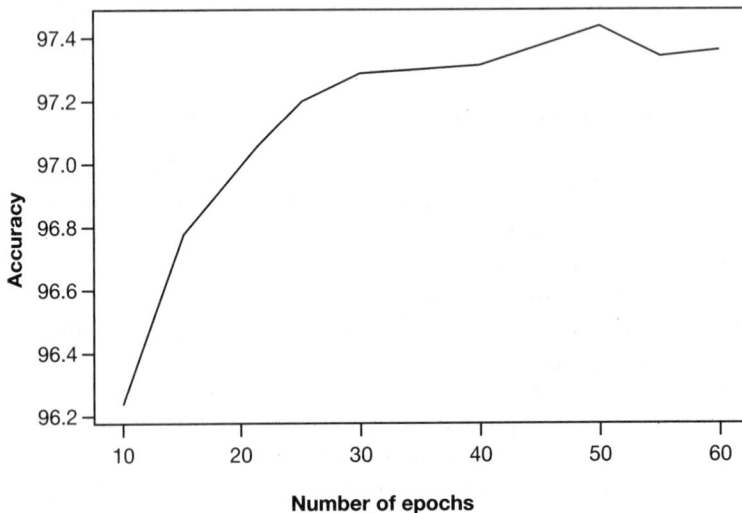

FIG. 9.23 Results for different number of epochs to show the need for early stopping

9.4 CONVOLUTIONAL NEURAL NETWORKS

So far, we have examined the MLP network with different number of layers. Some of the problems associated with MLP are:

- The sigmoid activation function can have a *vanishing or exploding gradient* problem. The gradient can become small and even zero when the weights are small and we encounter saturated neurons. This is the vanishing gradient problem. If we start with larger weights, then the gradient can explode while using backpropagation as it involves a product of many large values.
- The training data can be overfit if the number of hidden layers/neurons is large and the training data is small.
- Routinely backpropagation training was based on software simulations on slower machines. In order to make it work, researchers were using integer-valued weights to run the simulations faster.

We have considered different activation functions, weight initialization schemes and gradient-based optimizations earlier. We will now consider some of the popular dimensionality reduction techniques that pair well with deep neural networks. Specifically, we will discuss convolution and pooling.

9.4.1 Convolution

Convolution is a well-known operation in signal processing with applications in speech signals (one-dimensional) and images (two-dimensional). It is popular because convolution between two time-domain signals, like speech signals, can be realized by taking the product of the two signals in the frequency domain and transforming the resulting signal to time domain. A similar approach is used in the space domain underlying image signals.

Let s be a one-dimensional signal which is represented as a vector of size m. Let the convolution template or kernel K be a smaller size vector of size p. Consider the one-dimensional convolution that is popular in speech. It is shown in Fig. 9.24.

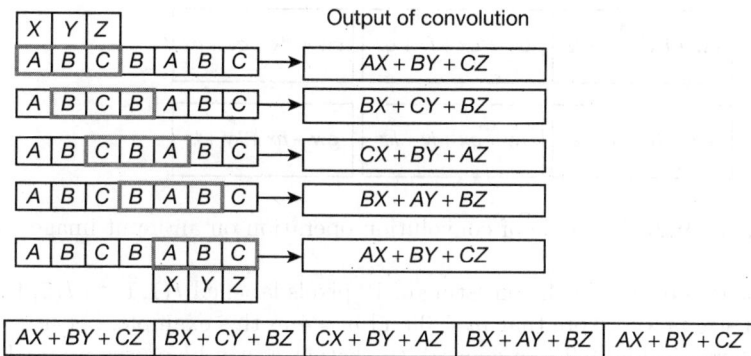

FIG. 9.24 One-dimensional convolution

Here, we have a one-dimensional sequence of data of 7 values. The kernel or filter or template is of size 3, having values X, Y and Z in the three locations. The filter is initially aligned with the first three entries in the one-dimensional sequence data; the three values are A, B, C in the data. We take a dot product between the vectors of size 3, that is, the kernel and the aligned part of the data of size 3. The output of the dot product is $AX + BY + CZ$, as shown in the convolution output at the top-right of the figure. The kernel now is aligned with three items in the data by shifting the kernel horizontally by one location. The output is $BX + CY + BZ$. Note that the kernel is the same but the aligned data entries are different now. This process goes on till the last entry in the kernel is aligned with the last entry in the data.

When the data sequence has 7 entries and the kernel has 3 entries, the possible shifts are 5 and the output for each shift is shown in the figure. In general, if the length of the data sequence is m and the length of the kernel is p, we will have $m - p + 1$ shifts/alignments and there will be one output value for each alignment. The sequence of the 5 output values is shown at the bottom of the figure.

The convolution output in one dimension, C_{OO}, that is, an array of size $(m - p + 1)$, is given by

$$C_{OO}(i) = f\left(\sum_{j=1}^{p} s(i + j - 1)K(j)\right),$$

for $i = 1, 2, \ldots, m - p + 1$.

The role of the kernel is to locate parts of the input s that match with the pattern present in K. Function f may be defined as $f(x) = 1$ if $x > \theta$, else $f(x) = 0$, where θ is a threshold. In general, it can be any activation function.

Let us illustrate the convolution operation in two dimensions using the example in Fig. 9.25.

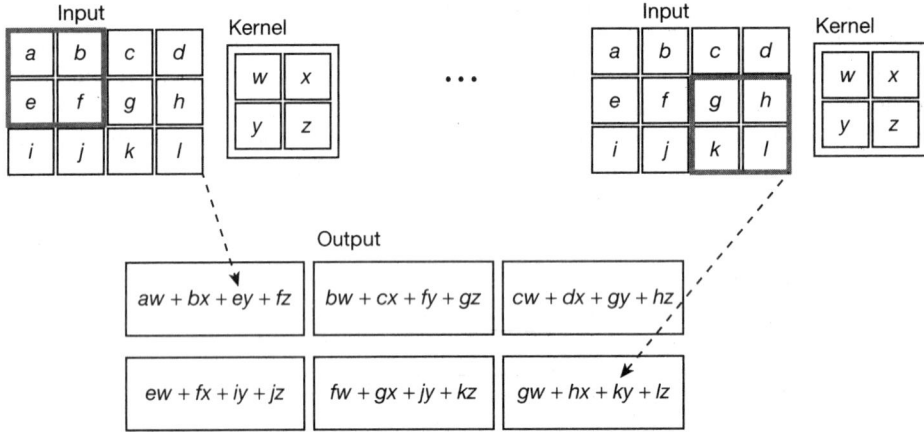

FIG. 9.25 Example of convolution operation on an input image

The input image is of size 3×4, consisting of 12 pixels labelled $I(1, 1)$ to $I(3, 4)$. In general, an image I can be of size $m \times n$. Note that $m = 3$ and $n = 4$ in this example. Let us consider a kernel of size 2×2, as shown in Fig. 9.25. In general, the kernel K can be of size $p \times q$. The convolution output in the two-dimensional case, C_{TO}, is given by

$$C_{TO}(i, j) = \sum_{k=1}^{p} \sum_{l=1}^{q} I(i + k - 1, j + l - 1)K(k, l),$$

for $i = 1, 2, \ldots, m - p + 1$ and $j = 1, 2, \ldots, n - q + 1$.

In the figure, we show two alignments between I and K corresponding to two different positions. They correspond to $C_{TO}(1,1)$ and $C_{TO}(2,3)$. The resulting output image given by C_{TO} is also shown in the figure. These kernels are typically used to capture various edges like horizontal lines or vertical lines in the input. It is popularly known as **mask** in image processing while **kernel/filter** is used in CNNs.

In a convolutional neural network (CNN), we will have multiple convolution layers. For example, we can have different kernels working on the same input image. One kernel may be looking for horizontal edges and the other may be looking for vertical edges, and so on. Each of the resulting outputs, obtained using a kernel on the image, may be viewed as a **feature map**. In a more generic setting, we will have multiple kernels, each looking for the presence of a different kind of feature in the input. It is possible to have more than one occurrence of a feature in the same input image. For example, it is possible for the input image to have a horizontal edge in the top-left portion and also in the bottom-right corner. Similarly, vertical edges occur in multiple parts of the image. So, using the same kernel, we may extract the respective features more than once.

In practice, we may have images that are much larger in size compared to the 3×4 input images shown in Figure 9.25. Also there can be multiple kernels, each looking for one or more occurrences of the feature embedded in it. Correspondingly, there can be several feature maps, one for each kernel. So, each hidden layer is made up of multiple feature maps that may be as many as the number of kernels employed.

In Fig. 9.25, we use a kernel that is specified beforehand. However, in a CNN, these kernels are learnt; learning here means learning the entries in the kernel. The kernel used in the figure is of size 2×2. That means we need to learn 4 weights. A bias may also have to be learnt. So, for each feature map, we need to learn only $pq + 1$ weights, where $p \times q$ is the size of the kernel and the extra 1 is to learn the bias term associated with the node in the hidden layer. This important characteristic of CNNs is called **weight sharing**.

Further, the value of $pq + 1$ is much smaller in practice than the size of the image given by $m \times n$. We have assumed that shifting of the kernel in Fig. 9.25, after each alignment and computing the dot product, is performed by one column horizontally or one row vertically (by one pixel) corresponding to the next alignment; in such a case, the **stride** is 1. We can have strides of length 2 or more in practice. The output of convolving the image I and the kernel K is C_{TO}. This is called a feature map; as we increase the stride, the size of the feature map decreases and calculation of the size is left as an exercise.

9.4.2 Padding Zero Rows and Columns

The feature map can be smaller than the input. However, if we want both the input and the feature map to be of the same size, we employ padding of zero rows and columns.

If we use a zero row at the top and a zero row at the bottom of the image and similarly one column of zeros to the left and another column to the right, then the $m \times n$ image will become $m + 2 \times n + 2$ sized as both the number of rows and columns have increased by 2. If the filter is of size 3×3 and stride is 1, the size of the resulting feature map will be $(m + 2 - 3 + 1) \times (n + 2 - 3 + 1)$, which is $m \times n$. So, the input and the feature map will have the same size after padding zero rows and columns as stated.

It is possible to show that if the image I is of size $m \times n$, if we use a kernel of size $p \times q$ and the stride is s, then the size of the feature map will be $\frac{m-p}{s} + 1 \times \frac{n-q}{s} + 1$. If padding is not used, we get a feature map of size $(m - 3 + 1) \times (n - 3 + 1) = (m - 2) \times (n - 2)$ when we use a filter of size 3×3 on an image of size $m \times n$.

EXAMPLE 2: Let us consider an image, I, of size 6×6 and a kernel, K, of size 3×3. Let

$$I = \begin{bmatrix} a_1 & a_2 & a_3 & a_4 & a_5 & a_6 \\ b_1 & b_2 & b_3 & b_4 & b_5 & b_6 \\ c_1 & c_2 & c_3 & c_4 & c_5 & c_6 \\ d_1 & d_2 & d_3 & d_4 & d_5 & d_6 \\ e_1 & e_2 & e_3 & e_4 & e_5 & e_6 \\ f_1 & f_2 & f_3 & f_4 & f_5 & f_6 \end{bmatrix} \quad \text{and} \quad K = \begin{bmatrix} p_1 & p_2 & p_3 \\ q_1 & q_2 & q_3 \\ r_1 & r_2 & r_3 \end{bmatrix}$$

Then, the feature map will be of the form

$$\begin{bmatrix} u_1 & u_2 & u_3 & u_4 \\ v_1 & v_2 & v_3 & v_4 \\ w_1 & w_2 & w_3 & w_4 \\ x_1 & x_2 & x_3 & x_4 \end{bmatrix}$$

where $u_1 = a_1 p_1 + a_2 p_2 + a_3 p_3 + b_1 q_1 + b_2 q_2 + b_3 q_3 + c_1 r_1 + c_2 r_2 + c_3 r_3$. Similarly, $x_4 = d_4 p_1 + d_5 p_2 + d_6 p_3 + e_4 q_1 + e_5 q_2 + f_4 r_1 + f_5 r_2 + f_6 r_3$. Other values can be obtained in a similar manner by considering the related alignments between I and K.

It is important to note that the resulting feature map is of size 4×4. If we want the feature map to be of size 6×6, that is, the same as the size of I, then we opt for zero padding. The resulting padded version of the image is

$$I_p = \begin{bmatrix} 0 & 0 & 0 & 0 & 0 & 0 & 0 & 0 \\ 0 & a_1 & a_2 & a_3 & a_4 & a_5 & a_6 & 0 \\ 0 & b_1 & b_2 & b_3 & b_4 & b_5 & b_6 & 0 \\ 0 & c_1 & c_2 & c_3 & c_4 & c_5 & c_6 & 0 \\ 0 & d_1 & d_2 & d_3 & d_4 & d_5 & d_6 & 0 \\ 0 & e_1 & e_2 & e_3 & e_4 & e_5 & e_6 & 0 \\ 0 & f_1 & f_2 & f_3 & f_4 & f_5 & f_6 & 0 \\ 0 & 0 & 0 & 0 & 0 & 0 & 0 & 0 \end{bmatrix}$$

The resulting feature map is

$$\begin{bmatrix} t_0 & t_1 & t_2 & t_3 & t_4 & t_5 \\ u_0 & u_1 & u_2 & u_3 & u_4 & u_5 \\ v_0 & v_1 & v_2 & v_3 & v_4 & v_5 \\ w_0 & w_1 & w_2 & w_3 & w_4 & w_5 \\ x_0 & x_1 & x_2 & x_3 & x_4 & x_5 \\ y_0 & y_1 & y_2 & y_4 & y_5 & y_6 \end{bmatrix},$$

where the values of $u_1, u_2, u_3, u_4, v_1, v_2, v_3, v_4, w_1, w_2, w_3, w_4, x_1, x_2, x_3, x_4$ will be the same as in the case with no padding. $t_0 = a_1 q_2 + a_2 q_3 + b_1 r_2 + b_2 r_3$. Similarly, one can work out the values of the other items. We will consider it as an exercise.

We can have convolution layers as hidden layers. Each convolution layer has input and output feature maps/hidden layers as shown in Fig. 9.26. We assume that each kernel moves across height (m) and width (n) of the input and the kernel has depth l. Such an entity is called a **tensor**.

Moving of the kernel is performed along height and width only. So, the output layer for a kernel will have size $m \times n$, that is, the same size as its input in height and width, and its depth will be 1. This forms a feature map. The corresponding output of convolution using o kernels will be a hidden layer with o feature maps corresponding to o kernels, as shown in the figure. So, the hidden layer or the output of convolution will have $o \times m \times n$ neurons.

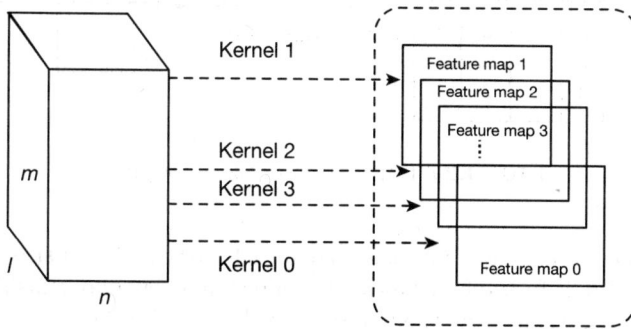

FIG. 9.26 Convolution layer with multiple feature maps in the hidden layer

The role of convolution is to reduce the number of trainable weights. We will examine it in an example.

EXAMPLE 3: Consider an input image of size 500×500 pixels. The total number of pixels is 250,000. So the input layer is of size 500×500. If we use 64 kernels of size 5×5 to get 64 feature maps, each of size 100×100 (stride $= 5$), then for each kernel, we require $5 \times 5 = 25$ weights; so, for 64 kernels we need to learn $64 \times 25 = 1600$ weights.

Instead if we use an MLP that is fully connected with one hidden layer of size 100×100, then the number of weights to be learnt will be $500 \times 500 \times 10^4 = 25 \times 10^{10}$ weights, ignoring any bias values.

In this example, the ratio of the number of weights to be learnt by MLP versus the number of weights to be learnt using convolution kernels is more than 10^8. This reduction occurs because of *weight sharing* in the case of convolution.

If there are N_i neurons in layer i and there are N_{i+1} neurons in layer $i+1$ in a network, then in the case of MLP, we need $N_{i+1} \times N_i$ weights because between a pair of neurons, one from the ith layer and the other from the $(i+1)^{th}$ layer, we require a connection and there are $N_{i+1} \times N_i$ such pairs. However, in the case of a kernel-based convolution, for each feature map, we have the same set of 5×5 kernel weights; note that the same kernel is used across the entire i^{th} layer.

9.4.3   Pooling to Reduce Dimensionality

In a CNN, there will be more than one convolution layer. Typically, after each convolution layer, there will be a pooling layer to further reduce the dimensionality. A pooling layer is obtained from the feature maps in the output of the previous convolution layer.

Let the size of each feature map be $r \times s$; so, the number of neurons in a feature map is rs. Pooling is performed by using a window of size $k \times k$, where k divides both r and s.

This is done by considering $k \times k$ neurons at a time in the feature map; the window is moved horizontally and vertically in a non-overlapping manner. In each window of k^2 neurons, the respective k^2 outputs are pooled to output one value that is stored in the corresponding location of the pooling layer. So, there is a reduction from k^2 to one using pooling, leading to a large reduction in dimensionality. The pooling operation may be explained using Fig. 9.27.

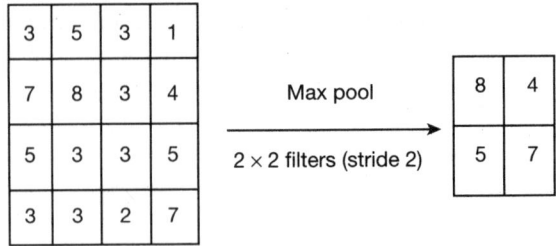

FIG. 9.27 Illustration of max pooling

Here, the input image is of size 4×4. It is split into 4 blocks, each of size 2×2 to suit the size of the window, that is, 2×2. In the first block, the values in a row-major fashion are 3, 5, 7 and 8 and by using max pooling, we get 8 as the maximum value which is stored in the top-left location of the output, as shown in the figure.

Similarly, the maximum value in the top-right block in the figure is 4 (out of 3, 1, 3, 4) which is stored in the top-right location of the output. The remaining two values at the bottom of the output are 5 (left) and 7(right). So, the example feature map of size 4×4 is reduced to 2×2 by using max pooling.

Max pooling is often used in CNNs to identify the most distinctive features of an object, such as its edges and corners. Instead of max pooling, one can use average pooling, as shown in Fig. 9.28. Average pooling is used for tasks such as image segmentation and object detection, where a more fine-grained representation of the input is required.

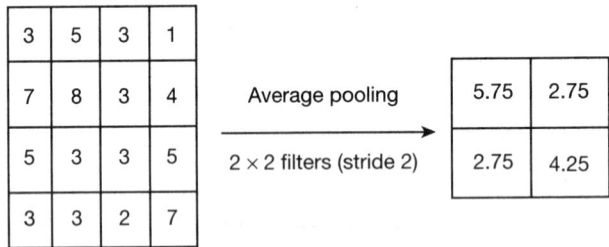

FIG. 9.28 Illustration of average pooling

The architecture of the CNN will consist of several convolution layers; after each convolution layer, there will be a pooling layer with the output of the feature maps in the layer forming the input to the pooling layer. The output of the pooling layer will be the input of the next convolution layer.

Typically, the final output layer of the CNN will be a fully connected layer that is connected to all the neurons in the previous layer. The reason is that we have a small number of units to deal with; so, we can afford to use a fully connected layer.

EXAMPLE 4: An example CNN is shown in Fig. 9.29.

The input to the CNN is a handwritten character of size 32×32. There are two convolution layers, each followed by a pooling layer. Here, for the first convolution layer, $S = 1$ means the stride is 1; $F = 5$ means the filter size is 5×5; $K = 10$ means there are 10 kernels and correspondingly 10 feature maps; $P = 0$ means there is no padding.

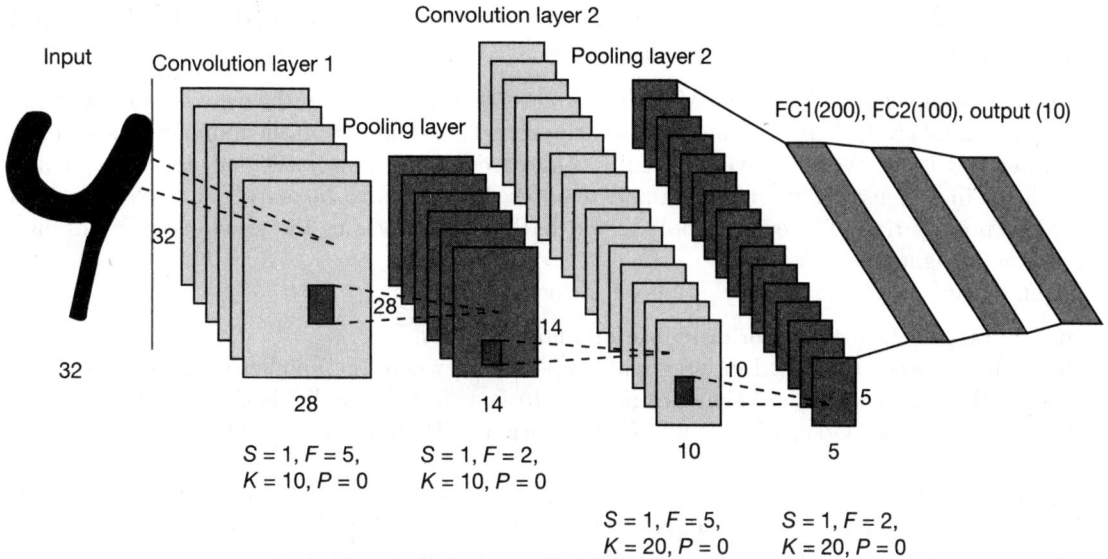

FIG. 9.29 Convolutional network with two convolution layers

The output of the second pooling layer is flattened and is connected to a fully connected layer of 200 neurons. The output of this fully connected layer is connected to another fully connected layer of 100 neurons. The output layer has 10 neurons as the number of classes/handwritten digits is 10.

We can work out the number of weights to be learnt in each layer. In the first convolution layer, there are 10 kernels, each of size 5×5. So, the number of weights to be learnt is $10 \times 5 \times 5 = 250$. It is as if we are convolving with a kernel that is a tensor of size $10 \times 5 \times 5$. The size of each feature map is 28×28 as we applied a $10 \times 5 \times 5$ filter on a 32×32 pixel image. We are not counting biases here.

Next, we have the first pooling layer where the filter size 2×2. So, each feature map of size 28×28 is scaled down to 14×14 as every 2×2 window results in a single value using max pooling. There are no weights to be learnt between the convolution and pooling layers. In the second convolution layer, we have 20 kernels and as many feature maps learnt from the 10 pooled layer outputs. So, the number of weights to be learnt per kernel is $10 \times 5 \times 5 = 250$. We have 20 kernels, so we require $20 \times 250 = 5000$ weights.

Each feature map is of size 10×10 as we had applied a filter of size $5 \times 5 \times 10$ on an input of size $10 \times 14 \times 14$. Here also, we are not counting the biases. The second pooling layer will not require any weights to be learnt. Its output will have 10 reduced representations each of size 5×5; this scaling occurs because of the max pooling window of size 2×2.

The output of the second pooling layer is input to a fully connected layer. Now the number of outputs of this pooling layer will be $20 \times 5 \times 5 = 500$. These 500 units are fully connected to a layer having 200 units. This requires 500×200 weights and 200 biases to be learnt. So, the total number weights to be learnt is 100, 200. The outputs of the 200 units will be fully connected to a layer having 100 neurons. This layer requires 200×100 weights and 100 biases, adding up to 20,100 weights to be learnt.

The output of the 100 units is connected to 10 output neurons requiring 1000 weights and 10 biases that adds up to 1010 weights to be learnt. Typically, the final layers will be fully connected as we can afford to have fully connected layers on the reduced dimensional data that is obtained after a sequence of convolution and pooling layers.

A CNN is implemented using an MLP. The main difference is that in the convolution layer, the number of weights to be learnt is dependent on the kernel size and not on the number of neurons in the subsequent layer. We can have some fully connected layers towards the end. Further, pooling layers helps in dimensionality reduction and requires no weights to be learnt. A CNN is trained using backpropagation. The error is propagated back from a layer to the previous layer through the relevant weights.

Some important properties of CNN are given below:

- It is a state-of-the-art tool for classification and prediction.
- It has been successfully used in large-scale applications where the number of training patterns and/or the dimensionality of the data is large. In fact, it works well when the training data is large. For example, GoogLeNet is a CNN that began with 22 layers and which has now grown to 152 layers.
- It became popular because of its applications in image processing, speech processing and natural language processing. One of the important outcomes is a variant that has become popular in network applications in the form of **graph convolutional net (GCN)**.
- For handwritten digit recognition, specifically for the MNIST data classification, a CNN architecture called LeNet-5 is used. It was proposed by Yann LeCun and it has five layers of trainable parameters. It is a simple CNN used extensively in MNIST data classification. It is shown in Fig. 9.30.

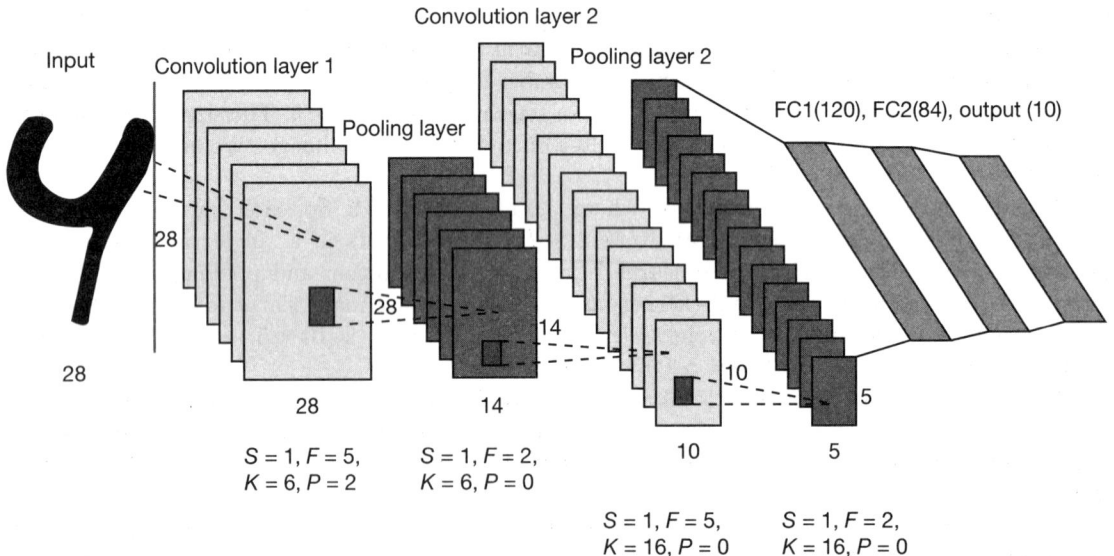

FIG. 9.30 Convolutional network: LeNet-5

- The results on the MNIST data set using LeNet-5 are:
  - Validation Accuracy: 98.75% (based on 5-fold cross validation)
  - Test Accuracy: 98.88%

9.5   RECURRENT NEURAL NETWORKS

In MLP and convolutional neural networks, the size of the input is fixed. For example, in MNIST digit classification, images of size 28×28 of the handwritten digits is given as input for classification of the digit images. Similarly, the number of outputs is also fixed. For a digit classification problem, we typically have 10 output nodes. Further, there is no dependence between two successive inputs presented at the input. However, there are several applications where we need a different set of requirements.

EXAMPLE 5:   Let us consider a simple network, as shown in Fig. 9.31.

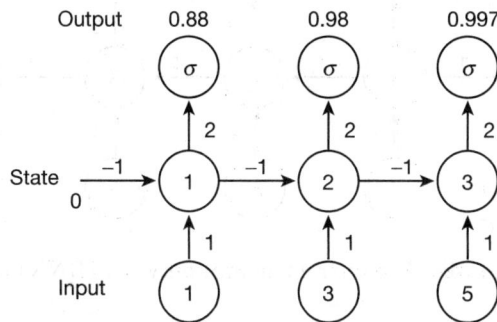

FIG. 9.31 A simple recurrent neural network

There is an input sequence of length 3; it is 1, 3, 5. There is a corresponding output sequence. The output sequence for the input sequence $1, 3, 5$ is $0.88, 0.98, 0.997$. We will see how it is computed shortly. There is an additional sequence called *state sequence* shown between the input and output. It is hidden.

Let us look at the first input $i(1)$, state $h(1)$ and output $o(1)$. The input is $i(1) = 1$ and there is a weight of value 1 between $i(p)$ and $h(p)$ for $p = 1, 2, 3$. We assume linear activation at these nodes. So, contribution of $i(1)$ to $h(1)$ is $1 \times 1 = 1$. The nodes $h(p)$ and $h(p+1)$ are linked by an edge with weight -1 for $p = 0, 1, 2$. Let the state at time 0, $h(0) = 0$; this is the initial state. It could be different from 0 in general. So, $h(1) = h(0) * (-1) + 1 * 1 = 1$. There is a weight of value 2 between $h(p)$ and $o(p)$ for $p = 1, 2, 3$. So, the activation input to unit $o(1)$ is $h(1) \times 2 = 2$.

A sigmoid activation is used at $o(1)$ to get $o(1) = \sigma(2) = \frac{1}{1+e^{-2}} = 0.88$ which is shown in the figure. Note that $o(2)$ can be computed by knowing $h(2)$ and multiplying it with weight 2 that is present between $h(2)$ and $o(2)$. $h(2) = h(1) \times (-1) + i(2) \times 1 = 1 * (-1) + 3 * 1 = 2$.

The activation input to the node representing the second output $o(2)$ is $h(2) * 2 = 4$. So, $o(2) = \sigma(4) = \frac{1}{1+e^{-4}} \approx 0.98$. Similarly, it is possible to show that $h(3) = 3$ for $i(3) = 5$ and $o(3) \approx 0.997$.

Some important properties of this network are given below:

* Note that the function executed at each time step is the same. The weight between the input and hidden state nodes is the same across the time steps. Similarly, the weight between the hidden state and the output is the same and the weight between two successive states is also the same and it is -1.
* Dependence is captured through the hidden states. For example, the state $h(p)$ depends upon $h(p-1)$ which in turn depends upon $h(p-2)$, and so on. So, dependence between states $h(p)$

and $h(1)$ is captured through states $h(2), h(3), \ldots, h(p-1)$. There is recursion and that is why the network is called a recurrent neural network.

- The number of inputs and outputs could vary. In Fig. 9.31, we have seen an input sequence of size 3 and correspondingly the output sequence is also of length 3. We can easily extend it to length 4 by adding one more step, as shown in Fig. 9.32.
- The fourth input is 8, the state is 5 and the output is 0.9999.

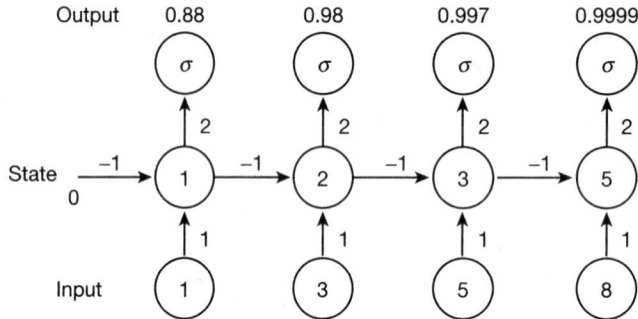

FIG. 9.32 A simple recurrent neural network (RNN) of length 4

Due to this flexibility, we can represent this network as shown in Fig. 9.33 in a compact manner.

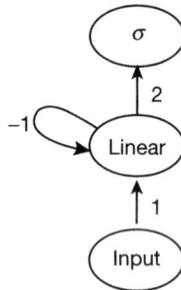

FIG. 9.33 A compact representation of the RNN

We want the same function and the same computation to take place at every time step. So, we have the same generic structure, as shown in Fig. 9.34, at each step with the same weights shared across steps. Specifically:

$h_i = f(ux_i + wh_{i-1} + b)$

$o_i = g(vh_i + c)$, where b and c are bias terms and weight vectors.

Further, f and g are activation functions at the hidden state and output nodes; b and c are bias terms; u, v and w are the appropriate weight vectors; and x_i, h_i and o_i are the input, state and output at time step i.

We need to learn w, u, v, b and c which are used across time steps. In the simple scenario we have considered in Example 5, w, u and v are assumed to be scalars. However, in most of the practical problems, they can be real vectors.

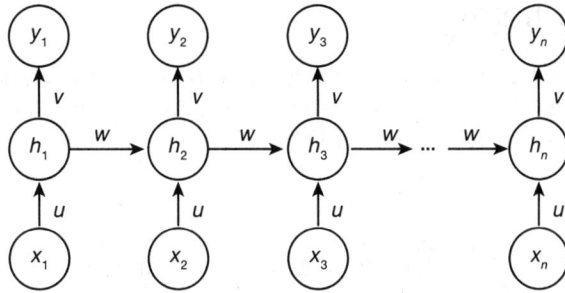

FIG. 9.34 Computation at each step is the same

The input vectors x_i are l-dimensional real. The state vectors h_i are k-dimensional real. The output vectors y_i are C-dimensional real. u is a $k \times l$ matrix. It is a linear transform that maps an l-dimensional vector to a k-dimensional vector.

w is a $k \times k$ matrix. It maps a state vector of size k to another state vector of size k. v is a matrix of size $C \times k$. It maps a state vector of size k to the output vector of size C.

There are several applications where the inputs are sequences or we have time series data. Recurrent neural networks are ideally suited to deal with such data.

Auto completion of a word: Here, a word is a sequence of characters. The problem is to predict the next character in the sequence based on the current and previous characters. In the case of the word *EAT*, when we input E, we will get the next character A as the output of the first element. Similarly, it has to predict the next character in the word and then stop, as shown in Fig. 9.35.

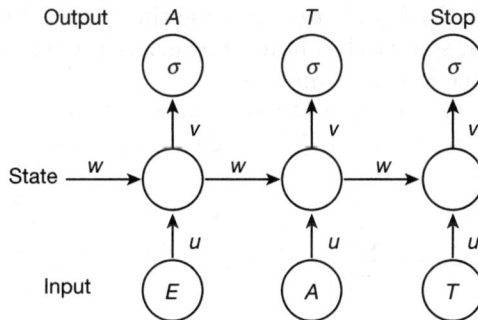

FIG. 9.35 Auto completion of successive characters in a word

Other important problems are weather prediction and stock market prediction. In weather prediction, we are given rainfall on different days and the problem is to predict rainfall on a day in the future. Here, input is a day and output is the amount of rainfall. Similarly, in stock market prediction, the input is a day and the output is the price of a share and the problem is to predict the stock price in the future.

Part of Speech (POS) tagging is an important activity in text analysis. Here, for each word in the input text, we would like to predict whether it is a noun, verb, adjective, etc. So, the input is a word and the output is the tag.

There are many other important sequence-based prediction problems.

9.5.1 Training an RNN

Training an RNN involves backpropagation of the error based on the chosen loss function. Due to its recursive nature, the gradient of the loss with respect to w requires some care, as shown next. Consider the word completion task shown in Fig. 9.36.

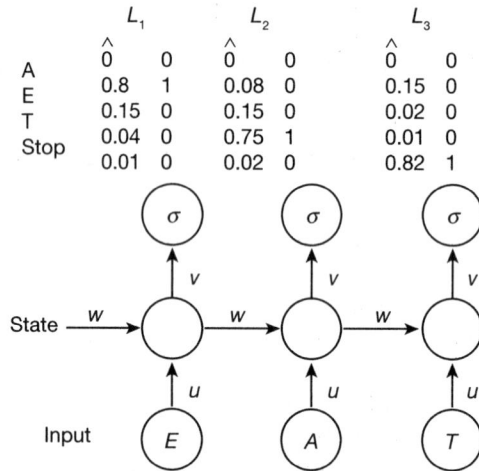

FIG. 9.36 Loss for auto completion

Note that o is the target output and $\hat{o}$ is the predicted output, assuming that we have only four characters in the alphabet, A, E, T and *stop*. At some time step i, let the predicted probabilities (soft max outputs) for the four symbols be obtained by $\hat{o}_i$ and the true or target output be specified by o_i. The overall loss is a function of o_is and $\hat{o}_i$s.

If the number of possible outputs at each time instance, like in auto completion or classification is small, we can use cross-entropy loss. If $\mathcal{L}_1$, $\mathcal{L}_2$, $\mathcal{L}_3$ are the losses at time steps 1, 2 and 3, the cross entropy loss is given by

$$\mathcal{L}_i = -\sum_{j=1}^{4} o_i(j) log\hat{o}_i(j), \ for \ i = 1, 2, 3$$

Note that only for one value of j, $o_i(j) = 1$; it is 0 for other values.

The total loss $\mathcal{L}$ is the sum of the losses at time steps 1, 2 and 3. So,

$$\mathcal{L} = \mathcal{L}_1 + \mathcal{L}_2 + \mathcal{L}_3$$

It is difficult to compute the gradient with respect to w because of inter-dependencies.

Let us consider the example in Fig. 9.36. Let us consider the gradient of $\mathcal{L}_3$ with respect to w. Using chain rule, we have

$$\frac{\partial \mathcal{L}_3}{\partial w} = \frac{\partial \mathcal{L}_3}{\partial h_3} \times \frac{\partial h_3}{\partial w}$$

The first partial derivative in the RHS is easy to compute. The second component offers some challenge as $h_3 = f(ux_3 + wh_2 + b)$; h_3 depends on h_2 through w. Further, h_2 is not a constant with respect to w. So,

$$\frac{\partial h_3}{\partial w} = \frac{\partial^s h_3}{\partial w} + \frac{\partial h_3}{\partial h_2} \times \frac{\partial h_2}{\partial w} = \frac{\partial^s h_3}{\partial w} + \frac{\partial h_3}{\partial h_2} \times \left(\frac{\partial^s h_2}{\partial w} + \frac{\partial h_2}{\partial h_1} \times \frac{\partial h_1}{\partial w} \right)$$

$$= \frac{\partial^s h_3}{\partial w} + \frac{\partial h_3}{\partial h_2} \times \left(\frac{\partial^s h_2}{\partial w} + \frac{\partial h_2}{\partial h_1} \times \left(\frac{\partial^s h_1}{\partial w} + \frac{\partial h_1}{\partial h_0} \times \frac{\partial h_0}{\partial w} \right) \right)$$

$$= \frac{\partial^s h_3}{\partial w} + \frac{\partial h_3}{\partial h_2} \times \left(\frac{\partial^s h_2}{\partial w} + \frac{\partial h_2}{\partial h_1} \times \frac{\partial^s h_1}{\partial w} \right),$$

where $\frac{\partial^s h_i}{\partial w}$ means take the partial derivative of h_i with respect to w assuming contributions of $h_{i-1}, h_{i-2}, \ldots, h_o$ to be constant, that is, viewing them as independent of w. The dependency is captured using the additional term as shown. This is called **backpropagation through time**.

Exploding and Vanishing Gradients

Let us consider the computation of $\frac{\partial h_k}{\partial h_1}$ and see how it can lead to a vanishing or exploding gradient.

$$\frac{\partial h_k}{\partial h_1} = \frac{\partial h_k}{\partial h_{k-1}} \times \frac{\partial h_{k-1}}{\partial h_{k-2}} \cdots \times \frac{\partial h_2}{\partial h_1}$$

Let us consider one term in the product, $\frac{\partial h_p}{\partial h_{p-1}}$. We know the activation $a_p = wh_{p-1}$ ignoring the bias term and $h_p = \sigma(a_p)$ assuming the activation function to be σ. Note that $\frac{\partial a_p}{\partial h_{p-1}} = w$. So, $\frac{\partial h_p}{\partial h_{p-1}} = \frac{\partial h_p}{\partial a_p} \times \frac{\partial a_p}{\partial h_{p-1}} = \frac{\partial h_p}{\partial a_p} \times w$.

Note that $a_p = (a_p(1), a_P(2), \ldots, a_p(l))$ and $h_p = (\sigma(a_p(1)), \sigma(a_P(2)), \ldots, \sigma(a_p(l)))$. So, $\frac{\partial h_p}{\partial a_p}$ is a matrix of size $l \times l$ where all the off-diagonal entries are 0. For example, the off-diagonal entry for row 1 and column 2 will be $\frac{\partial h_p(2)}{\partial a_p(1)} = 0$. So, $\frac{\partial h_p}{\partial a_p} = diag(\sigma'(a_p))$.

Now considering the norm of $\frac{\partial h_p}{\partial h_{p-1}} = $ norm of $\frac{\partial h_p}{\partial a_p} \times w = diag(\sigma'(a_p))w$. We know that $\sigma'(a_p) \leq \frac{1}{4}$ if we use sigmoid activation and it is ≤ 1 if it is tanh activation. So, $\sigma'(a_p) \leq c$, where $c = \frac{1}{4}$ or 1.

We know that the norm of $\frac{\partial h_p}{\partial h_{p-1}} \leq $ norm of $\frac{\partial h_p}{\partial a_p} \times w = diag(\sigma'(a_p))w$ using the Cauchy–Schwarz inequality. So, norm of $\frac{\partial h_p}{\partial h_{p-1}} \leq c \times norm(w)$ for $p = 2, \ldots, k$. So, norm of $\frac{\partial h_k}{\partial h_1} \leq \prod_{i=1}^{k} c\alpha = (c\alpha)^k$. So, if $c\alpha < 1$, then $\frac{\partial h_k}{\partial h_1}$ vanishes. If $c\alpha > 1$, then $\frac{\partial h_k}{\partial h_1}$ explodes.

One simple solution to this problem is to use **truncated backpropagation** where a fixed number of terms are used in the product.

Long Short-Term Memory (LSTM)

If the length of the sequence is large, there could be problems such as those mentioned below:

- Dependence between inputs is captured in RNNs through the hidden states. They capture, in some sense, the input history.
- Consider state h_k; it depends on h_{k-1}. As the value of k increases, the information at time step 1 or 2 will be completely erased from the state at k. This is due to information overload.

- In LSTM, we try to solve this problem by exercising additional control with the help of gates. There are several variants of LSTM; a popular version is gated recurrent unit (GRU).
- In GRU, there are two gates to control the information overload:

1. **Update Gate(z):** It determines how much of each component of the hidden state needs to be passed along into the future. It is given by

$$z_k = \sigma(u_z x_k + w_z h_{k-1})$$

So, we need to learn additional parameters w_z and u_z.

2. **Reset Gate(r):** It determines how much of the past knowledge to forget. It is given by $r_k = \sigma(u_r x_k + w_r h_{k-1})$. From this, we get $\tilde{h}_k = tanh(ux_k + r_k \odot wh_{k-1})$, where $\odot$ is the element-wise product. Using z_k and r_k, we get $h_k = z_k \odot h_{k-1} + (1 - z_k) \odot \tilde{h}_k$ from h_{k-1}.

A good property of GRUs is that they can deal with the vanishing gradients. However, it is possible that they face the problem of exploding gradients. This can be solved by **gradient clipping** or normalizing the gradients appropriately; that is, the direction of the gradient is preserved but the magnitude is controlled.

EXAMPLE 6: Let $h_{k-1} = \begin{pmatrix} 0.8 \\ 0 \\ 0.6 \\ 0 \\ 0 \\ 0.4 \end{pmatrix}$. Let w_z, u_z, w_r, u_r be zero vectors/matrices of appropriate sizes.

Let x_k be a zero vector. So, $z_k = \sigma(0, 0, 0, 0, 0, 0)^t = (0.5, 0.5, 0.5, 0.5, 0.5, 0.5)^t$
Similarly, $r_k = (0.5, 0.5, 0.5, 0.5, 0.5, 0.5)^t$ as u_r and w_r are zero matrices.

We can compute $\tilde{h}_k = \begin{pmatrix} 0.38 \\ 0 \\ 0.29 \\ 0 \\ 0 \\ 0.197 \end{pmatrix}$ So, $h_k = \begin{pmatrix} 0.4 \\ 0 \\ 0.3 \\ 0 \\ 0 \\ 0.2 \end{pmatrix} + \begin{pmatrix} 0.19 \\ 0 \\ 0.145 \\ 0 \\ 0 \\ 0.0985 \end{pmatrix} = \begin{pmatrix} 0.59 \\ 0 \\ 0.435 \\ 0 \\ 0 \\ 0.2985 \end{pmatrix}$

So, even if x_k is a zero vector, we can get a non-zero h_k.

9.5.2 Encoder–Decoder Models

There are several practical applications where two models are used to solve the problem. One model takes the input and returns a vector as output. This is the *encoding* part. Another model takes the output of the encoder and generates the output; this model is called the *decoder*. The output of the encoder is the input to the decoder.
The applications that exploit encoder–decoder models are given below:

- *Image captioning or annotation:* Given an image as input, the output is text describing objects in the image. Here, a CNN can be the encoder and an RNN the decoder.
- *Speech to text:* Given speech as input, the output is text corresponding to speech utterance. Here, an RNN is the encoder and another RNN is the decoder.
- *Machine translation:* Here, input is a sentence in some source language and the output is the translated text in the target language. Here, RNNs can be used in both encoding and decoding.

- *Image question answering*: Here, an image and a question in textual form is the input and the output is the answer in textual form. In this case, the encoder will have both a CNN and an RNN and the decoder will have an RNN.
- *Document summary generation*: Here, a document is the input and an abstract or summary of the document is the output. In this case, we can have RNNs for both encoding and decoding.

We now deal with a practical application of the encoder–decoder model. We use Australian weather data (from Kaggle). We use only a subset of columns and rows. We do not consider columns that have NaN values and consider only the location *Albury*. Month-wise average temperature values are predicted using a GRU encoder and decoder pair.

Data from January 2014 to September 2015, that is, for 21 months, is taken for training and the remaining data for testing. The true values and predicted values of the average temperature are shown in Fig. 9.37.

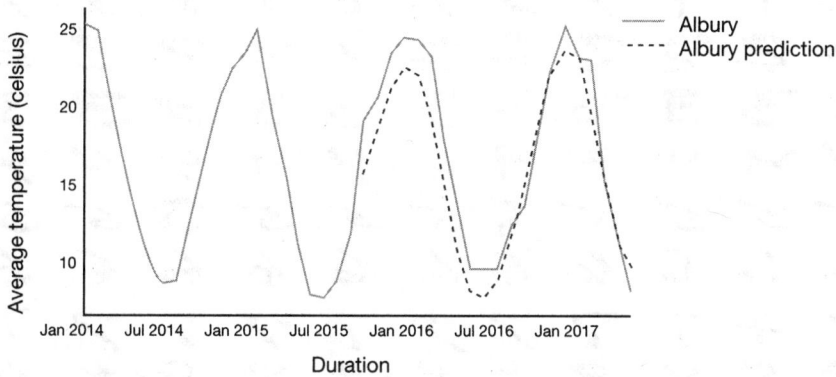

FIG. 9.37 Comparing predicted values with the ground truth (for colour figure, please see Colour Plate 2)

It is a regression problem. So, squared error loss is used. The loss values are plotted in Fig. 9.38.

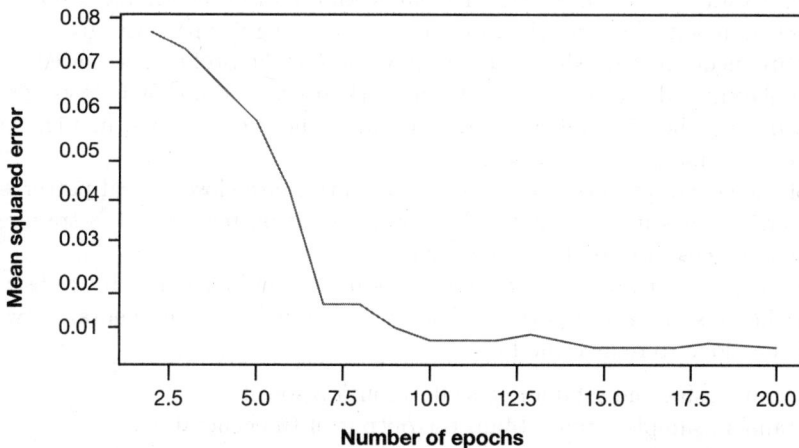

FIG. 9.38 Loss values

9.6 GENERATIVE ADVERSARIAL NETWORKS

We have seen, in the earlier chapters, some mechanisms to synthesize new patterns. We can generate new patterns using the centroid of the k nearest neighbors of each of the patterns. We can learn the underlying probability structure using the training data and then generate new patterns.

Suppose our interest is in sampling from the distribution and we are not particular about the form of the density function $P(x)$. Then we can generate new patterns. Suppose we are given some MNIST digit data for training, as shown in Fig. 9.39. We want to generate more images from this unknown distribution. It is possible that the unknown distribution is complex.

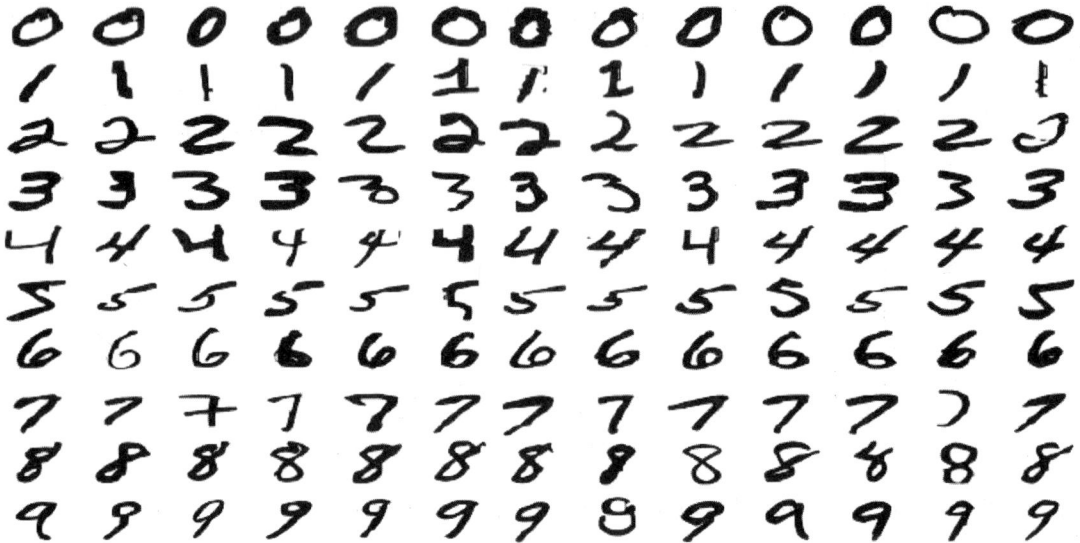

FIG. 9.39 Training data set: MNIST digits

A solution is to start with a sample from a simple known distribution and keep making changes to it, so that we end up with patterns that are difficult to distinguish from the given training images. These changes are incorporated using a neural network and the problem is viewed as a two-player game. The two players and the training of the network are carried out as follows: The two players are the generator and the discriminator. The overall structure is shown in Fig. 9.40. Both the generator and discriminator keep learning.

The role of the generator is to synthesize patterns that are close to real patterns. The role of the discriminator is to classify accurately whether the input pattern is real (a training pattern) or synthetic (a pattern generated by the generator).

The input to the generator is typically some kind of random noise. It is observed that the distribution of the noise is not important. Changing of weights of the neural network based on backpropagation is done to reduce the loss.

1. Obtain a sample using the random noise distribution specified.
2. Using the random sample input, obtain the output of the generator.
3. Obtain the output of the discriminator using the output of the generator as input.
4. Calculate gradient of the loss and backpropagate through the discriminator and the generator to obtain gradients to change the weights of the generator network.

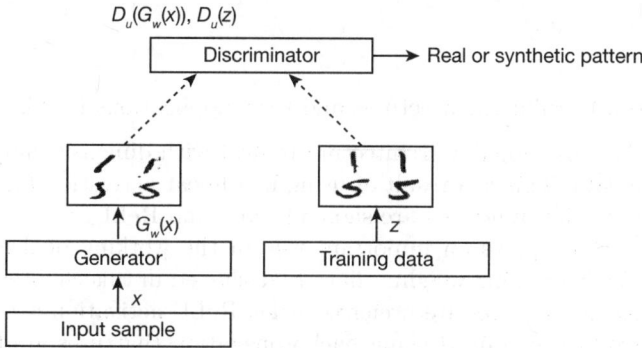

FIG. 9.40 Generative adversarial network

The discriminator and generator are trained for one or two epochs alternatively. The discriminator has two inputs: one from the generator given by $G_w(x)$, where x is the noise sample that is input to the generator and w is the weight matrix associated with the generator neural network, the other which is a sampled training image z.

So, there are two parts to the discriminator output: one is based on the generator's output which is processed by the discriminator to output $D_u(G_w(x))$ and the second part which is the ground truth given by $D_u(z)$, where z is a training pattern and u is the weight matrix of the discriminator network.

If the output of the discriminator is interpreted as the probability that the image input to it is real, then the discriminator would like to minimize $log(D_u(G_w(x)))$ and maximize $log(D_u(z))$. The generator would try to maximize $log(D_u(G_w(x)))$ or equivalently minimize $log(1 - D_u(G_w(x)))$, assuming an output 1 from the discriminator indicates that the input pattern is real and a 0 indicates that the input is synthetic. The combined objective is

$$\min_{w} \ \max_{u} \ [E_z \, log(D_u(z)) + E_x \, log(1 - D_u(G_w(x)))]$$

The first term in this objective is concerned only with the weights (u) of the discriminator network. The second term deals with the weights of both the networks as both w and u are involved. The discriminator would like to assign a 0 to $D_u(G_w(x))$ and thereby maximize the second term, and as an adversary, the generator would try to minimize the second term.

We show the results of the digits generated after 25 epochs in Fig. 9.41.

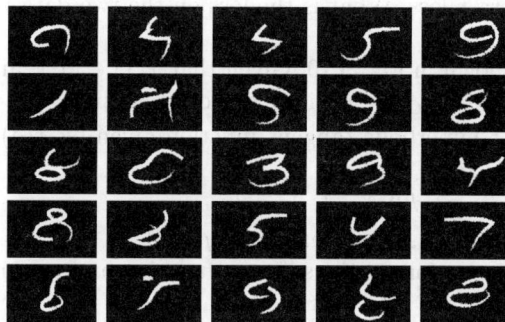

FIG. 9.41 Images generated after 25 epochs

SUMMARY

We have discussed deep learning architectures and their applications in this chapter.

- Autoencoder is the most popular architecture to deal with dimensionality reduction of non-linearly separable data. This is a result of using non-linear activation functions.
- The non-linear activation functions are sigmoid, tanh and ReLU.
- Initialization of weights plays an important role in the working of deep neural networks. Popular schemes for initializing weights that are discussed in this chapter are LeCun, Xavier and He. He and Xavier schemes are preferred when ReLU activation is used.
- Deep neural networks are trained using backpropagation that uses gradient descent (GD). The ADAM optimizer is the most popular scheme for optimization.
- CNN is the most popular architecture for dealing with image data. It employs weight sharing and pooling to reduce the dimensionality. We have considered a practical application of CNN on MNIST handwritten digit data.
- Sequential data can be processed by using recurrent neural networks. We have considered RNNs, LSTM and GRU, in this context and used them in some real-life prediction applications.
- GANs are popular in synthesizing or generating data. We have used GANs in handwritten digit data.
- A major problem with the neural networks is that they are opaque and explanation is a problem. Explainable AI is an active research topic. It is possible to explore how the hidden nodes represent various input patterns.

EXERCISES

1. Consider the gradient in an autoencoder given as

$$\nabla_{w_e}(\mathcal{L}(w)) = \frac{\partial \mathcal{L}(w)}{\partial \hat{x}} \times \frac{\partial \hat{x}}{\partial a} \times \frac{\partial a}{\delta h} \times \frac{\partial h}{\partial a_1} \times \frac{\partial a_1}{\partial w_e}.$$

   Find the expressions for each of the partial derivatives.

2. Consider the autoencoder given in Example 16 in Chapter 8. Let the weights be $w_1 = 1$ and $w_2 = 1$ and let the inputs be $\begin{pmatrix} 1 \\ 2 \end{pmatrix}$ and $\begin{pmatrix} 3 \\ 4 \end{pmatrix}$. Obtain the optimal values of w_3 and w_4 that minimize the squared error loss. What is the loss on the two data points using these optimal w_3 and w_4?

3. Consider the network shown in Fig. 9.10. Given that $x(1) \in [0, 1]$ and $x(2) \in [0, 1]$, show that updates to all of w_{AC} and w_{BC} are positive or both negative.

4. Let $f(a)$ be the tanh or hyperbolic tangent function given by $f(a) = \frac{e^a - e^{-a}}{e^a + e^{-a}}$. Show that its derivative is $f'(a) = 1 - f^2(a)$.

5. Show that $tanh(a) = 2sigmoid(2a) - 1$.

6. Consider the softmax function. Show that $\sum_{i=1}^{C} p_i = 1$.

7. Use the results from the section on the softmax function and assuming the loss to be

$$\sum_{i=1}^{C}(y_i - p_i)^2,$$

where y is the required or target output, and using the partial derivative of loss $\mathcal{L}$ to be

$$\frac{\partial \mathcal{L}}{\partial \hat{y}_s} = \sum_j \frac{\partial \mathcal{L}}{\partial p_j} \times \frac{\partial p_j}{\partial \hat{y}_s},$$

derive the form of $\frac{\partial \mathcal{L}}{\partial \hat{y}_s}$.

8. In the case of momentum-based updates, if $ud_0 = 0$, then show that $ud_{k+1} = \eta(\sum_{i=0}^{k} \alpha^{k-i} \nabla w_{i+1})$.

9. Consider the function $x^2 - log(x)$. Let $x_0 = 1.90377123$. Compute x_1 based on gradient descent on the function using a learning rate of 0.01.

10. Show that m_k in ADAM can be written as $m_k = (1 - dr) \sum_{i=1}^{k} dr^{k-i} \nabla w_i$.

11. Consider weight updates using SGD.

    a. In the case of SGD, the weights are updated using $w_{k+1} = w_k - \eta \nabla w_k$. Show that $w_0 - w_{k+1} = \eta \sum_{i=1}^{k} \nabla w_i$, where w_{k+1} is obtained after $k+1$ SGD-based updates to w_0.

    b. Use the result in (a) and the bound $max_i |\nabla w_i| < \delta$ to show that $|w_0 - w_{k+1}| < k\eta\delta$.

12. Let an image I be of size 9×9 with 81 pixels. If we use a kernel of size 3×3 on it, then what is the size of the resulting feature map if

    a. the stride is 1?

    b. the stride is 2?

    c. the stride is 3?

13. Consider Example 17 from Chapter 8. Obtain the values of $t_2, u_2, v_2, w_2, x_2, y_2$ in terms of entries of I_p and K.

14. Consider the LeNet-5 architecture shown in Fig. 9.30. Obtain the number of weights and biases to be learnt in each layer.

15. Consider the four-time-step RNN given in Fig. 9.32. Now add one more time stamp with input 5. Compute the output.

PRACTICAL EXERCISES

1. Download the MNIST handwritten digit data. There are 10 classes (corresponding to digits $0, 1, \ldots, 9$) and each digit is viewed as an imgae of size 28×28 (= 784) pixels; each pixel having values 0 to 255. There are around 6000 digit training patterns and around 1000 test patterns in each class and the class label is also provided for each of the digits. Visit `http://yann.lecun.com/exdb/mnist/` for more details.

   • **Task**: Build a convolution neural network based on the following:

     – conv1: Convolution layer having 2 feature detectors, with kernel size 3×3, and sigmoid as the activation function, with stride 1 and no padding.

     – Pool1: A max pooling layer with pool size 2×2 with stride 2.

     – conv2: Convolution layer having 2 feature detectors, with kernel size 3×3, and rectified linear unit as the activation function, with stride 1 and no padding.

     – Pool2: A max pooling layer with pool size 2×2 and stride 1.

     – FC1: Fully connected layer with 50 neurons and hyperbolic tangent as the activation function.

     – Output: Output layer containing 10 neurons, softmax as activation function.

- Fit the model with the training data set, with 20 epochs, mini-batch size of 200, and use the cross-entropy loss function.
- Obtain the training and validation accuracies and report.

Bibliography

- Srivastava, N et al. 2014. Dropout: A Simple Way to Prevent Neural Networks from Overfitting, *Journal of Machine Learning Research*, 15: 1929–1958.

- Khapra, M. 2018. Deep Learning, *NPTEL Course*, https://onlinecourses.nptel.ac.in/noc19_cs85/preview.

- Goodfellow, I, Bengio, Y and Courville, A. 2016. *Deep Learning*, MIT Press.

- Chollet, F. 2018. *Deep Learning with Python*, Manning Publications.

Colour Plate 2

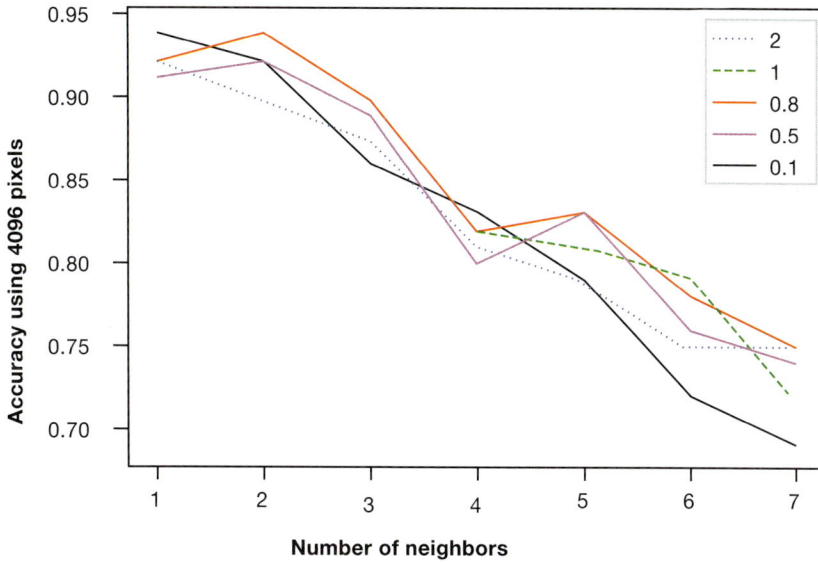

(a)

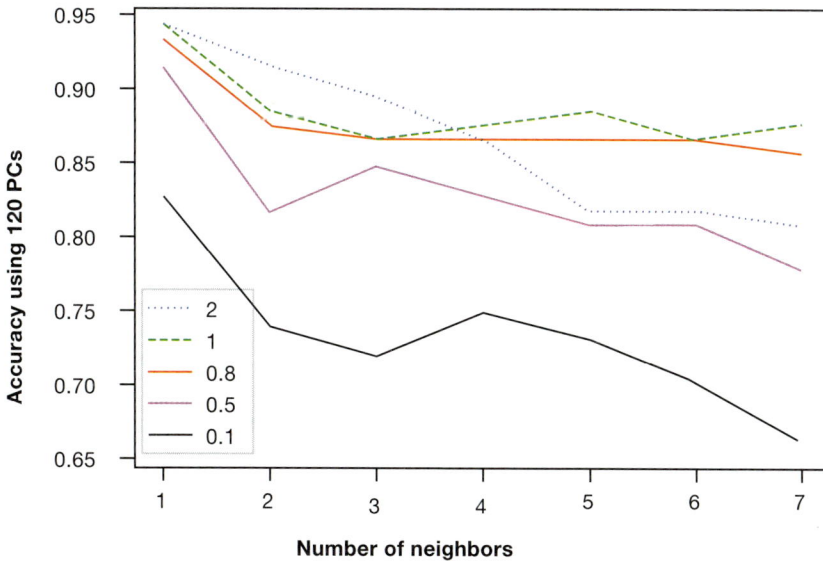

(b)

FIG. 6.3 Results on the Olivetti Face data set

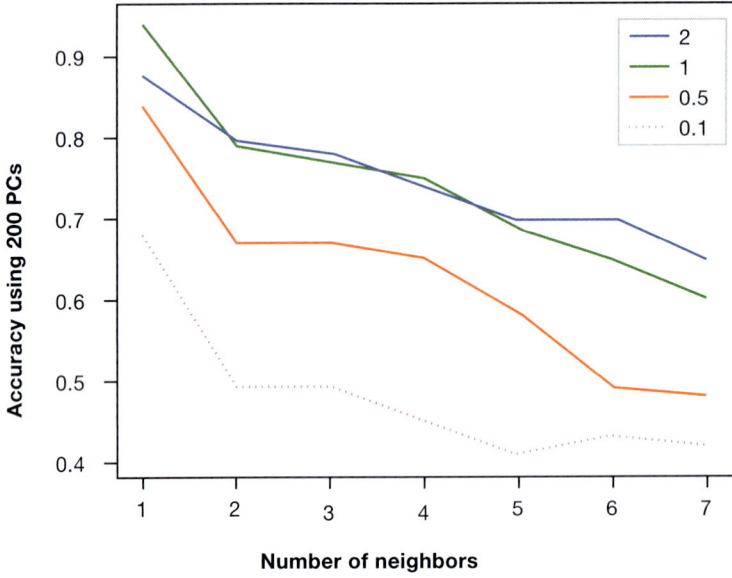

(a)

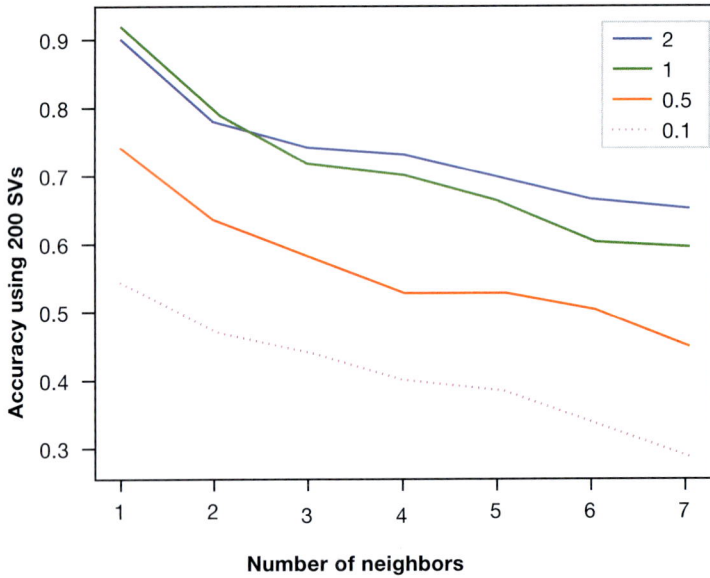

(b)

FIG. 6.4 Results on the Olivetti Face data set: 200 PCs and 200 SVs

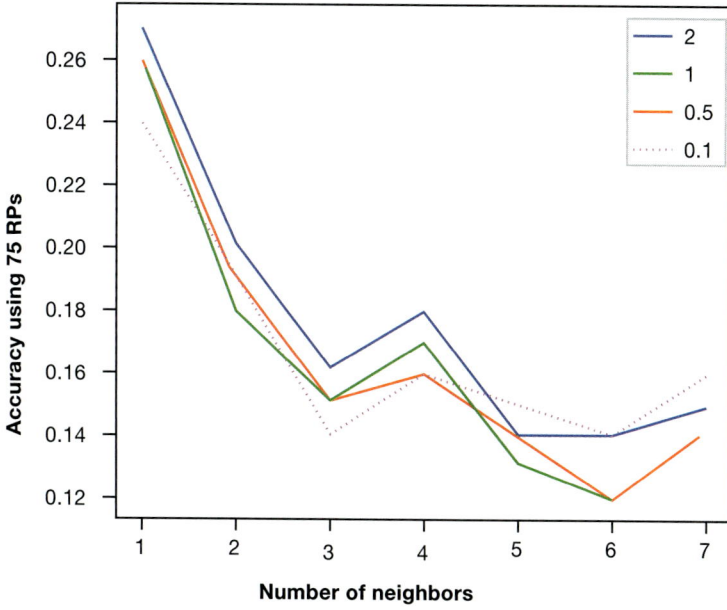

(a)

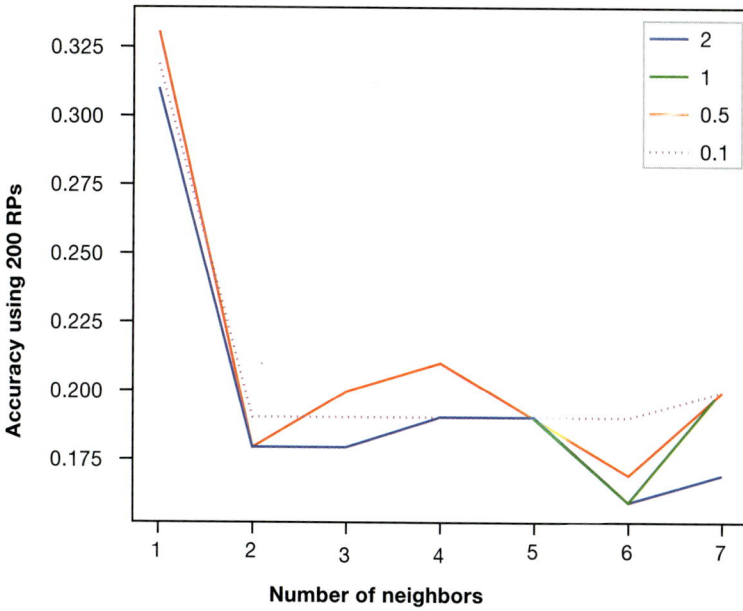

(b)

FIG. 6.5 Results on the Olivetti Face data set: 75 RPs and 200 RPs

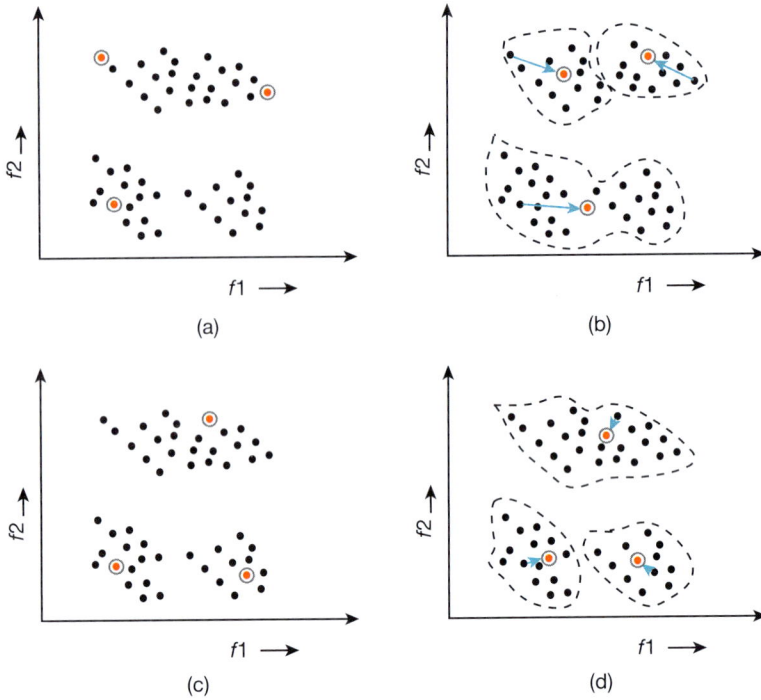

FIG. 7.13 Visualization of goodness of using k-means clustering for initialization

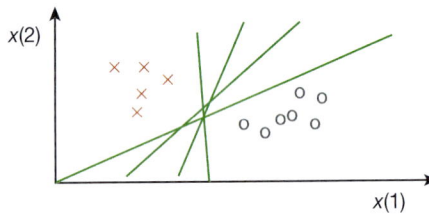

FIG. 8.4 Multiple decision boundaries

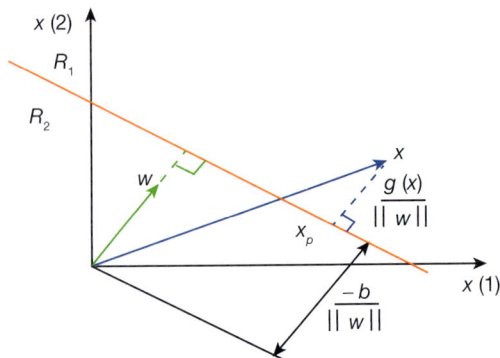

FIG. 8.5 Distance of a point from the decision boundary

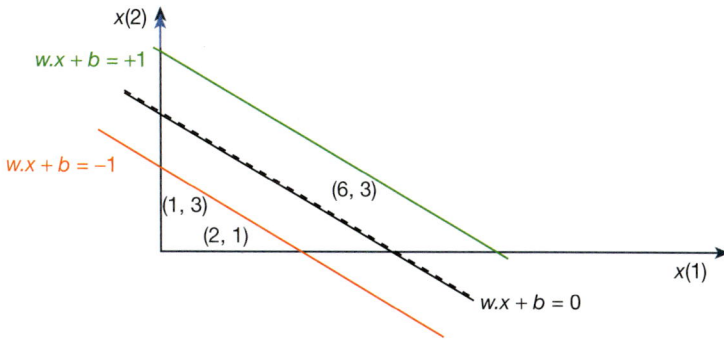

FIG. 8.7 An example to illustrate SVM

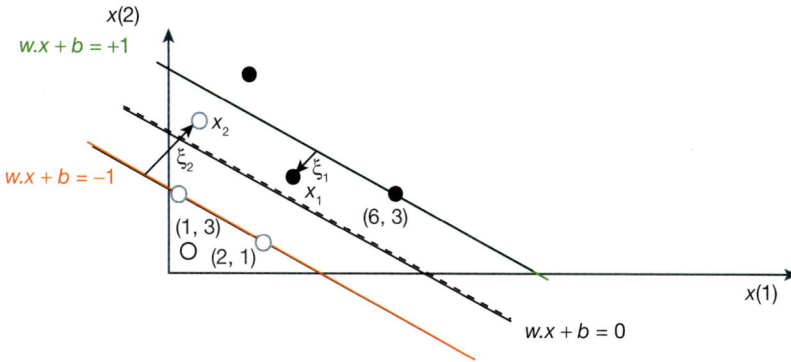

FIG. 8.8 An example to illustrate the margin violators

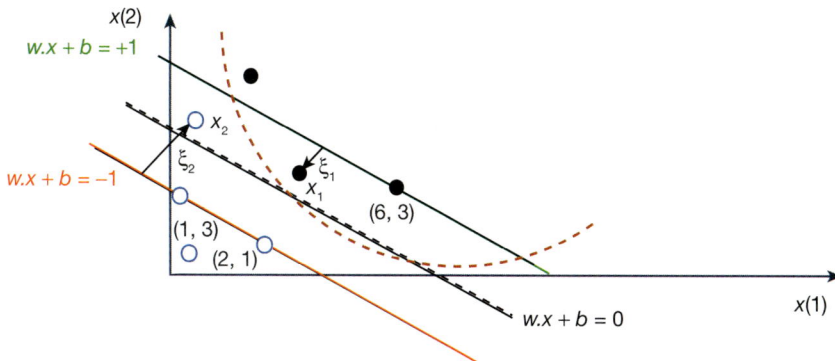

FIG. 8.9 An example to illustrate a non-linear SVM

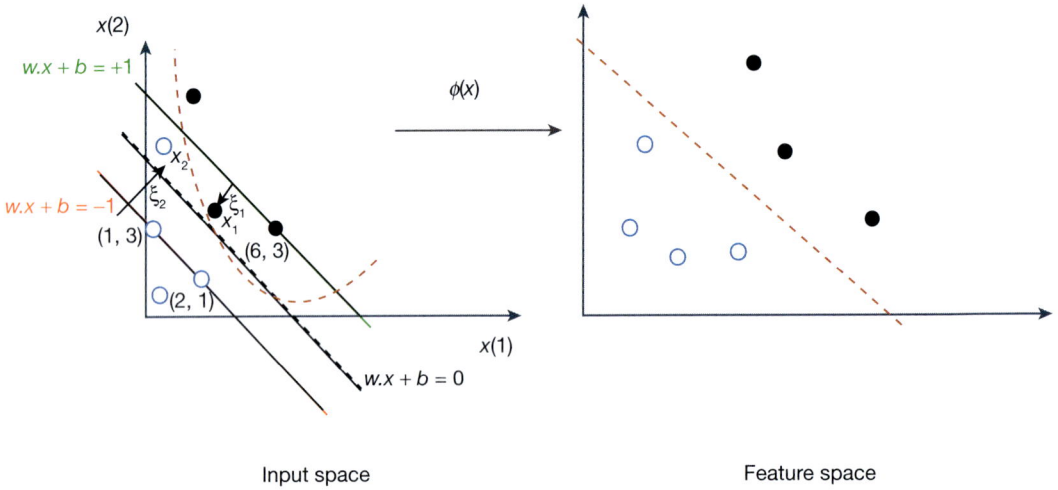

FIG. 8.10 Mapping the data to a high-dimensional feature space

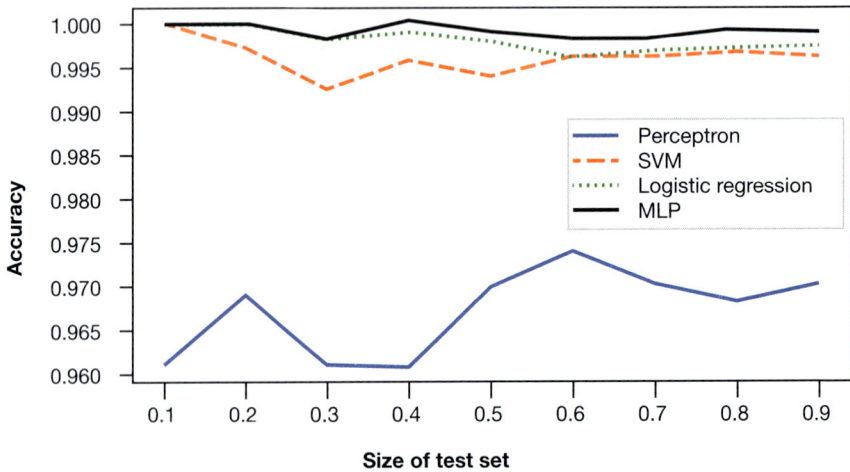

FIG. 8.17 Comparison of linear classifiers on the Digits data set

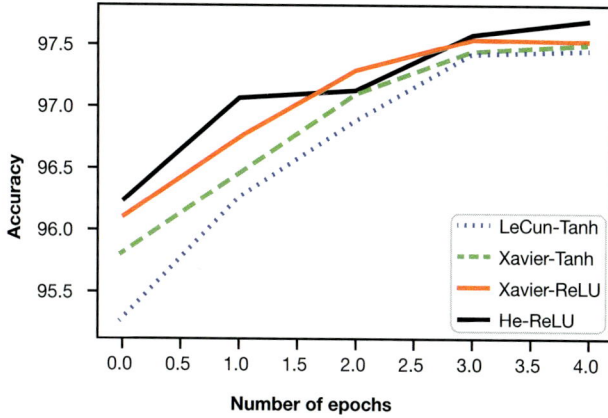

FIG. 9.14 Comparison of various weight initialization schemes

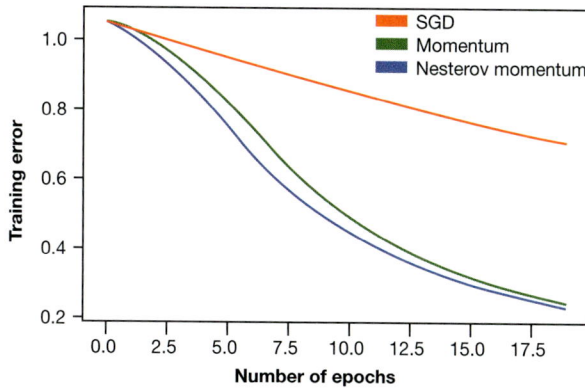

FIG. 9.16 Training error versus number of epochs on the Digits data set

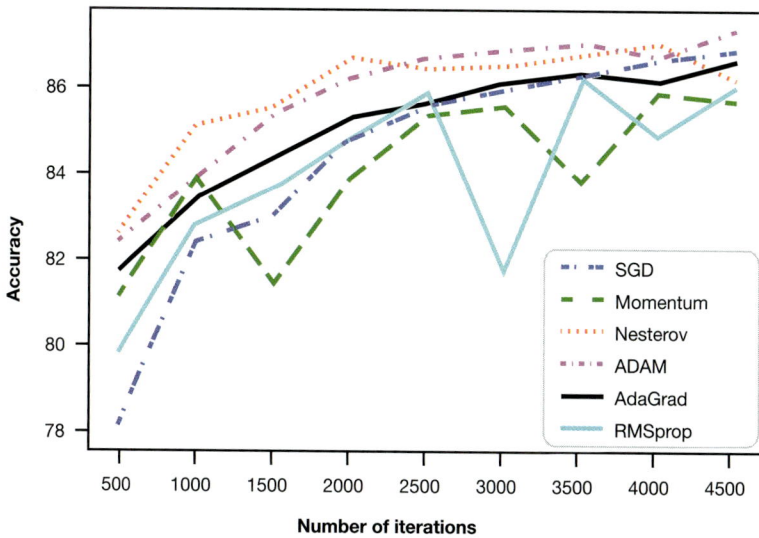

FIG. 9.18 Comparison of optimizers on the Fashion MNIST data set

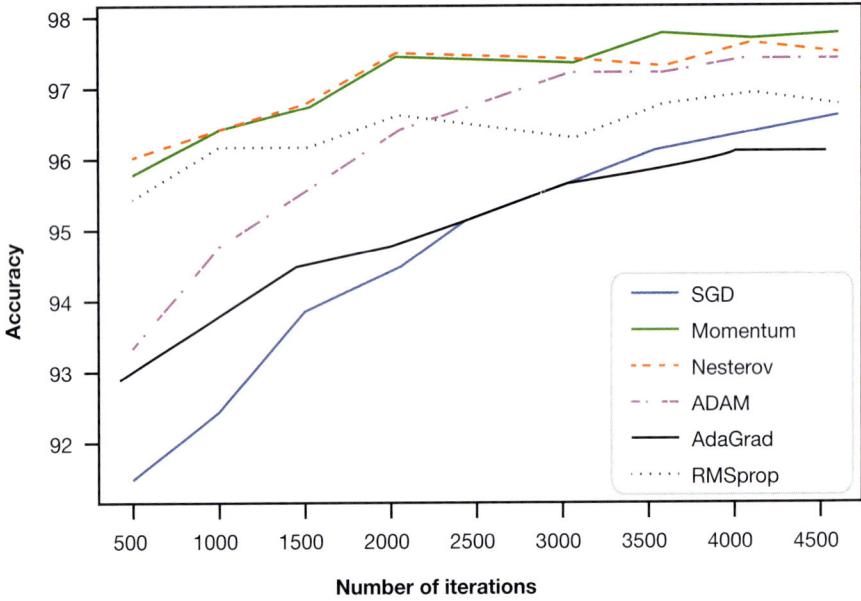

FIG. 9.19 Comparison of optimizers on the MNIST data set

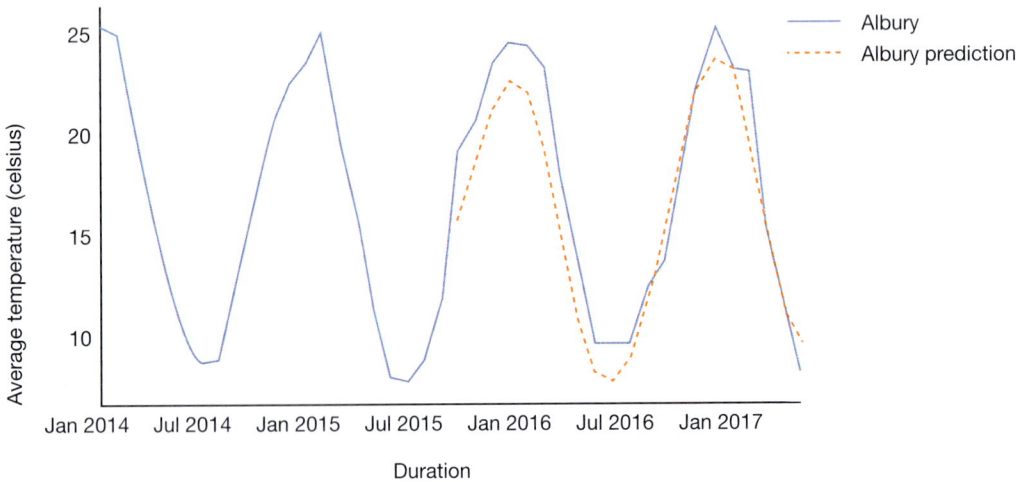

FIG. 9.37 Comparing predicted values with the ground truth

Conclusions

We have considered several important topics associated with ML in this book:

- **Feature engineering:** We have considered the need for dealing with missing values and outliers in Chapter 1. We have also considered the importance of choosing the right representation for efficient and effective ML in Chapters 1, 6 and 9; specifically we have considered feature selection methods, linear feature extraction schemes based on PCA, SVD and RP and non-linear feature extraction based on autoencoders.
- **Supervised learning:** Here, the ML model needs to predict the target value. In classification, the target is a class label drawn from a finite set of labels. In regression, the target is a real number. In both regression and classification, we are provided a set of labelled examples. We split the given set of examples into training, validation and test sets. There is no specific splitting ratio that is universally applicable. We can experiment with different splitting ratios for a given application to choose the best.
  The ML model is trained using the training examples and is fine-tuned using the validation patterns. The test patterns are used only for performance analysis; they should not be used in training. We have considered various ML models based on KNNs, DTC, SVM, NBC and NNs. All these schemes can be used for both classification and regression.
- **Unsupervised learning**: Here, we do not have class labels associated with the examples. We can use dimensionality reduction techniques on unlabelled examples also. In clustering, we partition the given set of examples into clusters. Note that we can cluster both labelled and unlabelled patterns. The need for clustering labelled patterns arises in the context of identifying subclasses in a class or outliers. We have considered the Leader clustering algorithm in Chapter 2 and a variety of partitional and hierarchical clustering algorithms in Chapter 7. We have considered both soft and hard clustering algorithms.

OBSERVATIONS AND CONCLUSIONS

We summarize them using the following important components of any ML system.

- **Search:** Search is an important activity carried out by any ML system. We need to search for the right ML model. Some important roles of search in ML are given below:
  - We need to search for the appropiate set of *parameter values*. One can broadly say that ML is parameter tuning. These parameters could be the number of clusters, number of hidden layers in a neural networks, the number of neighbors in KNN, etc.

- Search is important in selecting a model. Classifiers based on SVM and ANN can ideally deal with numerical attributes. On the other hand, DTC, NBC and models based on them can handle mixed type features.
- Fine-tuning the model involves search. Broadly we exploit the validation data to arrive at the best model. Note the relation to the bias–variance trade-off. Complex models will have more parameters to tune, larger number of hidden layers, higher-order non-linear decision boundaries, etc. The complexity of the chosen model will vary based on the application domain, number of features and number of patterns used in training.

- **Representation:** For successful ML, the most important component is the representation of the data and the model. Specifically, even on low-dimensional data, one can use most of the feature selection algorithms; mutual information-based features do well on high-dimensional applications including text analytics.

  We have observed that autoencoder-based features yield better classification accuracy compared to PCA or SVD-based features. Note that non-linear dimensionality reduction schemes based on autoencoder perform better than the ones based on linear feature extraction schemes like PCA and SVD if the classes are non-linearly separable.

  Note that encoder–decoder models that employ CNNs and RNNs also perform intrinsic dimensionality reduction. Studies show how the hidden layers/nodes capture some interesting features of the input in NN-based models. We have shown how the hidden nodes in the autoencoder capture input pixels.

  Representation and search go hand in hand in building and employing ML models. There are claims that DL systems learn the representations automatically and that they can work directly on the raw data in any application. There are interesting studies and practical results on representation learning using DL.

- **Large data sets:** Large data sets may either have a large number of patterns or features or both. Various algorithms will face difficulties while dealing with large data sets.

  - *High-dimensional data:* Models using KNN and DTC are not ideal for dealing with high-dimensional data. Models employing SVM and MLP are better suited, provided we have a large number of training patterns. The random forest model is also recommended to deal with high-dimensional data.
  - *Large number of patterns:* Models based on NBC and ANNs are recommended. It can be computationally prohibitive to use KNN.
  - *Large number of classes:* Models employing KNN and ANNs are recommended.
  - *Large unlabelled data sets:* Use of the divide-and-conquer clustering or an incremental algorithm like Leader is recommended.

- **Explainability:** It is well-known that the DL models based on ANNs are opaque compared to the ML models employing DTC, random forest or gradient boosting. Models based on DTC or frequent itemsets are ideally suited for explanation. Feature selection methods are suggested as they deal with a subset of the given features.

Bibliography

- Burges, CJC. 1998. A tutorial on support vector machines for pattern recognition, *Data Mining and Knowledge Discovery*, 2(2): 121–167.

- Schölkopf, B and Smola, AJ. 2001. *Learning with Kernels*, MIT Press.

- Rifkin, RM. 2008. Multiclass Classification, *Lecture Notes*, Spring08, MIT, USA.
- Witten, IH, Frank, E and Hall, MA. 2011. *Data Mining: Practical Machine Learning Tools and Techniques, Third Edition*, Morgan Kaufmann Publishers, Burlington.
- Murty, MN and Susheela Devi, V. 2015. *Introduction to Pattern Recognition and Machine Learning*, World Scientific Publishing Co. Pte. Ltd.: Singapore.
- Murty, MN and Manasvi, A. 2021. *Machine Learning in Social Networks: Embedding Nodes, Edges, Communities, and Graphs*, Springer Verlag, Singapore.
- Murty, MN and Avinash, M. 2023. *Representation in Machine Learning*, Springer Verlag, Singapore.

Appendix

Hints to Practical Exercises

Chapter 2

Practical Exercise Q1

```
[1]: #generate random 100 numbers between [0,1]
     import random
     datapoints=[]
     for i in range(100):
         n=random.random();
         n=round(n,3)
         datapoints.append(n)
     print(datapoints)
```

```
[0.133, 0.77, 0.579, 0.301, 0.939, 0.625, 0.83, 0.386, 0.192, 0.301, 0.529,
0.048, 0.006, 0.5, 0.892, 0.357, 0.633, 0.967, 0.114, 0.391, 0.674, 0.969,
0.787, 0.554, 0.55, 0.117, 0.028, 0.205, 0.374, 0.741, 0.708, 0.316, 0.355,
0.573, 0.541, 0.145, 0.092, 0.882, 0.892, 0.313, 0.761, 0.74, 0.195, 0.979,
0.708, 0.787, 0.816, 0.604, 0.958, 0.252, 0.462, 0.537, 0.097, 0.919, 0.811,
0.58, 0.407, 0.818, 0.064, 0.041, 0.563, 0.045, 0.436, 0.164, 0.713, 0.082,
0.349, 0.727, 0.858, 0.699, 0.417, 0.211, 0.912, 0.693, 0.077, 0.76, 0.14,
0.683, 0.551, 0.422, 0.446, 0.256, 0.71, 0.848, 0.996, 0.537, 0.36, 0.356,
0.358, 0.994, 0.561, 0.994, 0.493, 0.503, 0.166, 0.455, 0.303, 0.88, 0.834,
0.33]
```

```
[2]: from sklearn.neighbors import KNeighborsClassifier
     import numpy as np
     import matplotlib.pyplot as plt
```

```
[3]: targetresult=[]
     for data in datapoints:
         if(data<=0.5):
             targetresult.append(1)
         else:
             targetresult.append(2)
     print(targetresult)
```

```
[1, 2, 2, 1, 2, 2, 2, 1, 1, 1, 2, 1, 1, 1, 2, 1, 2, 2, 1, 1, 2, 2, 2, 2, 2, 1,
1, 1, 1, 2, 2, 1, 1, 2, 2, 1, 1, 2, 2, 1, 2, 2, 1, 2, 2, 2, 2, 2, 2, 1, 1, 2, 1,
2, 2, 2, 1, 2, 1, 1, 2, 1, 1, 1, 2, 1, 1, 2, 2, 2, 1, 1, 2, 2, 1, 2, 1, 2, 2, 1,
1, 1, 2, 2, 2, 2, 1, 1, 1, 2, 2, 2, 1, 2, 1, 1, 1, 2, 2, 1]
```

```
[16]:  train_data=datapoints[:50]
       train_result=targetresult[:50]
       test_data=datapoints[50:]
       test_result=targetresult[50:]
       #print(train_data)
       train_data=np.array(train_data).reshape(-1,1)
       print(train_data)
       test_data=np.array(test_data).reshape(-1,1)
       train_result=np.array(train_result)
       test_result=np.array(test_result)
```

```
[[0.133]
 [0.77 ]
 [0.579]
 [0.301]
 [0.939]
 [0.625]
 [0.83 ]
 [0.386]
 [0.192]
 [0.301]
 [0.529]
 [0.048]
 [0.006]
 [0.5  ]
 [0.892]
 [0.357]
 [0.633]
 [0.967]
 [0.114]
 [0.391]
 [0.674]
 [0.969]
 [0.787]
 [0.554]
 [0.55 ]
 [0.117]
 [0.028]
 [0.205]
 [0.374]
 [0.741]
 [0.708]
 [0.316]
 [0.355]
 [0.573]
 [0.541]
 [0.145]
 [0.092]
 [0.882]
 [0.892]
 [0.313]
 [0.761]
 [0.74 ]
 [0.195]
 [0.979]
 [0.708]
 [0.787]
 [0.816]
 [0.604]
 [0.958]
 [0.252]]
```

```
[[0.133]
 [0.77 ]
 [0.579]
 [0.301]
 [0.939]
 [0.625]
 [0.83 ]
 [0.386]
 [0.192]
 [0.301]
 [0.529]
 [0.048]
 [0.006]
 [0.5 ]
 [0.892]
 [0.357]
 [0.633]
 [0.967]
 [0.114]
 [0.391]
 [0.674]
 [0.969]
 [0.787]
 [0.554]
 [0.55 ]
 [0.117]
 [0.028]
 [0.205]
 [0.374]
 [0.741]
 [0.708]
 [0.316]
 [0.355]
 [0.573]
 [0.541]
 [0.145]
 [0.092]
 [0.882]
 [0.892]
 [0.313]
 [0.761]
 [0.74 ]
 [0.195]
 [0.979]
 [0.708]
 [0.787]
 [0.816]
 [0.604]
 [0.958]
 [0.252]] [1 2 2 1 2 2 2 1 1 1 2 1 1 1 2 1 2 2 1 1 2 2 2 2 2 1 1 1 1 2 2 1 1 2 2
 1 1
 2 2 1 2 2 1 2 2 2 2 2 2 1]
```

```
[21]: knn=KNeighborsClassifier(n_neighbors=1)
```

```
[22]: knn.fit(train_data,train_result)
```

```
[22]: KNeighborsClassifier(n\_neighbors=1)
```

```
[23]: print(knn.score(test_data,test_result))

      1.0
```

```
[24]: #predicting score for k= 1 to k=80
      score=[]
      kValue=[]
      for k in [1,2,3,4,5,20,30]:
          knn=KNeighborsClassifier(n_neighbors=k)
          knn.fit(train_data,train_result)
          score.append(knn.score(test_data,test_result))
          kValue.append(k)
```

```
[25]: plt.figure()
      plt.plot(kValue,score)
      plt.title('KNN Classfier accuracy - test data')
      plt.xlabel('K-value')
      plt.ylabel('accuracy')
      plt.show()
```

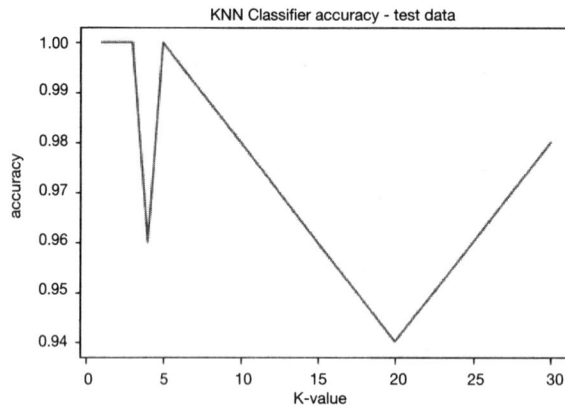

FIG. A.1 KNN classifier accuracy for test data (for colour figure, please see Colour Plate 3)

```
[26]: #predicting score for k= 1 to k=80
      score=[]
      kValue=[]
      for k in [1,2,3,4,5,20,30]:
          knn=KNeighborsClassifier(n_neighbors=k)
          knn.fit(train_data,train_result)
          score.append(knn.score(train_data,train_result))
          kValue.append(k)
```

```
[27]: plt.figure()
      plt.plot(kValue,score)
      plt.title('KNN Classfier accuracy - train data')
      plt.xlabel('K-value')
      plt.ylabel('accuracy')
      plt.show()
```

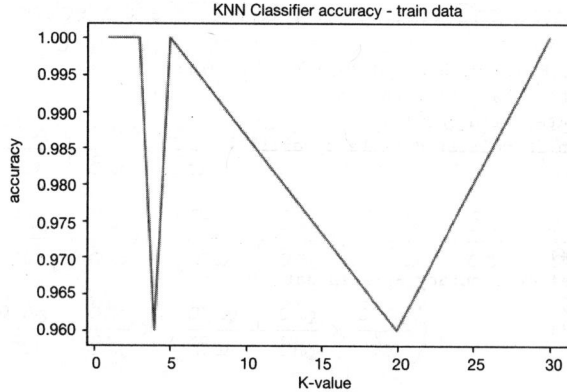

FIG. A.2 KNN classifier accuracy for training data (for colour figure, please see Colour Plate 3)

```
[28]: #predicting score for k= 1 to k=80
      score=[]
      kValue=[]
      for k in [1,2,3,4,5,20,30]:
          wknn=KNeighborsClassifier(n_neighbors=k,weights='distance')
          wknn.fit(train_data,train_result)
          score.append(wknn.score(test_data,test_result))
          kValue.append(k)
```

```
[29]: plt.figure()
      plt.plot(kValue,score)
      plt.title('WKNN Classfier accuracy - test data')
      plt.xlabel('K-value')
      plt.ylabel('accuracy')
      plt.show()
```

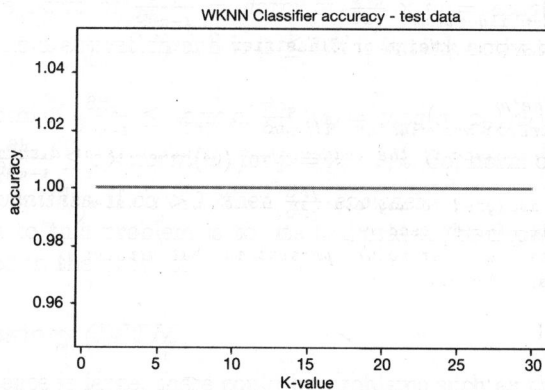

FIG. A.3 WKNN classifier accuracy for test data (for colour figure, please see Colour Plate 3)

```
[30]: from sklearn.neighbors import RadiusNeighborsClassifier
```

```
[31]: score=[]
      rValue=[]
      for i in [0.05,0.1,0.15,0.2,0.25,0.3,0.35,0.4,0.5,0.75]:
          rknn=RadiusNeighborsClassifier(radius=i)
          rknn.fit(train_data,train_result)
          score.append(rknn.score(test_data,test_result))
          rValue.append(i)
```

```
[32]: plt.figure()
      plt.plot(rValue,score)
      plt.title('rKNN Classfier accuracy - train data')
      plt.xlabel('R-value')
      plt.ylabel('accuracy')
      plt.show()
```

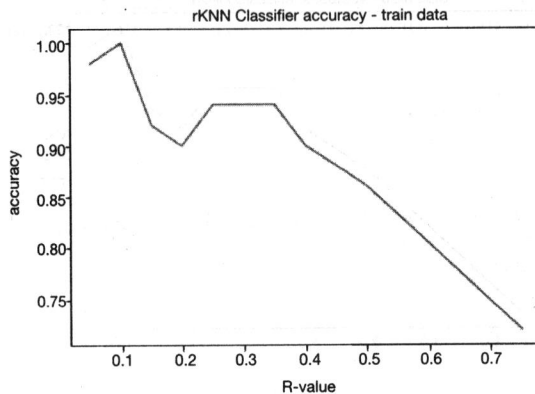

FIG. A.4 RKNN classifier accuracy for training data (for colour figure, please see Colour Plate 3)

Practical Exercise Q2

```
[33]: #clustering using leader algorithm
      from sklearn.neighbors import KNeighborsClassifier
```

```
[40]: #clustering based on theta
      #if no cluster found create new cluster with data point
      #if clusters are present check if the leader of the cluster is at disatance
       ↪less than theta
      #if data point is not assigned to any cluster create new cluster
      #clusters - list of all cluster leader
      #clusterPointMap- contain all datapoints present in that cluster
      def calculatePurity(data,theta):
          clusters=[]
          clusterPointsMap={}
          for i in data:
              assingned=0
              if(len(clusters)==0):
                  clusters.append(i)
                  clusterPointsMap[i]=[i]
                  continue
              for j in clusters:
                  if(abs(i-j)<=theta):
```

```
                        clusterPointsMap[j].append(i)
                        assingned=1
                        break
            if(assingned==0):
                clusters.append(i)
                clusterPointsMap[i]=[i]
    #calculating purity of clusters
    #for each cluster count number of points belong to class1 and number of
↪points belong to class2
    #calculate purity as sum(max(class1count,class2count)) for all clusters
↪divided by number of data points
    purity=0
    for j in clusterPointsMap:
        class1count=0
        class2count=0
        for data in clusterPointsMap[j]:
            if(data<=0.5):
                class1count+=1
            else:
                class2count+=1
        purity+=max(class1count,class2count)
    return purity/120.0
```

Practical Exercise Q3

```
[41]:  theta=0.01
       purity=[]
       theta_values=[]
       while(theta<1):
           purity.append(calculatePurity(datapoints,theta))
           theta_values.append(theta)
           theta+=0.01
```

```
[42]:  plt.figure()
       plt.scatter(theta_values,purity)
       plt.title('leader algorithm clustering - data')
       plt.xlabel('theta')
       plt.ylabel('purity')
       plt.show()
```

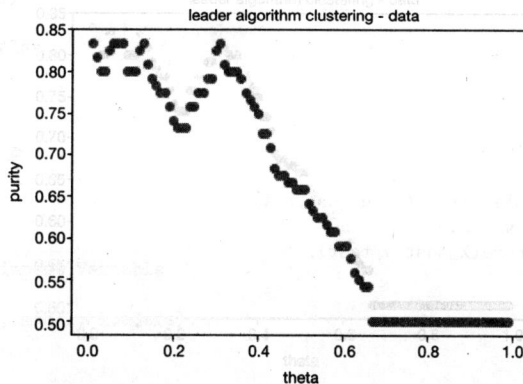

FIG. A.5 Purity of Leader algorithm clustering (for colour figure, please see Colour Plate 3)

Practical Exercise Q4

```
[44]: %matplotlib inline
      from sklearn.datasets import load_digits
      from sklearn.neighbors import KNeighborsClassifier
      from sklearn.neighbors import NearestNeighbors
      import numpy as np
      import torch
      import collections
      from sklearn.neighbors import KNeighborsClassifier
      from sklearn.model_selection import train_test_split
      from sklearn.linear_model import Perceptron
      from sklearn import svm
      from mlxtend.frequent_patterns import fpgrowth
      import pandas as pd
      import matplotlib.pyplot as plt
      import warnings
      warnings.filterwarnings('ignore')
      # import warnings
      # warnings.filterwarnings('error', category=DeprecationWarning)
```

```
[45]: #pip install mlxtend
```

```
[46]: digits = load_digits()
      x = digits.data
      y = digits.target
      image_data=[]
      for i in x:
          image_data.append(i.flatten())
      len(image_data)
```

```
[46]: 1797
```

```
[47]: for i in range(len(image_data)):
          for j in range(len(image_data[0])):
              if(image_data[i][j]<8):
                  image_data[i][j]=0
              else:
                  image_data[i][j]=1
```

```
[48]: X_train,X_test,y_train,y_test=train_test_split(image_data,y, train_size = 0.10
       ↪, stratify = y)
      len(X_train)
```

```
[48]: 179
```

```
[49]: k=[1, 3, 5, 10, 20]
      acc=[]
      for i in k:
          neigh = KNeighborsClassifier(n_neighbors=i)
          neigh.fit(X_train, y_train)
          acc.append(neigh.score(X_test,y_test))
```

```
[50]:  %matplotlib inline
       import matplotlib.pyplot as plt
       plt.plot(k, acc, color = 'green')
       plt.xlabel('Value of k')
       plt.ylabel('Testing Accuracy')
```

```
[50]:  Text(0, 0.5, 'Testing Accuracy')
```

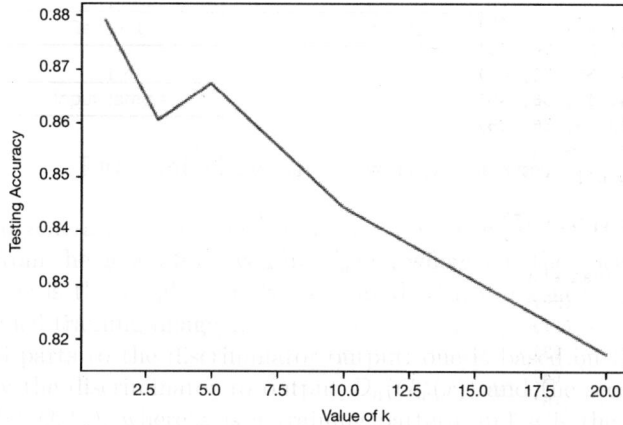

FIG. A.6 Testing accuracy (for colour figure, please see Colour Plate 3)

```
[51]:  df = pd.DataFrame(X_train, columns = digits.feature_names)
       df2 = pd.DataFrame(X_test, columns = digits.feature_names)
```

```
[52]:  fdf=df.append(df2)
```

```
[59]:  fdf.keys
       len(df)
```

```
[59]:  179
```

```
[54]:  fpgrowth(df, min_support=0.1)
```

```
[54]:           support              itemsets
       0        0.877095                 (59)
       1        0.865922                 (11)
       2        0.837989                  (3)
       3        0.832402                  (4)
       4        0.698324                 (18)
       ...       ...                     ...
       299572   0.100559          (17, 18, 10)
       299573   0.100559      (11, 17, 18, 59)
       299574   0.100559      (10, 17, 18, 59)
       299575   0.100559      (10, 17, 18, 11)
       299576   0.100559  (10, 11, 17, 18, 59)

       [299577 rows x 2 columns]
```

```
[55]: fpgrowth(df, min_support=0.3)
```

```
[55]:         support           itemsets
      0      0.877095               (59)
      1      0.865922               (11)
      2      0.837989                (3)
      3      0.832402                (4)
      4      0.698324               (18)
      ...        ...                 ...
      3080   0.301676   (11, 37, 36, 45)
      3081   0.312849   (11, 59, 36, 45)
      3082   0.301676    (4, 37, 36, 45)
      3083   0.301676    (4, 11, 36, 45)
      3084   0.318436    (59, 3, 36, 45)

      [3085 rows x 2 columns]
```

```
[56]: fpgrowth(df, min_support=0.5)
```

```
[56]:         support       itemsets
      0      0.877095           (59)
      1      0.865922           (11)
      2      0.837989            (3)
      3      0.832402            (4)
      4      0.698324           (18)
      ..         ...            ...
      158    0.513966   (11, 60, 53)
      159    0.508380       (4, 45)
      160    0.508380      (11, 45)
      161    0.508380      (59, 45)
      162    0.530726      (60, 45)

      [163 rows x 2 columns]
```

```
[57]: fpgrowth(df, min_support=0.7)
```

```
[57]:         support    itemsets
      0      0.877095        (59)
      1      0.865922        (11)
      2      0.837989         (3)
      3      0.832402         (4)
      4      0.810056        (60)
      5      0.737430        (12)
      6      0.787709    (59, 11)
      7      0.821229     (59, 3)
      8      0.748603     (11, 3)
      9      0.731844  (11, 59, 3)
      10     0.720670     (59, 4)
      11     0.715084     (11, 4)
      12     0.709497    (11, 60)
[ ]:
```

```
[ ]:
```

Chapter 3

Practical Exercise Q1: Task 1

```
[13]: import numpy as np
      from numpy import mean
      from numpy import std
      import matplotlib.pyplot as plt
      from sklearn.utils import Bunch
      from sklearn.datasets import fetch_olivetti_faces
      from sklearn.tree import DecisionTreeClassifier
      from sklearn import tree
      from sklearn.model_selection import train_test_split
      from sklearn.model_selection import cross_val_score
```

```
[19]: dataset = fetch_olivetti_faces()
      X = dataset.data
      y = dataset.target
      X_train, X_test, y_train, y_test = train_test_split (X, y,test_size = 0.1,
          ↪random_state=0)
```

```
[22]: clf = DecisionTreeClassifier(random_state=0)
      path = clf.cost_complexity_pruning_path(X_train, y_train)
      ccp_alphas, impurities = path.ccp_alphas, path.impurities
```

```
[23]: fig, ax = plt.subplots()
      ax.plot(ccp_alphas[:-1], impurities[:-1], marker="o", drawstyle="steps-post")
      ax.set_xlabel("effective alpha",fontname="Calibri", size=15,fontweight="bold" )
      ax.set_ylabel("total impurity of leaves",fontname="Calibri",
          ↪size=15,fontweight="bold")
      ax.set_title("Total Impurity vs effective alpha for training set")
```

```
[23]: Text(0.5, 1.0, 'Total Impurity vs effective alpha for training set')
```

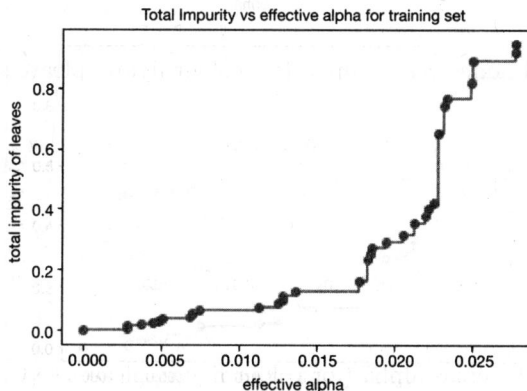

Fig. A.7 Total impurity versus effective alpha for training data (for colour figure, please see Colour Plate 3)

```
[24]: clfs = []
      for ccp_alpha in ccp_alphas:
          clf = DecisionTreeClassifier(random_state=0, ccp_alpha=ccp_alpha)
          clf.fit(X_train, y_train)
          clfs.append(clf)
      print(
          "Number of nodes in the last tree is: {} with ccp_alpha: {}".format(
              clfs[-1].tree_.node_count, ccp_alphas[-1]
          )
      )
```

Number of nodes in the last tree is: 1 with ccp_alpha: 0.027707671957671987

```
[25]: clfs = clfs[:-1]
      ccp_alphas = ccp_alphas[:-1]

      node_counts = [clf.tree_.node_count for clf in clfs]
      depth = [clf.tree_.max_depth for clf in clfs]
      fig, ax = plt.subplots(2, 1)
      ax[0].plot(ccp_alphas, node_counts, marker="o", drawstyle="steps-post")
      ax[0].set_xlabel("alpha",fontname="Calibri", size=15,fontweight="bold")
      ax[0].set_ylabel("number of nodes",fontname="Calibri",
        ↪size=15,fontweight="bold")
      ax[0].set_title("Number of nodes vs alpha")
      ax[1].plot(ccp_alphas, depth, marker="o", drawstyle="steps-post")
      ax[1].set_xlabel("alpha",fontname="Calibri", size=15,fontweight="bold")
      ax[1].set_ylabel("depth of tree",fontname="Calibri", size=15,fontweight="bold")
      ax[1].set_title("Depth vs alpha")
      fig.tight_layout()
```

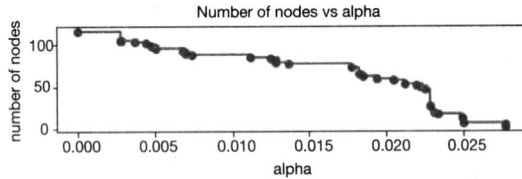

FIG. A.8 Number of nodes versus alpha (for colour figure, please see Colour Plate 3)

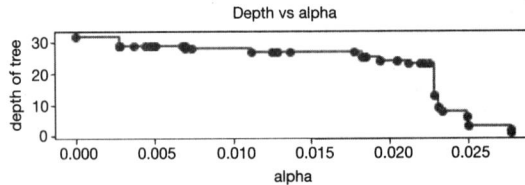

FIG. A.9 Depth versus alpha (for colour figure, please see Colour Plate 3)

```
[26]: train_scores = [clf.score(X_train, y_train) for clf in clfs]
      test_scores = [clf.score(X_test, y_test) for clf in clfs]

      fig, ax = plt.subplots()
```

```
ax.set_xlabel("alpha",fontname="Calibri", size=15,fontweight="bold")
ax.set_ylabel("accuracy",fontname="Calibri", size=15,fontweight="bold")
ax.set_title("Accuracy vs alpha for training and testing sets")
ax.plot(ccp_alphas, train_scores, marker="o", label="train",
  ↪drawstyle="steps-post")
ax.plot(ccp_alphas, test_scores, marker="o", label="test",
  ↪drawstyle="steps-post")
ax.legend()
plt.show()
```

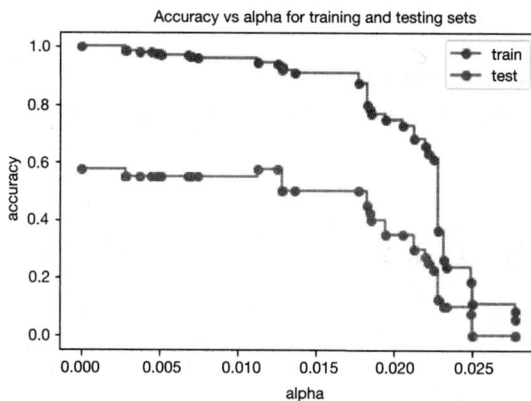

Fig. A.10 Accuracy versus alpha for training and testing data (for colour figure, please see Colour Plate 3)

```
[27]: clf_DTC = DecisionTreeClassifier(criterion = 'gini', random_state=0,
        ↪ccp_alpha=0.011)
      scores = cross_val_score(clf_DTC, X, y, cv = 10,scoring='accuracy', n_jobs=-1)
      print('The accuracy using Gini index is ','%.3f (%.3f)' % (mean(scores),
        ↪std(scores)))
      scores
```

The accuracy using Gini index is 0.595 (0.077)

```
[27]: array([0.7  , 0.575, 0.575, 0.6  , 0.55 , 0.6  , 0.525, 0.725, 0.45 ,
      0.65 ])
```

```
[28]: clf_DTC = DecisionTreeClassifier(criterion = 'entropy', random_state=0,
        ↪ccp_alpha=0.011)
      scores = cross_val_score(clf_DTC, X, y, cv = 10,scoring='accuracy', n_jobs=-1)
      print('The accuracy using Entropy is ','%.3f (%.3f)' % (mean(scores),
        ↪std(scores)))
      scores
```

The accuracy using Entropy is 0.532 (0.062)

```
[28]: array([0.625, 0.65 , 0.55 , 0.425, 0.5  , 0.525, 0.525, 0.525, 0.475,
      0.525])
```

```
[29]: clf_DTC = DecisionTreeClassifier(criterion = 'gini', random_state=0,
        ↪ccp_alpha=0.011)
      scores = cross_val_score(clf_DTC, X, y, cv = 10,scoring='accuracy', n_jobs=-1)
      print('The accuracy using Gini index is ','%.3f (%.3f)' % (mean(scores),
```

```
    ↪std(scores)))
    scores
```

The accuracy using Gini index is 0.595 (0.077)

[29]: array([0.7 , 0.575, 0.575, 0.6 , 0.55 , 0.6 , 0.525, 0.725, 0.45 ,
       0.65])

[30]:
```
clf_DTC = DecisionTreeClassifier(criterion = 'entropy')
scores = cross_val_score(clf_DTC, X, y, cv = 10,scoring='accuracy', n_jobs=-1)
print('The accuracy using Entropy is ','%.3f (%.3f)' % (mean(scores),
    ↪std(scores)))
scores
```

The accuracy using Entropy is 0.560 (0.067)

[30]: array([0.575, 0.575, 0.625, 0.425, 0.6 , 0.575, 0.5 , 0.675, 0.5 ,
       0.55])

[ ]:

Practical Exercise Q1: Task 2

[1]:
```
import numpy as np
from numpy import mean
from numpy import std
import matplotlib.pyplot as plt
from sklearn.datasets import fetch_olivetti_faces
from sklearn.utils import Bunch
from sklearn.ensemble import RandomForestClassifier
from sklearn.model_selection import cross_val_score
from sklearn.model_selection import LeaveOneOut
```

[2]:
```
dataset = fetch_olivetti_faces()
```

[3]:
```
dataset = fetch_olivetti_faces()
X_train = dataset.data
y_train = dataset.target
```

[6]:
```
Xtr = dataset.data
cv = 10
n = [0.2, 0.4,0.6]
for i in n:
    clf_RF = RandomForestClassifier(n_estimators = 50, random_state=0,
    ↪max_features=i, n_jobs = -1)
    scores = cross_val_score(clf_RF, X_train, y_train, cv =
    ↪cv,scoring='accuracy', n_jobs=-1)
    print('The accuracy using 50 decision trees with {}% features is '.
    ↪format(i*100),'%.3f (%.3f)' % (mean(scores), std(scores)))
```

The accuracy using 50 decision trees with 20.0% features is 0.925 (0.057)
The accuracy using 50 decision trees with 40.0% features is 0.902 (0.051)
The accuracy using 50 decision trees with 60.0% features is 0.920 (0.047)

[ ]:

Practical Exercise Q1: Task 3

```
[1]:  import numpy as np
      from numpy import mean
      from numpy import std
      import matplotlib.pyplot as plt
      from sklearn.datasets import fetch_olivetti_faces
      from xgboost import XGBClassifier
      from sklearn.model_selection import train_test_split
      from sklearn.model_selection import cross_val_score
      from sklearn.metrics import accuracy_score
```

```
[2]:  dataset = fetch_olivetti_faces()
      dataset = fetch_olivetti_faces()
      X = dataset.data
      y = dataset.target
```

```
[3]:  n = [50,100]
      for i in n:
          clf_xgb = XGBClassifier(n_estimators=i, use_label_encoder=False,
      ↪eval_metric='mlogloss')
          scores = cross_val_score(clf_xgb, X, y, cv = 10,scoring='accuracy',
      ↪n_jobs=-1)
          print('The accuracy with number of trees = {} is '.format(i),'%.3f (%.3f)'
      ↪% (mean(scores), std(scores)), '{}'.format(scores))
```

```
The accuracy with number of trees = 50 is 0.825 (0.062) [0.85 0.85 0.85 0.8
0.825 0.85 0.725 0.95 0.725 0.825]
The accuracy with number of trees = 100 is 0.825 (0.062) [0.85 0.85 0.85 0.8
0.825 0.85 0.725 0.95 0.725 0.825]
```

```
[ ]:
```

Chapter 4

Practical Exercise Q1

```
[7]:  import numpy as np
      from numpy import mean
      from numpy import std
      import matplotlib.pyplot as plt
      from sklearn.datasets import fetch_olivetti_faces
      from sklearn.utils import Bunch
      from sklearn.model_selection import cross_val_score
      from sklearn.naive_bayes import GaussianNB
```

```
[8]:  dataset = fetch_olivetti_faces()
      X = dataset.data
      y = dataset.target
```

```
[9]:  clf_NBC = GaussianNB()
      for i in range(5,11,5):
          scores = cross_val_score(clf_NBC, X, y, cv = i,scoring='accuracy',
      ↪n_jobs=-1)
          print('The accuracy with kfolds = {} is '.format(i),'%.3f (%.3f)' %
      ↪(mean(scores), std(scores)), '{}'.format(scores))
```

```
The accuracy with kfolds = 5 is  0.882 (0.027) [0.9     0.9125 0.9     0.85    0.85
]
The accuracy with kfolds = 10 is  0.920 (0.047) [0.975 0.975 0.925 0.925 0.9
0.925 0.85  0.95  0.825 0.95 ]
```

[ ]:

Chapter 6

Practical Exercise Q1: Task 1

```
[2]:  import numpy as np
      import pandas as pd
      from sklearn import neighbors
      from sklearn.preprocessing import StandardScaler
      from sklearn.metrics import accuracy_score
      import matplotlib.pyplot as plt
      from sklearn.model_selection import train_test_split as tts
      from sklearn.datasets import fetch_olivetti_faces as faces
      from sklearn.datasets import load_digits as digits
      from sklearn.neighbors import NearestNeighbors
      from sklearn.datasets import load_breast_cancer as LBC
```

```
[11]: data = LBC()
      X = data['data']
      y = data['target']
      # from sklearn.preprocessing import StandardScaler
      # X = StandardScaler().fit_transform(X)
      X_train,X_test,y_train,y_test = tts(X,y)
      type(data)
```

```
[11]: sklearn.utils._bunch.Bunch
```

```
[146]: import matplotlib.pyplot as plt
       from sklearn import random_projection
       x = np.random.rand(30, 12)
       transformer = random_projection.GaussianRandomProjection(n_components =12)
       x = transformer.fit_transform(x)
       R = np.matmul(X,x)
       from sklearn.model_selection import train_test_split as tts
       X_train,X_test,y_train,y_test = tts(
           R,y
       )
       R
```

```
[146]: array([[-3050.21084671, -536.36313165, -2051.20760158, ...,
          -1637.23044529, -810.55551211, -3004.93597166],
         [-3257.18763399, -548.77385214, -2181.9648471 , ...,
          -1682.46596365, -936.25348679, -3096.01531926],
         [-2960.56284756, -481.39306351, -2000.57665057, ...,
          -1511.63949953, -861.37387988, -2767.75588841],
         ...,
         [-2065.23399096, -304.43566497, -1401.65869207, ...,
          -1031.53342766, -615.66629288, -1909.52440135],
         [-3154.10400578, -496.312506  , -2132.77534013, ...,
          -1609.82910724, -912.98716832, -2960.1002104 ],
         [ -548.65028133,  -52.08782406,  -384.52326817, ...,
          -262.71353945, -163.46136591, -507.46548506]]])
```

```
[12]: acc_dict = {}
      for j in [2, 1, 0.5, 0.1]:

          acc_dict[j]=[]

          def mydist(x1, x2):
                  return pow(sum(pow(abs(x1 - x2), j)), (1 / j))

          for i in list(range(1,26,2)): #K parameters

              model = neighbors.KNeighborsClassifier(i, algorithm='ball_tree',
                              metric='pyfunc', metric_params="func":mydist)
              #c = KNeighborsClassifier(n_neighbors=i, metric='my_distance')
              model.fit(X_train,y_train)
              y_pred = model.predict(X_test)
                  #print("Y test is ", y_test, "y predicted is", y_pred)
              acc_test = accuracy_score(y_test, y_pred)
              #acc_tra = accuracy_score(y_train, y_pr)
              #print("# neighbours K is", i, "R is", j, "train Accuracy:", acc_tra)

              #print("# neighbours K is", i, "R is", j, "test Accuracy:", acc_test)
              acc_dict[j].append((i,acc_test))

          print(j, len(acc_dict), len(acc_dict[j]))

      2 1 13
      1 2 13
      0.5 3 13
      0.1 4 13
```

```
[13]: colors = ['blue', 'green', 'red', 'magenta', 'black', 'cyan']

      for i, r in enumerate(acc_dict.keys()):
          x,y = list(zip(*acc_dict[r]))
          plt.plot(x,y,label=r,color=colors[i])
      plt.title('KNNC with Random Projections: Breast Cancer Data')
      plt.legend()
      plt.xlabel('Number of Neighbors')
      plt.ylabel('Accuracy')
      plt.show()
```

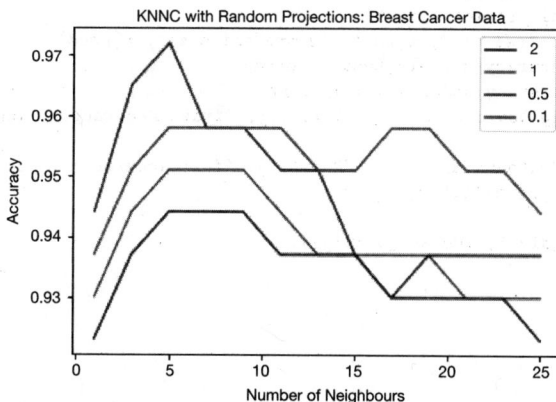

FIG. A.11 Accuracy of KNN classifier with random projections for the Breast Cancer data set (for colour figure, please see Colour Plate 3)

Practical Exercise Q1: Task 2

```
[19]: from sklearn.decomposition import TruncatedSVD
      X = data['data']
      y = data['target']
      svd = TruncatedSVD(12)
      X = svd.fit_transform(X)
```

```
[20]: from sklearn.metrics import accuracy_score
      X_train, X_test, y_train, y_test = tts(X, y)
```

```
[21]: # for i in list((1,25,2)): #K parameters
      #     for j in [2, 1,0.5,0.1]:
      #         def mydist(x1, x2):
      #             return pow(sum(pow(abs(x1 - x2), j)), (1 / j))
      #         model = neighbors.KNeighborsClassifier(i, algorithm='ball_tree',
      #                     metric='pyfunc', metric_params="func":mydist)
      #         #c = KNeighborsClassifier(n_neighbors=i, metric='my_distance')
      #         model.fit(X_train,y_train)
      #         y_pred = model.predict(X_test)
      #             #print("Y test is ", y_test, "y predicted is", y_pred)
      #         acc_test = accuracy_score(y_test, y_pred)
      #         #print("# neighbours K is", i, "R is", j, "train Accuracy:", acc_tra)
      ↪
      #         print("# neighbours K is", i, "R is", j, "test Accuracy:", acc_test)
```

```
[22]: acc_dict = {}
      for j in [2, 1, 0.5, 0.1]:

          acc_dict[j]=[]

          def mydist(x1, x2):
                  return pow(sum(pow(abs(x1 - x2), j)), (1 / j))

          for i in list(range(1,25,2)): #K parameters

              model = neighbors.KNeighborsClassifier(i, algorithm='ball_tree',
                          metric='pyfunc', metric_params="func":mydist)
              #c = KNeighborsClassifier(n_neighbors=i, metric='my_distance')
              model.fit(X_train,y_train)
              y_pred = model.predict(X_test)
                  #print("Y test is ", y_test, "y predicted is", y_pred)
              acc_test = accuracy_score(y_test, y_pred)
              #acc_tra = accuracy_score(y_train, y_pr)
              #print("# neighbours K is", i, "R is", j, "train Accuracy:", acc_tra)
      ↪
              #print("# neighbours K is", i, "R is", j, "test Accuracy:", acc_test)
              acc_dict[j].append((i,acc_test))

          print(j, len(acc_dict), len(acc_dict[j]))
```

```
2 1 12
1 2 12
0.5 3 12
0.1 4 12
```

```
[23]:  colors = ['blue', 'green', 'red', 'magenta', 'black', 'cyan']

       for i, r in enumerate(acc_dict.keys()):
           x,y = list(zip(*acc_dict[r]))
           plt.plot(x,y,label=r,color=colors[i])
       plt.title('KNNC with Singular Vectors: Breast Cancer Data')
       plt.legend()
       plt.xlabel('Number of Neighbors')
       plt.ylabel('Accuracy')
       plt.show()
```

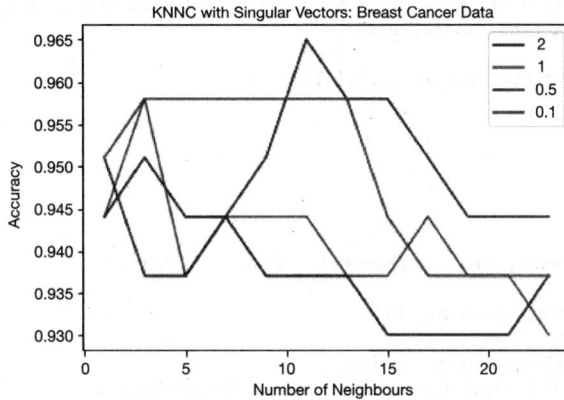

FIG. A.12 Accuracy of KNN classifier with singular vectors for the Breast Cancer data set (for colour figure, please see Colour Plate 3)

```
[ ]:
```

Practical Exercise Q1: Task 3

```
[14]: from sklearn.decomposition import PCA
      pca = PCA(n_components=12)
      X = data['data']
      y = data['target']
      from sklearn.preprocessing import StandardScaler
      X = StandardScaler().fit_transform(X)
      Xtr,Xts,y_train,y_test = tts(
          X,y
      )
      X_train = pca.fit_transform(Xtr)
      X_test = pca.fit_transform(Xts)
```

```
[15]: acc_dict = {}
      for j in [2, 1, 0.5, 0.1]:

          acc_dict[j]=[]

          def mydist(x1, x2):
                  return pow(sum(pow(abs(x1 - x2), j)), (1 / j))
```

```
      for i in list(range(1,25,2)): #K parameters

          model = neighbors.KNeighborsClassifier(i, algorithm='ball_tree',
                        metric='pyfunc', metric_params="func":mydist)
          #c = KNeighborsClassifier(n_neighbors=i, metric='my_distance')
          model.fit(X_train,y_train)
          y_pred = model.predict(X_test)
              #print("Y test is ", y_test, "y predicted is", y_pred)
          acc_test = accuracy_score(y_test, y_pred)
          #acc_tra = accuracy_score(y_train, y_pr)
          #print("# neighbours K is", i, "R is", j, "train Accuracy:", acc_tra)
↪
          #print("# neighbours K is", i, "R is", j, "test Accuracy:", acc_test)
          acc_dict[j].append((i,acc_test))

      print(j, len(acc_dict), len(acc_dict[j]))

2 1 12
1 2 12
0.5 3 12
0.1 4 12
```

```
[16]: colors = ['blue', 'green', 'red', 'magenta', 'black', 'cyan']

      for i, r in enumerate(acc_dict.keys()):
          x,y = list(zip(*acc_dict[r]))
          plt.plot(x,y,label=r,color=colors[i])
      plt.title('KNNC with Principal Components: Breast Cancer Data')
      plt.legend()
      plt.xlabel('Number of Neighbors')
      plt.ylabel('Accuracy')
      plt.show()
```

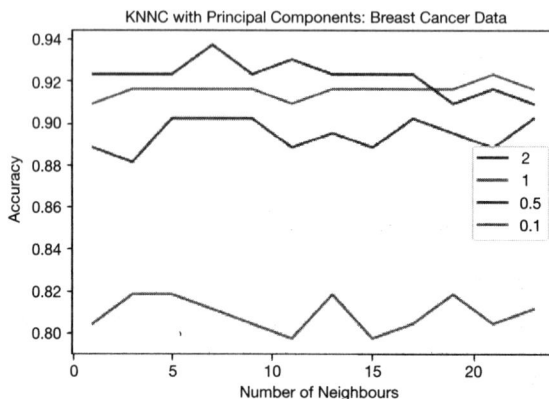

FIG. A.13 Accuracy of KNN classifier with principal components for the Breast Cancer data set (for colour figure, please see Colour Plate 3)

Chapter 7

Practical Exercise Q1: Task 1

```
[23]: from sklearn.datasets import fetch_olivetti_faces
      from sklearn.neighbors import KNeighborsClassifier
      from sklearn.model_selection import train_test_split
      import numpy as np
      import warnings
      warnings.filterwarnings('ignore')
      from sklearn import tree
      from statistics import mean
      from sklearn.cluster import KMeans
      import matplotlib.pyplot as plt
      from sklearn.neighbors import NearestCentroid
      from collections import Counter
      from sklearn.naive_bayes import GaussianNB
      from sklearn.naive_bayes import BernoulliNB
      from sklearn.datasets import load_breast_cancer as LBC
```

```
[24]: bcdata = LBC()
      data = bcdata.data
      target = bcdata.target
      data
```

```
[24]: array([[1.799e+01, 1.038e+01, 1.228e+02, ..., 2.654e-01, 4.601e-01,
              1.189e-01],
             [2.057e+01, 1.777e+01, 1.329e+02, ..., 1.860e-01, 2.750e-01,
              8.902e-02],
             [1.969e+01, 2.125e+01, 1.300e+02, ..., 2.430e-01, 3.613e-01,
              8.758e-02],
             ...,
             [1.660e+01, 2.808e+01, 1.083e+02, ..., 1.418e-01, 2.218e-01,
              7.820e-02],
             [2.060e+01, 2.933e+01, 1.401e+02, ..., 2.650e-01, 4.087e-01,
              1.240e-01],
             [7.760e+00, 2.454e+01, 4.792e+01, ..., 0.000e+00, 2.871e-01,
              7.039e-02]])
```

```
[25]: X_train_list=[]
      X_test_list=[]
      y_train_list=[]
      y_test_list=[]
      for j in range(10):
          X_train,X_test,y_train,y_test=train_test_split(data,target, train_size = 0.
      →8 , stratify = target)
          X_train_list.append(X_train)
          X_test_list.append(X_test)
          y_train_list.append(y_train)
          y_test_list.append(y_test)
```

```
[26]: def kmeans(feature_mat,p):
          result_mat=np.empty((len(feature_mat),p))
          kma = KMeans(n_clusters=p, n_init="auto").fit(feature_mat.T)
          centers= kma.cluster_centers_
          centers=centers.T
          return centers
```

```
[27]: kmeans_res={}
      K=[12,20,30]
      arr=[]
      for k in K:
          X_train_list_single=[]
          X_test_list_single=[]
          for i in range(len(X_train_list)):
              cluster_array=np.append(X_train_list[i],X_test_list[i],axis=0)
              arr=kmeans(cluster_array,k)
              X_train_list_single.append(arr[:455])
              X_test_list_single.append(arr[455:])
              #print("done")
          kmeans_res[k]=[X_train_list_single,X_test_list_single]
```

```
[28]: len(arr[0])
```

```
[28]: 30
```

```
[29]: fin_acc={}
      x=[1,2,3,4,5,6,7,8,9,10]
      for k_value in K:
          acc=[]
          train_test_data=kmeans_res[k_value]
          for i in range(len(train_test_data[0])):
              clf = GaussianNB()
              clf.fit(train_test_data[0][i], y_train_list[i])
              acc.append(clf.score(train_test_data[1][i],y_test_list[i]))
              print(acc)
          fin_acc[k_value]=acc
      f = plt.figure()
      f.set_figwidth(20)
      f.set_figheight(6)
      for i in fin_acc:
          plt.plot(x,fin_acc[i], label = i)
      plt.legend()
      plt.ylabel("accuracy")
      plt.show()
```

```
[0.9385964912280702]
[0.9385964912280702, 0.9473684210526315]
[0.9385964912280702, 0.9473684210526315, 0.9298245614035088]
[0.9385964912280702, 0.9473684210526315, 0.9298245614035088, 0.9122807017543859]
[0.9385964912280702, 0.9473684210526315, 0.9298245614035088, 0.9122807017543859,
0.9298245614035088]
[0.9385964912280702, 0.9473684210526315, 0.9298245614035088, 0.9122807017543859,
0.9298245614035088, 0.9649122807017544]
[0.9385964912280702, 0.9473684210526315, 0.9298245614035088, 0.9122807017543859,
0.9298245614035088, 0.9649122807017544, 0.9210526315789473]
[0.9385964912280702, 0.9473684210526315, 0.9298245614035088, 0.9122807017543859,
0.9298245614035088, 0.9649122807017544, 0.9210526315789473, 0.9298245614035088]
[0.9385964912280702, 0.9473684210526315, 0.9298245614035088, 0.9122807017543859,
0.9298245614035088, 0.9649122807017544, 0.9210526315789473, 0.9298245614035088,
0.9210526315789473]
[0.9385964912280702, 0.9473684210526315, 0.9298245614035088, 0.9122807017543859,
0.9298245614035088, 0.9649122807017544, 0.9210526315789473, 0.9298245614035088,
0.9210526315789473, 0.9298245614035088]
[0.9210526315789473]
[0.9210526315789473, 0.956140350877193]
[0.9210526315789473, 0.956140350877193, 0.9385964912280702]
[0.9210526315789473, 0.956140350877193, 0.9385964912280702, 0.9385964912280702]
```

```
[0.9210526315789473, 0.956140350877193, 0.9385964912280702, 0.9385964912280702,
0.956140350877193]
[0.9210526315789473, 0.956140350877193, 0.9385964912280702, 0.9385964912280702,
0.956140350877193, 0.9736842105263158]
[0.9210526315789473, 0.956140350877193, 0.9385964912280702, 0.9385964912280702,
0.956140350877193, 0.9736842105263158, 0.9385964912280702]
[0.9210526315789473, 0.956140350877193, 0.9385964912280702, 0.9385964912280702,
0.956140350877193, 0.9736842105263158, 0.9385964912280702, 0.9385964912280702]
[0.9210526315789473, 0.956140350877193, 0.9385964912280702, 0.9385964912280702,
0.956140350877193, 0.9736842105263158, 0.9385964912280702, 0.9385964912280702,
0.9473684210526315]
[0.9210526315789473, 0.956140350877193, 0.9385964912280702, 0.9385964912280702,
0.956140350877193, 0.9736842105263158, 0.9385964912280702, 0.9385964912280702,
0.9473684210526315, 0.9298245614035088]
[0.9122807017543859]
[0.9122807017543859, 0.956140350877193]
[0.9122807017543859, 0.956140350877193, 0.9473684210526315]
[0.9122807017543859, 0.956140350877193, 0.9473684210526315, 0.9385964912280702]
[0.9122807017543859, 0.956140350877193, 0.9473684210526315, 0.9385964912280702,
0.956140350877193]
[0.9122807017543859, 0.956140350877193, 0.9473684210526315, 0.9385964912280702,
0.956140350877193, 0.9736842105263158]
[0.9122807017543859, 0.956140350877193, 0.9473684210526315, 0.9385964912280702,
0.956140350877193, 0.9736842105263158, 0.9473684210526315]
[0.9122807017543859, 0.956140350877193, 0.9473684210526315, 0.9385964912280702,
0.956140350877193, 0.9736842105263158, 0.9473684210526315, 0.9473684210526315]
[0.9122807017543859, 0.956140350877193, 0.9473684210526315, 0.9385964912280702,
0.956140350877193, 0.9736842105263158, 0.9473684210526315, 0.9473684210526315,
0.9298245614035088]
[0.9122807017543859, 0.956140350877193, 0.9473684210526315, 0.9385964912280702,
0.956140350877193, 0.9736842105263158, 0.9473684210526315, 0.9473684210526315,
0.9298245614035088, 0.9298245614035088]
```

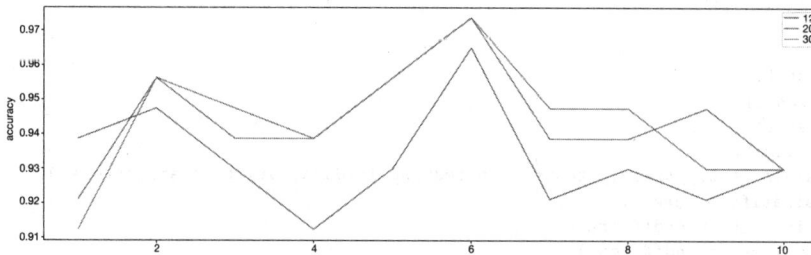

FIG. A.14 Output of Practical Exercise Q1: Task 1 (for colour figure, please see Colour Plate 3)

[ ]:

[ ]:

Practical Exercise Q1: Task 2

```
[15]: from sklearn.datasets import fetch_olivetti_faces
      from sklearn.neighbors import KNeighborsClassifier
      from sklearn.model_selection import train_test_split
      import numpy as np
      from sklearn import tree
      from statistics import mean
      from sklearn.cluster import kmeans_plusplus
      import matplotlib.pyplot as plt
      from sklearn.neighbors import NearestCentroid
      from collections import Counter
      from sklearn.naive_bayes import GaussianNB
      from sklearn.naive_bayes import BernoulliNB
      from sklearn.datasets import load_breast_cancer as LBC
```

```
[16]: bcdata = LBC()
      data = bcdata.data
      target = bcdata.target
      data
```

```
[16]: array([[1.799e+01, 1.038e+01, 1.228e+02, ..., 2.654e-01, 4.601e-01,
              1.189e-01],
             [2.057e+01, 1.777e+01, 1.329e+02, ..., 1.860e-01, 2.750e-01,
              8.902e-02],
             [1.969e+01, 2.125e+01, 1.300e+02, ..., 2.430e-01, 3.613e-01,
              8.758e-02],
             ...,
             [1.660e+01, 2.808e+01, 1.083e+02, ..., 1.418e-01, 2.218e-01,
              7.820e-02],
             [2.060e+01, 2.933e+01, 1.401e+02, ..., 2.650e-01, 4.087e-01,
              1.240e-01],
             [7.760e+00, 2.454e+01, 4.792e+01, ..., 0.000e+00, 2.871e-01,
              7.039e-02]])
```

```
[17]: X_train_list=[]
      X_test_list=[]
      y_train_list=[]
      y_test_list=[]
      for j in range(10):
          X_train,X_test,y_train,y_test=train_test_split(data,target, train_size = 0.
      ↪8 , stratify = target)
          X_train_list.append(X_train)
          X_test_list.append(X_test)
          y_train_list.append(y_train)
          y_test_list.append(y_test)
```

```
[18]: def kmeansplus(feature_mat,K):
          result_mat=np.empty((len(feature_mat),K))
          centers, indices = kmeans_plusplus(feature_mat.T, n_clusters=K)
          centers=centers.T
          return centers
```

```
[19]: kmeans_plus_res={}
      K=[12,20,30]
      arr=[]
      for k_value in K:
          X_train_list_single=[]
          X_test_list_single=[]
```

```
    for i in range(len(X_train_list)):
        cluster_array=np.append(X_train_list[i],X_test_list[i],axis=0)
        arr=kmeansplus(cluster_array,k_value)
        X_train_list_single.append(arr[:455])
        X_test_list_single.append(arr[455:])
        #print("done")
    kmeans_plus_res[k_value]=[X_train_list_single,X_test_list_single]
```

[20]: `len(arr[0])`

[20]: 30

[21]:
```
fin_acc={}
x=[1,2,3,4,5,6,7,8,9,10]
for k_value in K:
    acc=[]
    train_test_data=kmeans_plus_res[k_value]
    for i in range(len(train_test_data[0])):
        clf = GaussianNB()
        clf.fit(train_test_data[0][i], y_train_list[i])
        acc.append(clf.score(train_test_data[1][i],y_test_list[i]))
        print(acc)
    fin_acc[k_value]=acc
f = plt.figure()
f.set_figwidth(20)
f.set_figheight(6)
for i in fin_acc:
    plt.plot(x,fin_acc[i], label = i)
plt.legend()
plt.ylabel("accuracy")
plt.show()
```

```
[0.9210526315789473]
[0.9210526315789473, 0.9298245614035088]
[0.9210526315789473, 0.9298245614035088, 0.9122807017543859]
[0.9210526315789473, 0.9298245614035088, 0.9122807017543859, 0.9649122807017544]
[0.9210526315789473, 0.9298245614035088, 0.9122807017543859, 0.9649122807017544,
0.9298245614035088]
[0.9210526315789473, 0.9298245614035088, 0.9122807017543859, 0.9649122807017544,
0.9298245614035088, 0.9649122807017544]
[0.9210526315789473, 0.9298245614035088, 0.9122807017543859, 0.9649122807017544,
0.9298245614035088, 0.9649122807017544, 0.9385964912280702]
[0.9210526315789473, 0.9298245614035088, 0.9122807017543859, 0.9649122807017544,
0.9298245614035088, 0.9649122807017544, 0.9385964912280702, 0.9210526315789473]
[0.9210526315789473, 0.9298245614035088, 0.9122807017543859, 0.9649122807017544,
0.9298245614035088, 0.9649122807017544, 0.9385964912280702, 0.9210526315789473,
0.9210526315789473]
[0.9210526315789473, 0.9298245614035088, 0.9122807017543859, 0.9649122807017544,
0.9298245614035088, 0.9649122807017544, 0.9385964912280702, 0.9210526315789473,
0.9210526315789473, 0.9298245614035088]
[0.9649122807017544]
[0.9649122807017544, 0.9298245614035088]
[0.9649122807017544, 0.9298245614035088, 0.9385964912280702]
[0.9649122807017544, 0.9298245614035088, 0.9385964912280702, 0.9298245614035088]
[0.9649122807017544, 0.9298245614035088, 0.9385964912280702, 0.9298245614035088,
0.9385964912280702]
[0.9649122807017544, 0.9298245614035088, 0.9385964912280702, 0.9298245614035088,
0.9385964912280702, 0.9736842105263158]
```

```
[0.9649122807017544, 0.9298245614035088, 0.9385964912280702, 0.9298245614035088,
0.9385964912280702, 0.9736842105263158, 0.956140350877193]
[0.9649122807017544, 0.9298245614035088, 0.9385964912280702, 0.9298245614035088,
0.9385964912280702, 0.9736842105263158, 0.956140350877193, 0.9298245614035088]
[0.9649122807017544, 0.9298245614035088, 0.9385964912280702, 0.9298245614035088,
0.9385964912280702, 0.9736842105263158, 0.956140350877193, 0.9298245614035088,
0.9385964912280702]
[0.9649122807017544, 0.9298245614035088, 0.9385964912280702, 0.9298245614035088,
0.9385964912280702, 0.9736842105263158, 0.956140350877193, 0.9298245614035088,
0.9385964912280702, 0.956140350877193]
[0.9736842105263158]
[0.9736842105263158, 0.9210526315789473]
[0.9736842105263158, 0.9210526315789473, 0.9298245614035088]
[0.9736842105263158, 0.9210526315789473, 0.9298245614035088, 0.9298245614035088]
[0.9736842105263158, 0.9210526315789473, 0.9298245614035088, 0.9298245614035088,
0.9385964912280702]
[0.9736842105263158, 0.9210526315789473, 0.9298245614035088, 0.9298245614035088,
0.9385964912280702, 0.956140350877193]
[0.9736842105263158, 0.9210526315789473, 0.9298245614035088, 0.9298245614035088,
0.9385964912280702, 0.956140350877193, 0.9649122807017544]
[0.9736842105263158, 0.9210526315789473, 0.9298245614035088, 0.9298245614035088,
0.9385964912280702, 0.956140350877193, 0.9649122807017544, 0.9473684210526315]
[0.9736842105263158, 0.9210526315789473, 0.9298245614035088, 0.9298245614035088,
0.9385964912280702, 0.956140350877193, 0.9649122807017544, 0.9473684210526315,
0.956140350877193]
[0.9736842105263158, 0.9210526315789473, 0.9298245614035088, 0.9298245614035088,
0.9385964912280702, 0.956140350877193, 0.9649122807017544, 0.9473684210526315,
0.956140350877193, 0.9649122807017544]
```

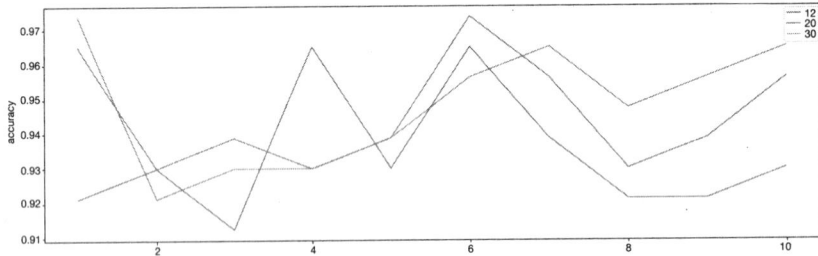

FIG. A.15 Output of Practical Exercise Q1: Task 2 (for colour figure, please see Colour Plate 3)

[ ]:

[ ]:

Chapter 8

Practical Exercise Q1

```
[1]: from sklearn import datasets
     from sklearn.datasets import load_digits
     import pandas as pd
     import numpy as np
     from sklearn.model_selection import train_test_split
     from sklearn.linear_model import Perceptron
     from sklearn import svm
     from sklearn.linear_model import LogisticRegression as lr
     from sklearn.neural_network import MLPClassifier as mlp
     import matplotlib.pyplot as plt
     X, y = load_digits(return_X_y=True)
     X_train,X_test,y_train,y_test = train_test_split(X,y,test_size=0.2)
     import warnings
     warnings.filterwarnings("ignore")
```

```
[2]: data = datasets.fetch_olivetti_faces()
     facedata = pd.DataFrame(data.data)
     facename = data.target
     X=np.array(facedata)
     y=np.array(facename)
     X_train, X_test, y_train, y_test = train_test_split(X,y)
     X.shape
```

```
[2]: (400, 4096)
```

```
[3]: from sklearn.linear_model import Perceptron
     from sklearn import svm
     from sklearn.linear_model import LogisticRegression as lr
     from sklearn.neural_network import MLPClassifier as mlp
```

```
[4]: clfp = Perceptron()
     clfp.fit(X_train, y_train)
     clfp.score(X_test, y_test)
```

```
[4]: 0.87
```

```
[5]: clfs = svm.SVC()
     clfs.fit(X_train,y_train)
     clfs.score(X_test,y_test)
```

```
[5]: 0.85
```

```
[6]: clfl = lr()
     clfl.fit(X_train,y_train)
     clfl.score(X_test,y_test)
```

```
[6]: 0.94
```

```
[7]: clfmlp=mlp()
     clfmlp.fit(X_train,y_train)
```

```
[7]: MLPClassifier()
```

```
[8]: clfmlp.score(X_test,y_test)
```

```
[8]: 0.57
```

```
[9]: acc_dict = {}
     for j in [0.1,0.2,0.3,0.4,0.5,0.6,0.7,0.8,0.9]:
         acc_dict[j]=[]
         X_train,X_test,y_train,y_test = train_test_split(X,y,test_size=j)
         #clfp.fit(X_train,y_train)
         acc = clfp.score(X_test,y_test)
         acc_dict[j].append((1,acc))
         acc= clfs.score(X_test,y_test)
         acc_dict[j].append((2,acc))
         acc= clfl.score(X_test,y_test)
         acc_dict[j].append((3,acc))
         acc= clfmlp.score(X_test,y_test)
         acc_dict[j].append((4,acc))
```

```
[10]: colors = ['blue', 'green', 'red', 'black']
      c1 = []
      c2 = []
      c3 = []
      c4 = []
      for i,r in enumerate(acc_dict.keys()):
          x,y = list(zip(*acc_dict[r]))
          #print (x,y)
          c1.append(y[0])
          c2.append(y[1])
          c3.append(y[2])
          c4.append(y[3])
      print(c1)
      print(y)
```

```
[0.9, 0.95, 0.9333333333333333, 0.9375, 0.93, 0.9416666666666667, 0.925,
0.93125, 0.9361111111111111]
(0.9361111111111111, 0.9583333333333334, 0.9861111111111112, 0.7416666666666667)
```

```
[11]: xlist = [0.1,0.2,0.3,0.4,0.5,0.6,0.7,0.8,0.9]
      plt.plot(xlist, c1, label='perceptron', color = 'blue')
      plt.plot(xlist,c2, label = 'svm', color = 'red')
      plt.plot(xlist,c3, label = 'logistic regression', color = 'green')
      plt.plot(xlist,c4, label='mlp', color = 'black')
      plt.title('Linear Classifiers on Olivetti Faces Data',fontname="Calibri",
        ↪size=15,fontweight="bold")
      plt.xlabel('Size of Test Set',fontname="Calibri", size=15,fontweight="bold" )
      plt.ylabel('Accuracy',fontname="Calibri", size=15,fontweight="bold")
      plt.legend()
      plt.show()
```

```
findfont: Font family ['Calibri'] not found. Falling back to DejaVu Sans.
```

Linear Classifiers on Olivetti Faces Data

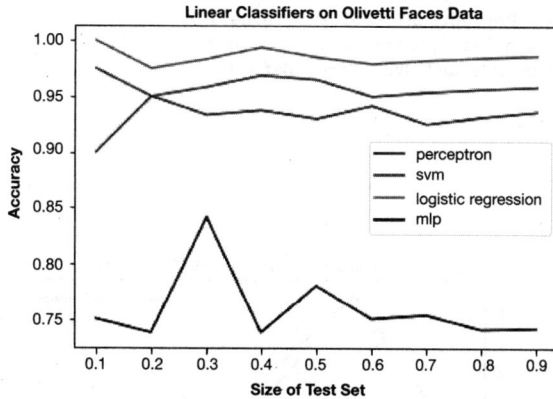

FIG. A.16 Accuracy of linear classifiers for the Olivetti Face data set (for colour figure, please see Colour Plate 3)

```
[90]: X=np.array(facedata)
      y=np.array(facename)
      X.shape
```

```
[90]: (400, 4096)
```

```
[91]: from sklearn.decomposition import PCA
      from sklearn.preprocessing import StandardScaler
      X = StandardScaler().fit_transform(X)
      pca = PCA(n_components=400)
      from sklearn.decomposition import TruncatedSVD
```

```
[92]: svd = TruncatedSVD(400)
      XS = svd.fit_transform(X)
      X = pca.fit_transform(X)
```

```
[93]: X.shape
```

```
[93]: (400, 400)
```

```
[94]: modelp = Perceptron()
      cp = []
      cs = []
      for i in [1,2,3,4,5]:
          xtp, xttp, ytp, yttp = train_test_split(X, y, test_size = i/10)
          clfpp = modelp.fit(xtp,ytp)
          accp = clfpp.score(xttp,yttp)
          cp.append(accp)
          xts,xtts,yts,ytts = train_test_split(XS,y, test_size=i/10)
          clfps = modelp.fit(xts,yts)
          accs = clfps.score(xtts,ytts)
          cs.append(accs)
      print(cp)
      print(cs)
```

```
[0.875, 0.925, 0.8833333333333333, 0.83125, 0.835]
[0.875, 0.925, 0.8916666666666667, 0.90625, 0.835]
```

```
[95]:  plt.plot(cp, label='pca-Perceptron', color = 'blue')
       plt.plot(cs, label = 'svd-Perceptron', color = 'red')
       plt.title('Perceptron on PCs and SVs: Faces Data')
       plt.legend()
       plt.xlabel('Test Size*10')
       plt.ylabel('Accuracy')
       plt.show()
```

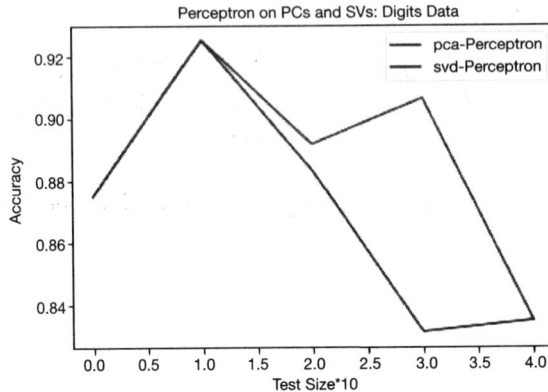

FIG. A.17 Accuracy of perceptron on principal components and singular vectors for the Digits data set (for colour figure, please see Colour Plate 3)

```
[96]:  models = svm.SVC()
       cp = []
       cs = []
       for i in [1,2,3,4,5]:
           xtp, xttp, ytp, yttp = train_test_split(X, y, test_size = i/10)
           clfsp = models.fit(xtp,ytp)
           accp = clfsp.score(xttp,yttp)
           cp.append(accp)
           xts,xtts,yts,ytts = train_test_split(XS,y, test_size=i/10)
           clfss = models.fit(xts,yts)
           accs = clfss.score(xtts,ytts)
           cs.append(accs)
       print(cp)
       print(cs)

       [0.95, 0.925, 0.825, 0.875, 0.62]
       [0.925, 0.85, 0.8666666666666667, 0.78125, 0.765]
```

```
[97]:  plt.plot(cp, label='pca-SVM', color = 'blue')
       plt.plot(cs, label = 'svd-SVM', color = 'red')
       plt.title('SVM on PCs and SVs: Faces Data')
       plt.legend()
       plt.xlabel('Test Size*10')
       plt.ylabel('Accuracy')
       plt.show()
```

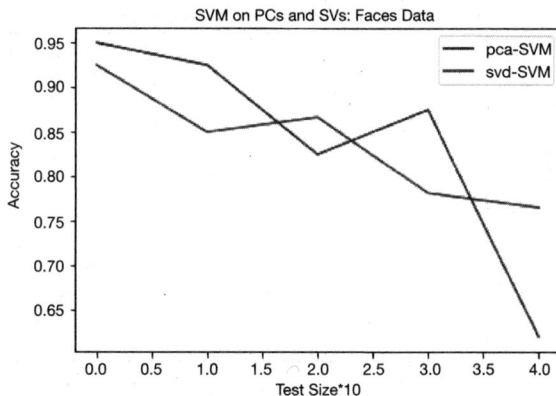

FIG. A.18 Accuracy of support vector machine on principal components and singular vectors for the Olivetti Face data set (for colour figure, please see Colour Plate 3)

```
[98]: modell = lr()
      cp = []
      cs = []
      for i in [1,2,3,4,5]:
          xtp, xttp, ytp, yttp = train_test_split(X, y, test_size = i/10)
          clflp = modell.fit(xtp,ytp)
          accp = clflp.score(xttp,yttp)
          cp.append(accp)
          xts,xtts,yts,ytts = train_test_split(XS,y, test_size=i/10)
          clfls = modell.fit(xts,yts)
          accs = clfls.score(xtts,ytts)
          cs.append(accs)
      print(cp)
      print(cs)

      [1.0, 0.9625, 0.9416666666666667, 0.95625, 0.925]
      [0.95, 0.9875, 0.925, 0.93125, 0.925]
```

```
[99]: plt.plot(cp, label='pca-LR', color = 'blue')
      plt.plot(cs, label = 'svd-LR', color = 'red')
      plt.title('LR on PCs and SVs: Faces Data')
      plt.legend()
      plt.xlabel('Test Size*10')
      plt.ylabel('Accuracy')
      plt.show()
```

```
[100]: modelm = mlp()
       cp = []
       cs = []
       for i in [1,2,3,4,5]:
           xtp, xttp, ytp, yttp = train_test_split(X, y, test_size = i/10)
           clfmp = modelm.fit(xtp,ytp)
           accp = clfmp.score(xttp,yttp)
           cp.append(accp)
           xts,xtts,yts,ytts = train_test_split(XS,y, test_size=i/10)
           clfms = modelm.fit(xts,yts)
           accs = clfms.score(xtts,ytts)
```

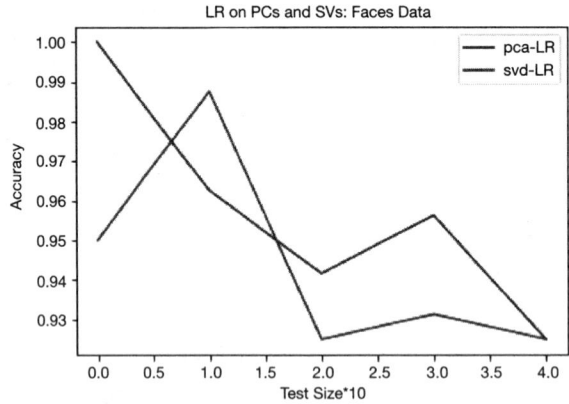

FIG. A.19 Accuracy of linear regression on principal components and singular vectors for the Olivetti Face data set (for colour figure, please see Colour Plate 3)

```
        cs.append(accs)
print(cp)
print(cs)

[0.725, 0.6125, 0.65, 0.60625, 0.475]
[0.825, 0.7, 0.5916666666666667, 0.6, 0.44]
```

```
[101]:  plt.plot(cp, label='pca-MLP', color = 'blue')
        plt.plot(cs, label = 'svd-MLP', color = 'red')
        plt.title('MLP on PCs and SVs: Faces Data')
        plt.legend()
        plt.xlabel('Test Size*10')
        plt.ylabel('Accuracy')
        plt.show()
```

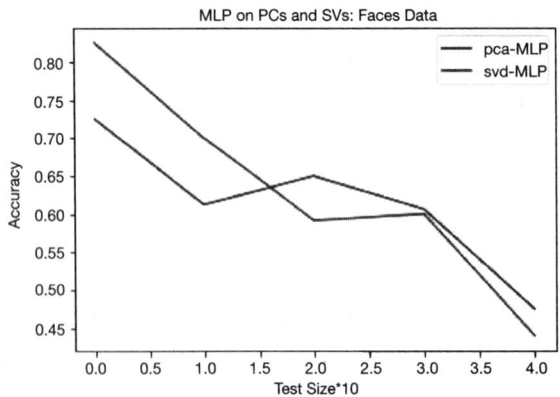

FIG. A.20 Accuracy of multi-layer perceptron on principal components and singular vectors for the Olivetti Face data set (for colour figure, please see Colour Plate 3)

```
[113]:  X=np.array(facedata)
        y=np.array(facename)
        X.shape
```

```
[113]:  (400, 4096)
```

```
[114]:  from sklearn.decomposition import PCA
        from sklearn.preprocessing import StandardScaler
        X = StandardScaler().fit_transform(X)
        pca = PCA(n_components=400)
        from sklearn.decomposition import TruncatedSVD
        svd = TruncatedSVD(400)
        XS = svd.fit_transform(X)
        X = pca.fit_transform(X)
        XS.shape
```

```
[114]:  (400, 400)
```

```
[115]:  acc_dict = {}
        for j in [0.1,0.2,0.3,0.4,0.5,0.6,0.7,0.8,0.9]:
            acc_dict[j]=[]
            X_train,X_test,y_train,y_test = train_test_split(X,y,test_size=j)
            #clfp.fit(X_train,y_train)
            acc = clfpp.score(X_test,y_test)
            acc_dict[j].append((1,acc))
            acc= clfsp.score(X_test,y_test)
            acc_dict[j].append((2,acc))
            acc= clflp.score(X_test,y_test)
            acc_dict[j].append((3,acc))
            acc= clfmp.score(X_test,y_test)
            acc_dict[j].append((4,acc))
```

```
[117]:  colors = ['blue', 'green', 'red', 'black']
        c1 = []
        c2 = []
        c3 = []
        c4 = []
        for i,r in enumerate(acc_dict.keys()):
            x,p = list(zip(*acc_dict[r]))
            #print (x,y)
            c1.append(p[0])
            c2.append(p[1])
            c3.append(p[2])
            c4.append(p[3])
        print(c1)
        print(p)
```

```
        [0.95, 0.9125, 0.9166666666666666, 0.9, 0.935, 0.9083333333333333,
        0.9214285714285714, 0.915625, 0.9222222222222223]
        (0.9222222222222223, 0.8722222222222222, 0.9666666666666667, 0.7305555555555555)
```

```
[118]:  xlist = [0.1,0.2,0.3,0.4,0.5,0.6,0.7,0.8,0.9]
        plt.plot(xlist, c1, label='perceptron', color = 'blue')
        plt.plot(xlist,c2, label = 'svm', color = 'red')
        plt.plot(xlist,c3, label = 'logistic regression', color = 'green')
        plt.plot(xlist,c4, label='mlp', color = 'black')
        plt.title('PCs and Linear Classifier Faces Data',fontname="Calibri",
          ↪size=12,fontweight="bold")
        plt.xlabel('Size of Test Set',fontname="Calibri", size=15,fontweight="bold" )
        plt.ylabel('Accuracy',fontname="Calibri", size=15,fontweight="bold")
        plt.legend()
        plt.show()
```

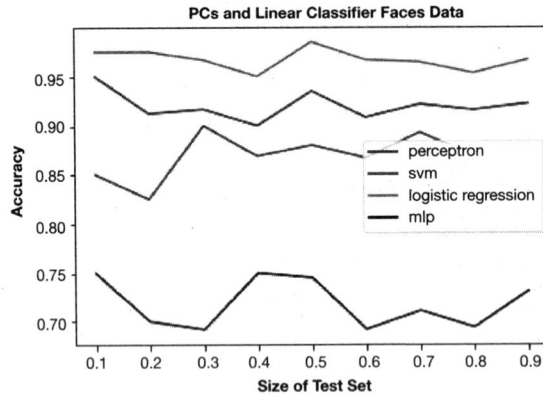

FIG. A.21 Accuracy of principal components and linear classifiers for the Olivetti Face data set (for colour figure, please see Colour Plate 3)

```
[121]: acc_dict = {}
       for j in [0.1,0.2,0.3,0.4,0.5,0.6,0.7,0.8,0.9]:
           acc_dict[j]=[]
           X_train,X_test,y_train,y_test = train_test_split(XS,y,test_size=j)
           #clfp.fit(X_train,y_train)
           acc = clfps.score(X_test,y_test)
           acc_dict[j].append((1,acc))
           acc= clfss.score(X_test,y_test)
           acc_dict[j].append((2,acc))
           acc= clfls.score(X_test,y_test)
           acc_dict[j].append((3,acc))
           acc= clfms.score(X_test,y_test)
           acc_dict[j].append((4,acc))
```

```
[122]: colors = ['blue', 'green', 'red', 'black']
       c1 = []
       c2 = []
       c3 = []
       c4 = []
       for i,r in enumerate(acc_dict.keys()):
           x,p = list(zip(*acc_dict[r]))
           #print (x,y)
           c1.append(p[0])
           c2.append(p[1])
           c3.append(p[2])
           c4.append(p[3])
       print(c1)
       print(p)
```

```
       [0.975, 0.8875, 0.8916666666666667, 0.93125, 0.915, 0.9, 0.9107142857142857,
       0.903125, 0.9111111111111111]
       (0.9111111111111111, 0.8888888888888888, 0.9583333333333334, 0.7166666666666667)
```

```
[123]: xlist = [0.1,0.2,0.3,0.4,0.5,0.6,0.7,0.8,0.9]
       plt.plot(xlist, c1, label='perceptron', color = 'blue')
       plt.plot(xlist,c2, label = 'svm', color = 'red')
       plt.plot(xlist,c3, label = 'logistic regression', color = 'green')
```

```
plt.plot(xlist,c4, label='mlp', color = 'black')
plt.title('SVs and Linear Classifier Faces Data',fontname="Calibri",
    ↪size=12,fontweight="bold")
plt.xlabel('Size of Test Set',fontname="Calibri", size=15,fontweight="bold" )
plt.ylabel('Accuracy',fontname="Calibri", size=15,fontweight="bold")
plt.legend()
plt.show()
```

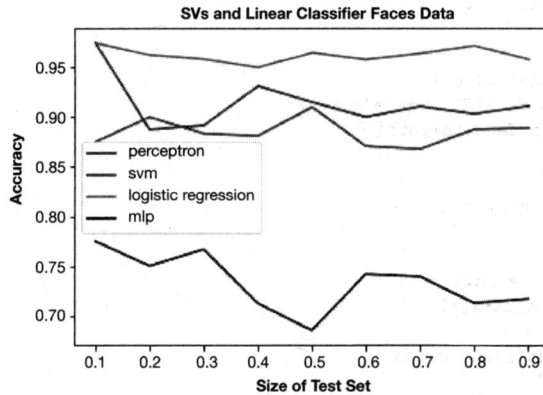

FIG. A.22 Accuracy of singular vectors and linear classifiers for the Olivetti Face data set (for colour figure, please see Colour Plate 3)

[ ]:

Chapter 9

Practical Exercise Q1

```
[1]: import torch
     from torchvision import datasets
     from torchvision.transforms import ToTensor
     from torch.utils.data import DataLoader


     # Device configuration
     device = torch.device('cuda' if torch.cuda.is_available() else 'cpu')
     device
```

```
[1]: device(type='cpu')
```

```
[2]: train_data = datasets.MNIST(
         root = 'data',
         train = True,
         transform = ToTensor(),
         download = True,
     )
     test_data = datasets.MNIST(
         root = 'data',
         train = False,
```

```
        transform = ToTensor()
)

print(train_data)
print(test_data)
print(train_data.data.size())
print(train_data.targets.size())

print(test_data.data.size())
print(test_data.targets.size())


import matplotlib.pyplot as plt
plt.imshow(train_data.data[0], cmap='gray')
plt.title('%i' % train_data.targets[0])
plt.show()

figure = plt.figure(figsize=(10, 8))
cols, rows = 10, 5
j = 0
for i in range(1, cols * rows + 1):
    img, label = train_data[j]
    j=j+1
    figure.add_subplot(rows, cols, i)
    plt.title(label)
    plt.axis("off")
    plt.imshow(img.squeeze(), cmap="gray")
plt.show()
```

```
Dataset MNIST
    Number of datapoints: 60000
    Root location: data
    Split: Train
    StandardTransform
Transform: ToTensor()
Dataset MNIST
    Number of datapoints: 10000
    Root location: data
    Split: Test
    StandardTransform
Transform: ToTensor()
torch.Size([60000, 28, 28])
torch.Size([60000])
torch.Size([10000, 28, 28])
torch.Size([10000])
```

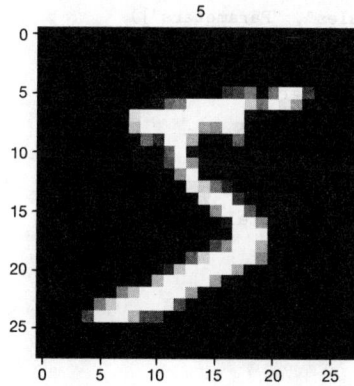

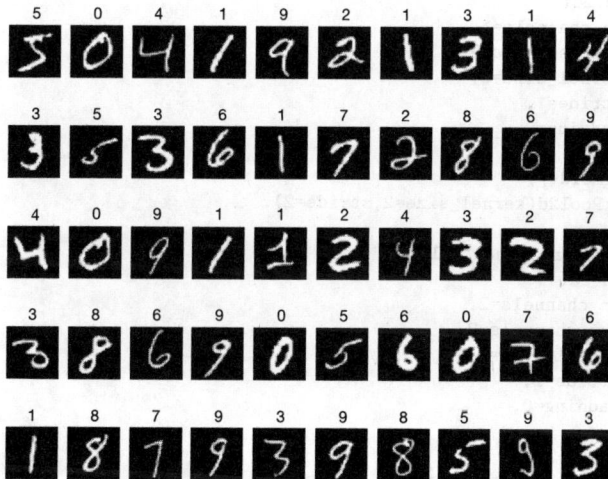

FIG. A.23 MNIST handwritten digits data set

```
[3]:  loaders = {
          'train' : DataLoader(train_data,
                                          batch_size=200,
                                          shuffle=True,
                                          num_workers=1),
          'test' : DataLoader(test_data,
                                          batch_size=200,
                                          shuffle=True,
                                          num_workers=1),
      }
      loaders
```

```
[3]: {'train': <torch.utils.data.dataloader.DataLoader at 0x7f7eae64d6d0>,
      'test': <torch.utils.data.dataloader.DataLoader at 0x7f7eae6a75b0>}
```

```
[4]:  from prettytable import PrettyTable

      def count_parameters(model):
```

```
        table = PrettyTable(["Modules", "Parameters"])
        total_params = 0
        for name, parameter in model.named_parameters():
            if not parameter.requires_grad: continue
            params = parameter.numel()
            table.add_row([name, params])
            total_params+=params
        print(table)
        print(f"Total Trainable Params: total_params")
        return total_params
```

```
[5]:  import torch.nn as nn
      class CNN(nn.Module):
          def __init__(self):
              super(CNN, self).__init__()
              self.conv1 = nn.Sequential(
                  nn.Conv2d(
                      in_channels=1,
                      out_channels=2,
                      kernel_size=3,
                      stride=1,
                      padding=0,
                  ),
                  nn.Sigmoid(),
                  nn.MaxPool2d(kernel_size=2,stride=2),
              )
              self.conv2 = nn.Sequential(
                  nn.Conv2d(
                      in_channels=2,
                      out_channels=2,
                      kernel_size=3,
                      stride=1,
                      padding=0,
                  ),
                  nn.ReLU(),
                  nn.MaxPool2d(kernel_size=2,stride=2),
              )
              # fully connected layer, output 10 classes
              self.FC1 = nn.Sequential(nn.Linear(2 * 5 * 5, 50),nn.Tanh())
              self.out = nn.Sequential(nn.Linear( 50, 10),nn.Softmax(dim=1))

          def forward(self, x):
              x = self.conv1(x)
              x = self.conv2(x)
              # flatten the output of conv2 to (batch_size, 2 * 5 * 5)
              x = x.view(x.size(0), -1)
              x = self.FC1(x)
              output = self.out(x)
              return output, x # return x for visualization

      cnn = CNN()
      print(cnn)


      count_parameters(cnn)

      loss_func = nn.CrossEntropyLoss()
      loss_func
```

```
from torch import optim
optimizer = optim.Adam(cnn.parameters(), lr = 0.01)
optimizer
```

```
CNN(
  (conv1): Sequential(
    (0): Conv2d(1, 2, kernel_size=(3, 3), stride=(1, 1))
    (1): Sigmoid()
    (2): MaxPool2d(kernel_size=2, stride=2, padding=0, dilation=1,
ceil_mode=False)
  )
  (conv2): Sequential(
    (0): Conv2d(2, 2, kernel_size=(3, 3), stride=(1, 1))
    (1): ReLU()
    (2): MaxPool2d(kernel_size=2, stride=2, padding=0, dilation=1,
ceil_mode=False)
  )
  (FC1): Sequential(
    (0): Linear(in_features=50, out_features=50, bias=True)
    (1): Tanh()
  )
  (out): Sequential(
    (0): Linear(in_features=50, out_features=10, bias=True)
    (1): Softmax(dim=1)
  )
)
+----------------+------------+
|    Modules     | Parameters |
+----------------+------------+
| conv1.0.weight |     18     |
|  conv1.0.bias  |     2      |
| conv2.0.weight |     36     |
|  conv2.0.bias  |     2      |
|  FC1.0.weight  |    2500    |
|   FC1.0.bias   |     50     |
|  out.0.weight  |    500     |
|   out.0.bias   |     10     |
+----------------+------------+
Total Trainable Params: 3118
```

```
[5]: Adam (
     Parameter Group 0
         amsgrad: False
         betas: (0.9, 0.999)
         capturable: False
         differentiable: False
         eps: 1e-08
         foreach: None
         fused: None
         lr: 0.01
         maximize: False
         weight\_decay: 0
     )
```

```
[6]: from torch.autograd import Variable
     num_epochs = 20
     def train(num_epochs, model, loaders):

         acc_dict = []
         loss_dict = []
```

```
model.train()

# Train the model
total_step = len(loaders['train'])

for epoch in range(num_epochs):
    acc1 = 0
    total = 0
    total_loss = 0
    iter = 0
    for i, (images, labels) in enumerate(loaders['train']):

        iter = iter+1
        # gives batch data, normalize x when iterate train_loader
        b_x = Variable(images)        # batch x
        b_y = Variable(labels)        # batch y
        output = model(b_x)[0]
        loss = loss_func(output, b_y)

        # clear gradients for this training step
        optimizer.zero_grad()

        # backpropagation, compute gradients
        loss.backward()
        # apply gradients
        optimizer.step()

        pred_y = torch.max(output, 1)[1].data.squeeze()
        accuracy = (pred_y == labels).sum().item() / float(labels.size(0))
        acc1 = acc1 + accuracy
        total = total + float(labels.size(0))
        total_loss = total_loss + loss.item()



        if (i+1) % 100 == 0:
            print ('Epoch [{}/{}], Step [{}/{}], Loss: {:.4f}, Accuracy: {:.
↪4f}'
                        .format(epoch + 1, num_epochs, i + 1, total_step, loss.
↪item(),accuracy))
                #if (i+1) % 300 == 0 :
                    #acc_dict.append(accuracy)
                    #loss_dict.append(loss.item())
                pass

    epoch_accuracy = acc1/iter
    epoch_loss = total_loss/iter
    acc_dict.append(epoch_accuracy)
    loss_dict.append(epoch_loss)
    print ('Epoch [{}/{}], Loss: {:.4f}, Accuracy: {:.4f}'
                .format(epoch + 1, num_epochs,
↪epoch_loss,epoch_accuracy))
    pass


pass
return (acc_dict,loss_dict)
```

```
[7]: acc_dict, loss_dict = train(num_epochs, cnn, loaders)

     print(acc_dict)
     print(loss_dict)
```

```
Epoch [1/20], Step [100/300], Loss: 1.7602, Accuracy: 0.7200
Epoch [1/20], Step [200/300], Loss: 1.6384, Accuracy: 0.8350
Epoch [1/20], Step [300/300], Loss: 1.6376, Accuracy: 0.8350
Epoch [1/20], Loss: 1.7749, Accuracy: 0.6956
Epoch [2/20], Step [100/300], Loss: 1.6345, Accuracy: 0.8300
Epoch [2/20], Step [200/300], Loss: 1.6261, Accuracy: 0.8450
Epoch [2/20], Step [300/300], Loss: 1.6174, Accuracy: 0.8550
Epoch [2/20], Loss: 1.6034, Accuracy: 0.8650
Epoch [3/20], Step [100/300], Loss: 1.6363, Accuracy: 0.8300
Epoch [3/20], Step [200/300], Loss: 1.5772, Accuracy: 0.8950
Epoch [3/20], Step [300/300], Loss: 1.5950, Accuracy: 0.8750
Epoch [3/20], Loss: 1.5874, Accuracy: 0.8790
Epoch [4/20], Step [100/300], Loss: 1.5883, Accuracy: 0.8900
Epoch [4/20], Step [200/300], Loss: 1.5841, Accuracy: 0.8800
Epoch [4/20], Step [300/300], Loss: 1.5923, Accuracy: 0.8750
Epoch [4/20], Loss: 1.5762, Accuracy: 0.8891
Epoch [5/20], Step [100/300], Loss: 1.5410, Accuracy: 0.9150
Epoch [5/20], Step [200/300], Loss: 1.5648, Accuracy: 0.8950
Epoch [5/20], Step [300/300], Loss: 1.5662, Accuracy: 0.9100
Epoch [5/20], Loss: 1.5721, Accuracy: 0.8931
Epoch [6/20], Step [100/300], Loss: 1.5684, Accuracy: 0.9000
Epoch [6/20], Step [200/300], Loss: 1.5837, Accuracy: 0.8750
Epoch [6/20], Step [300/300], Loss: 1.5686, Accuracy: 0.8900
Epoch [6/20], Loss: 1.5666, Accuracy: 0.8980
Epoch [7/20], Step [100/300], Loss: 1.5582, Accuracy: 0.9100
Epoch [7/20], Step [200/300], Loss: 1.5582, Accuracy: 0.9100
Epoch [7/20], Step [300/300], Loss: 1.5637, Accuracy: 0.9000
Epoch [7/20], Loss: 1.5640, Accuracy: 0.9008
Epoch [8/20], Step [100/300], Loss: 1.5588, Accuracy: 0.9050
Epoch [8/20], Step [200/300], Loss: 1.5736, Accuracy: 0.8850
Epoch [8/20], Step [300/300], Loss: 1.5631, Accuracy: 0.9050
Epoch [8/20], Loss: 1.5610, Accuracy: 0.9027
Epoch [9/20], Step [100/300], Loss: 1.5628, Accuracy: 0.9000
Epoch [9/20], Step [200/300], Loss: 1.5525, Accuracy: 0.9000
Epoch [9/20], Step [300/300], Loss: 1.5227, Accuracy: 0.9500
Epoch [9/20], Loss: 1.5553, Accuracy: 0.9092
Epoch [10/20], Step [100/300], Loss: 1.5370, Accuracy: 0.9200
Epoch [10/20], Step [200/300], Loss: 1.5707, Accuracy: 0.8900
Epoch [10/20], Step [300/300], Loss: 1.5766, Accuracy: 0.8900
Epoch [10/20], Loss: 1.5556, Accuracy: 0.9083
Epoch [11/20], Step [100/300], Loss: 1.5530, Accuracy: 0.9150
Epoch [11/20], Step [200/300], Loss: 1.5338, Accuracy: 0.9350
Epoch [11/20], Step [300/300], Loss: 1.5752, Accuracy: 0.8850
Epoch [11/20], Loss: 1.5518, Accuracy: 0.9120
Epoch [12/20], Step [100/300], Loss: 1.6073, Accuracy: 0.8500
Epoch [12/20], Step [200/300], Loss: 1.5477, Accuracy: 0.9150
Epoch [12/20], Step [300/300], Loss: 1.5677, Accuracy: 0.8900
Epoch [12/20], Loss: 1.5497, Accuracy: 0.9140
Epoch [13/20], Step [100/300], Loss: 1.5537, Accuracy: 0.9100
Epoch [13/20], Step [200/300], Loss: 1.5314, Accuracy: 0.9250
Epoch [13/20], Step [300/300], Loss: 1.5301, Accuracy: 0.9400
Epoch [13/20], Loss: 1.5465, Accuracy: 0.9170
Epoch [14/20], Step [100/300], Loss: 1.5511, Accuracy: 0.9050
Epoch [14/20], Step [200/300], Loss: 1.5371, Accuracy: 0.9250
Epoch [14/20], Step [300/300], Loss: 1.5665, Accuracy: 0.8950
```

```
Epoch [14/20], Loss: 1.5445, Accuracy: 0.9188
Epoch [15/20], Step [100/300], Loss: 1.5670, Accuracy: 0.8950
Epoch [15/20], Step [200/300], Loss: 1.5580, Accuracy: 0.9050
Epoch [15/20], Step [300/300], Loss: 1.5542, Accuracy: 0.9050
Epoch [15/20], Loss: 1.5432, Accuracy: 0.9205
Epoch [16/20], Step [100/300], Loss: 1.5633, Accuracy: 0.8950
Epoch [16/20], Step [200/300], Loss: 1.5381, Accuracy: 0.9400
Epoch [16/20], Step [300/300], Loss: 1.5768, Accuracy: 0.8800
Epoch [16/20], Loss: 1.5403, Accuracy: 0.9233
Epoch [17/20], Step [100/300], Loss: 1.5315, Accuracy: 0.9350
Epoch [17/20], Step [200/300], Loss: 1.5686, Accuracy: 0.8900
Epoch [17/20], Step [300/300], Loss: 1.5482, Accuracy: 0.9100
Epoch [17/20], Loss: 1.5372, Accuracy: 0.9263
Epoch [18/20], Step [100/300], Loss: 1.5490, Accuracy: 0.9200
Epoch [18/20], Step [200/300], Loss: 1.5353, Accuracy: 0.9300
Epoch [18/20], Step [300/300], Loss: 1.5437, Accuracy: 0.9200
Epoch [18/20], Loss: 1.5357, Accuracy: 0.9275
Epoch [19/20], Step [100/300], Loss: 1.5619, Accuracy: 0.8950
Epoch [19/20], Step [200/300], Loss: 1.5499, Accuracy: 0.9050
Epoch [19/20], Step [300/300], Loss: 1.5282, Accuracy: 0.9300
Epoch [19/20], Loss: 1.5363, Accuracy: 0.9273
Epoch [20/20], Step [100/300], Loss: 1.5272, Accuracy: 0.9350
Epoch [20/20], Step [200/300], Loss: 1.5354, Accuracy: 0.9250
Epoch [20/20], Step [300/300], Loss: 1.5287, Accuracy: 0.9300
Epoch [20/20], Loss: 1.5345, Accuracy: 0.9290
[0.6956333333333339, 0.8649666666666667, 0.8789666666666662, 0.8890666666666663,
0.8930666666666666, 0.8979666666666666, 0.9007666666666667, 0.9027499999999996,
0.9092, 0.908333333333332, 0.9119999999999989, 0.9140166666666671,
0.9169666666666675, 0.9187666666666666, 0.9204666666666664, 0.9232666666666666,
0.9263000000000003, 0.9274666666666669, 0.9272500000000002, 0.9290000000000002]
[1.7749135692914326, 1.603350172440211, 1.5874224253495535, 1.576213575998942,
1.5720923511187235, 1.566618199745814, 1.5639670741558076, 1.5610192465782164,
1.555258759657542, 1.5555630211035412, 1.5517947101593017, 1.5497142906983694,
1.5465335675080618, 1.5445477374394734, 1.5431940452257793, 1.540329338312149,
1.5372106997172037, 1.5357297921180726, 1.5363101089000701, 1.534504676659902]
```

```
[8]:  def plot_model_history(acc_dict,loss_dict,model_name) :
          plt.figure(figsize=(15, 15))
          plt.plot(loss_dict, label='Train Loss')
          plt.xlabel('Epoch', fontsize=18)
          plt.ylabel('Loss', fontsize=16)
          plt.legend()
          plt.title(model_name+' Training Loss')
          plt.show()

          plt.figure(figsize=(15, 15))
          plt.plot(acc_dict, label='Train Accuracy')
          plt.xlabel('Epoch', fontsize=18)
          plt.ylabel('Accuracy', fontsize=16)
          plt.legend()
          plt.title(model_name+' Training Accuracy')
          plt.show()


[9]:  plot_model_history(acc_dict,loss_dict,'CNN')
```

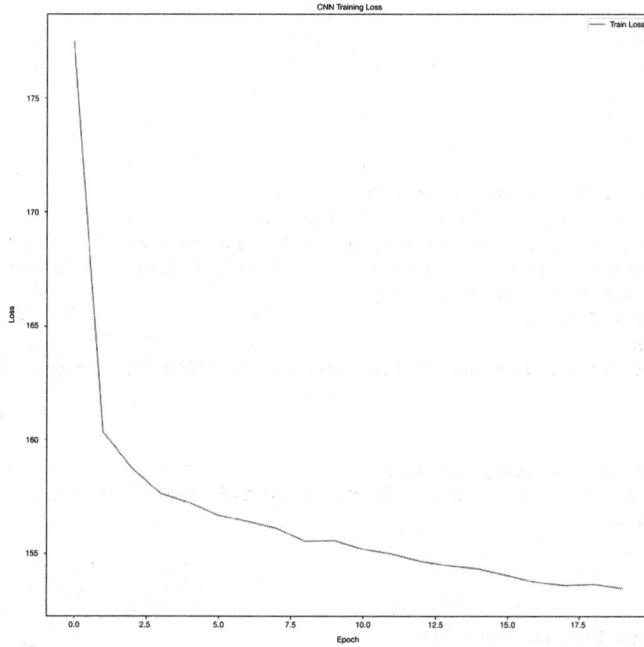

FIG. A.24 CNN training loss (for colour figure, please see Colour Plate 3)

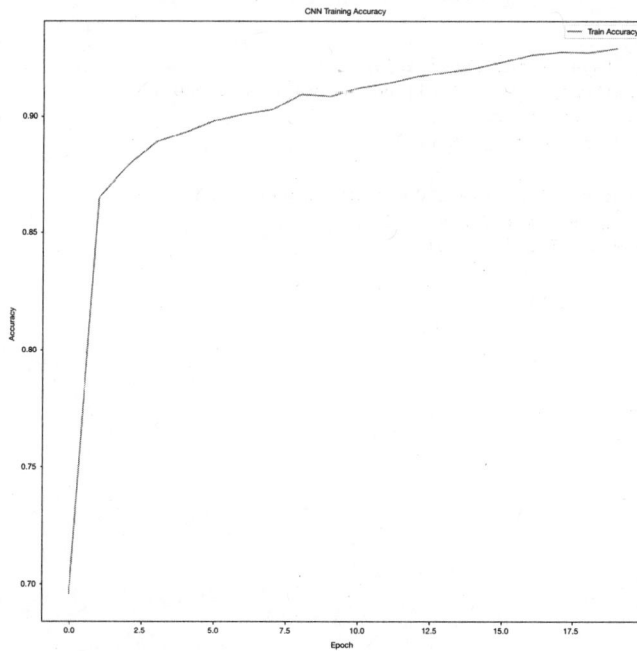

FIG. A.25 CNN training accuracy (for colour figure, please see Colour Plate 3)

```
[10]: def test(model):
          # Test the model
          model.eval()
          with torch.no_grad():
              correct = 0
              acc_sum = 0
              iter= 0
              for images, labels in loaders['test']:
                  test_output, last_layer = model(images)
                  pred_y = torch.max(test_output, 1)[1].data.squeeze()
                  accuracy = (pred_y == labels).sum().item() / float(labels.size(0))
                  acc_sum = acc_sum + accuracy
                  iter = iter +1
                  pass
                  #print('Test Accuracy of the model on the 10000 test images: %.2f'
       ↪% accuracy)
                  pass

              final_accuracy = (acc_sum/iter)
          print('Test Accuracy of the model on the 10000 test images: %.2f' %
       ↪final_accuracy)
```

```
[11]: test(cnn)

      sample = next(iter(loaders['test']))
      imgs, lbls = sample


      actual_number = lbls[:10].numpy()
      actual_number


      test_output, last_layer = cnn(imgs[:10])
      pred_y = torch.max(test_output, 1)[1].data.numpy().squeeze()
      print(f'Prediction number: pred_y')
      print(f'Actual number: actual_number')

      Test Accuracy of the model on the 10000 test images: 0.93
      Prediction number: [5 9 2 6 7 2 2 4 2 5]
      Actual number: [5 9 2 6 7 2 3 4 2 5]
```

```
[ ]:
```

Colour Plate 3

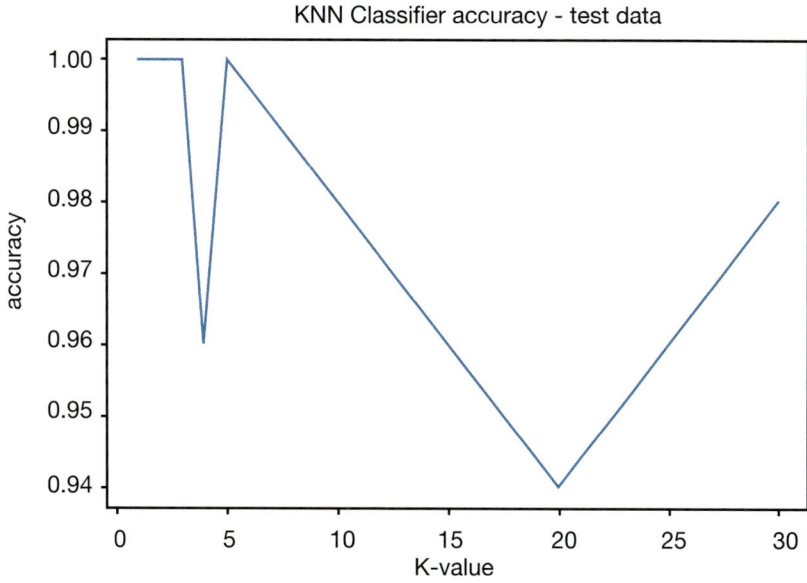

FIG. A.1 KNN classifier accuracy for test data

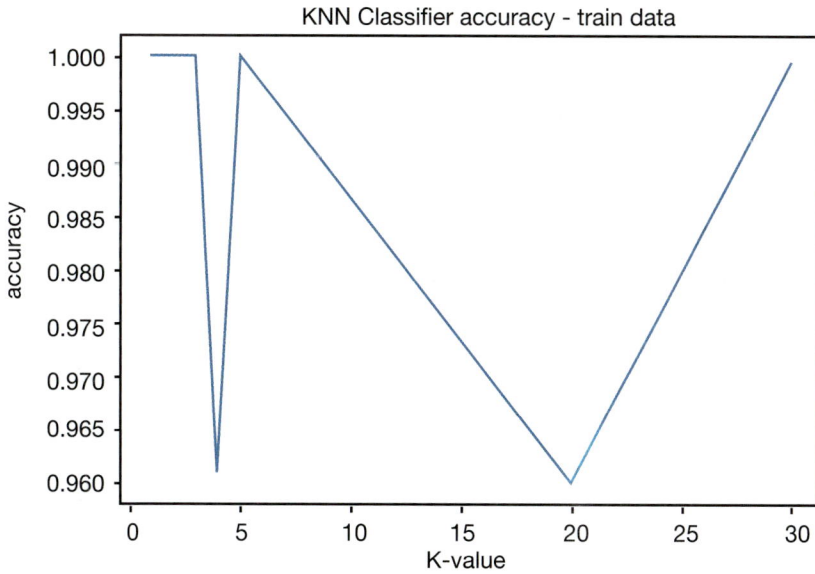

FIG. A.2 KNN classifier accuracy for training data

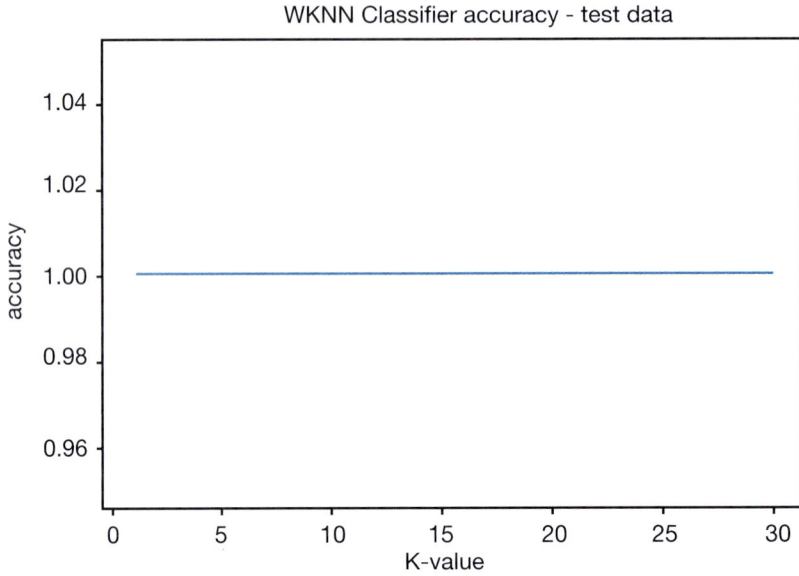

Fig. A.3 WKNN classifier accuracy for test data

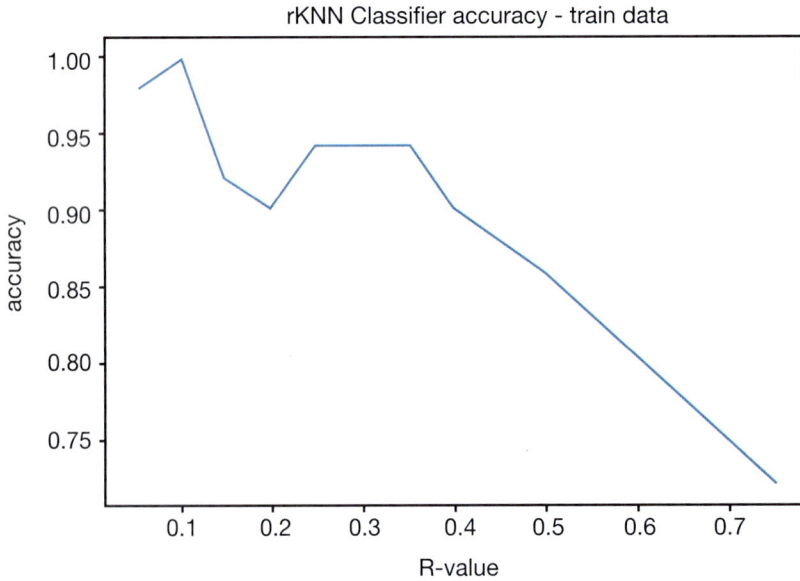

Fig. A.4 RKNN classifier accuracy for training data

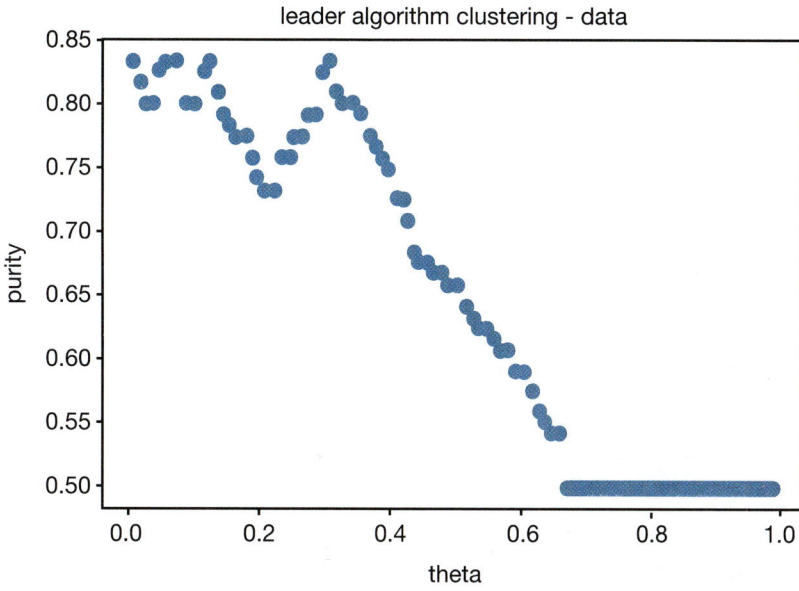

FIG. A.5 Purity of Leader algorithm clustering

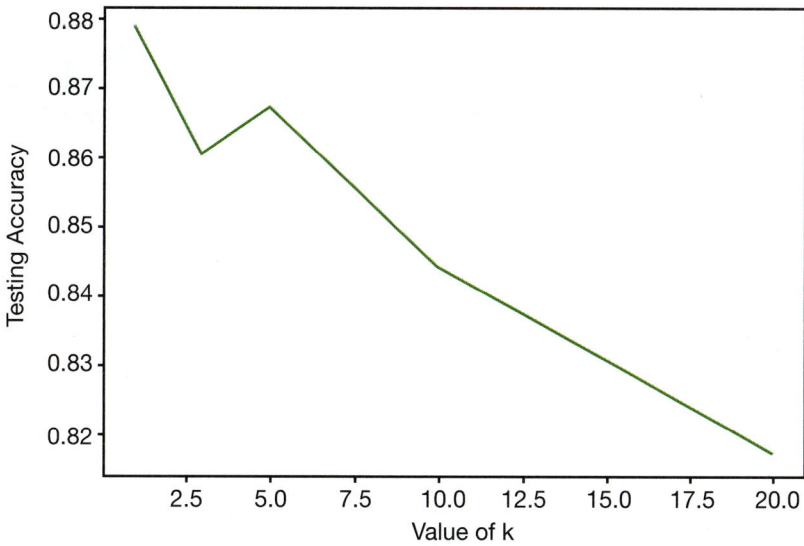

FIG. A.6 Testing accuracy

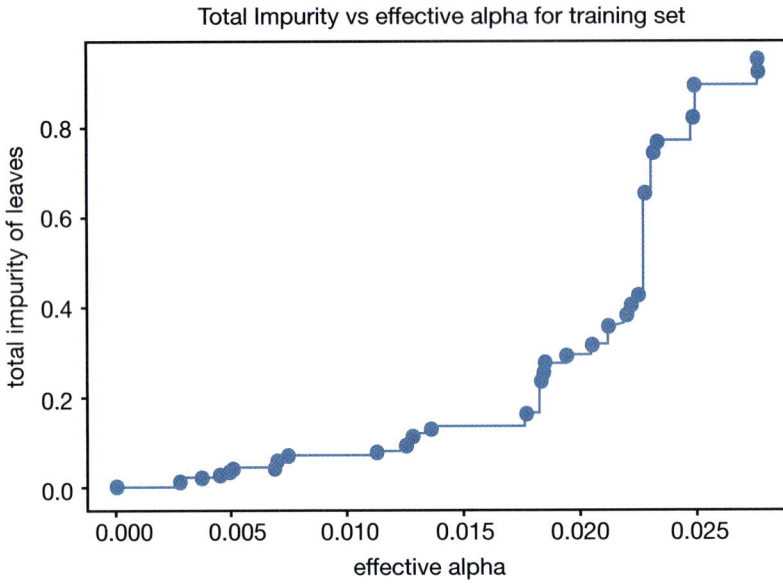

FIG. A.7 Total impurity versus effective alpha for training data

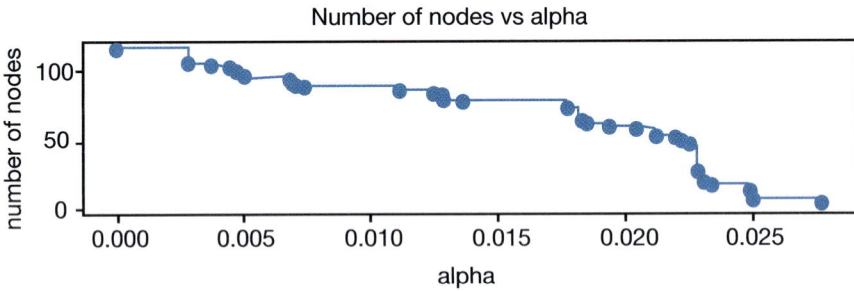

FIG. A.8 Number of nodes versus alpha

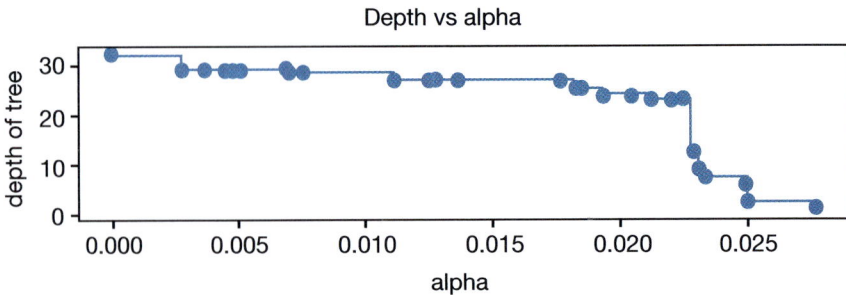

FIG. A.9 Depth versus alpha

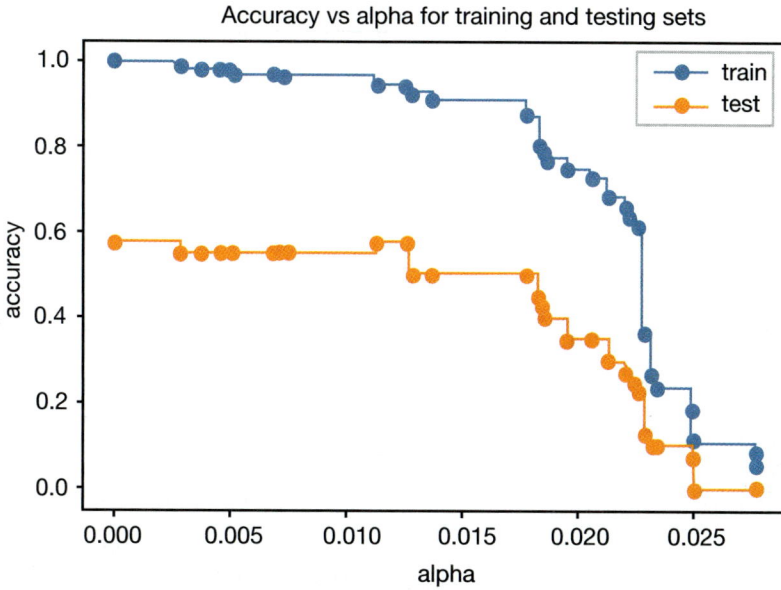

FIG. A.10 Accuracy versus alpha for training and testing data

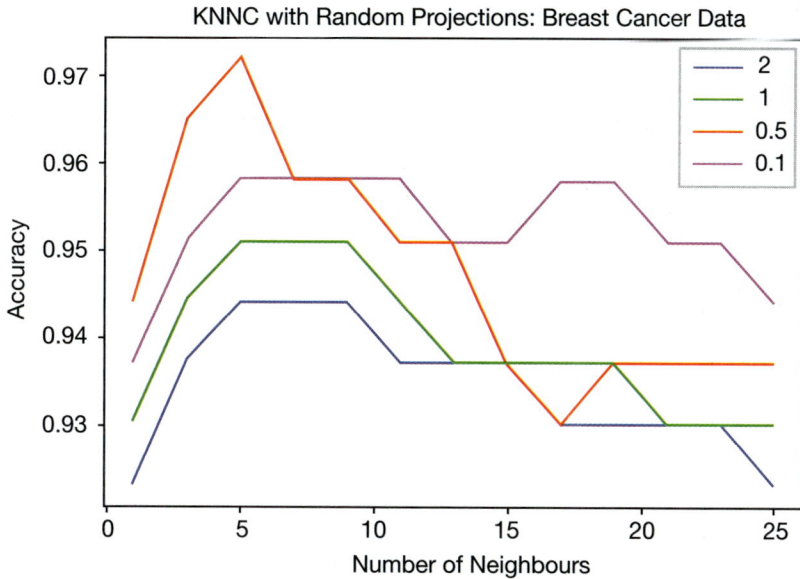

FIG. A.11 Accuracy of KNN classifier with random projections for the Breast Cancer data set

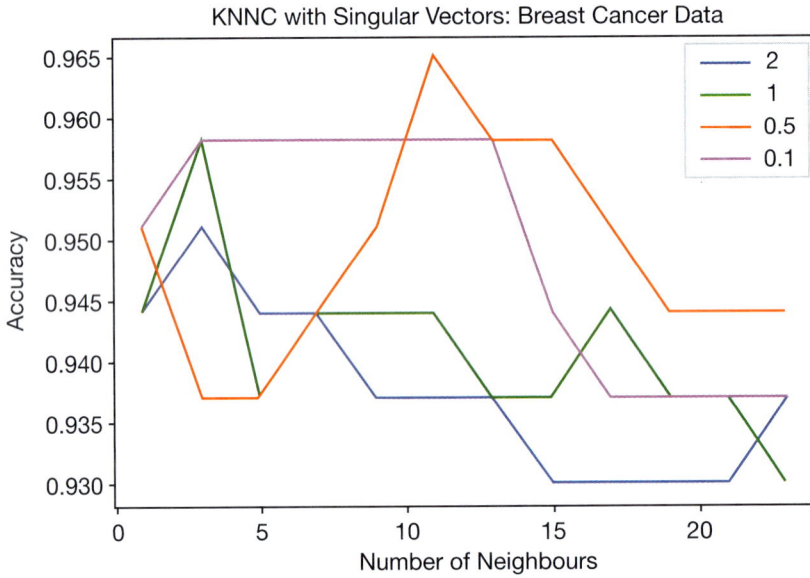

FIG. A.12 Accuracy of KNN classifier with singular vectors for the Breast Cancer data set

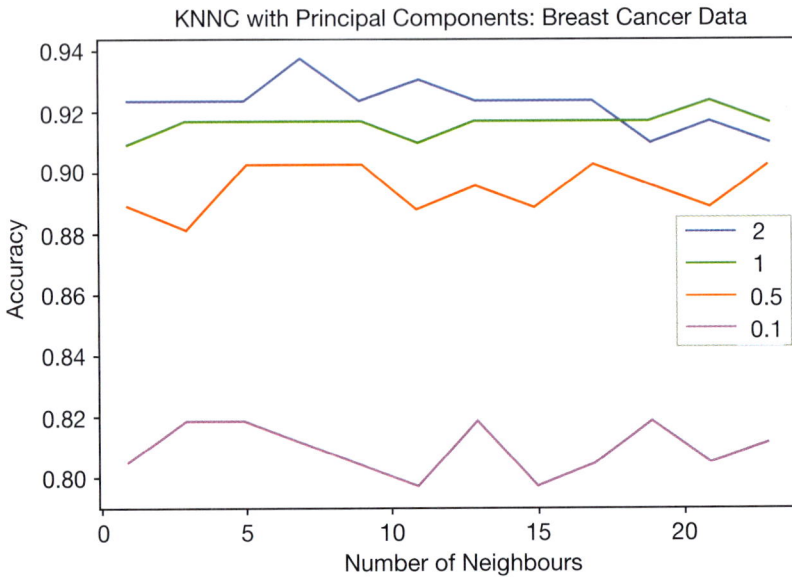

FIG. A.13 Accuracy of KNN classifier with principal components for the Breast Cancer data set

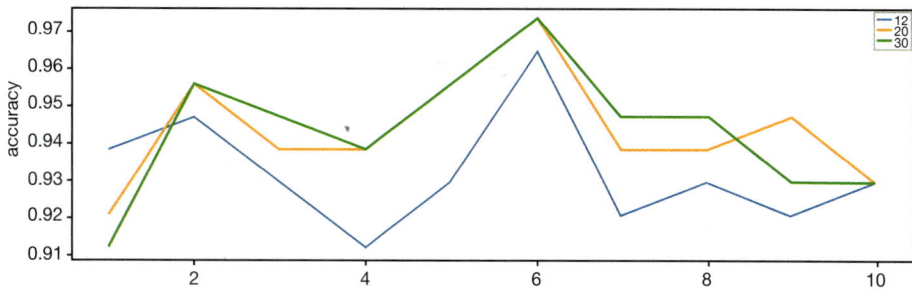

FIG. A.14 Output of Practical Exercise Q1: Task 1

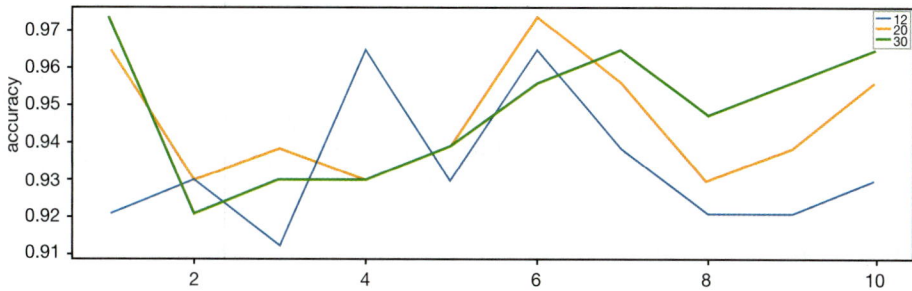

FIG. A.15 Output of Practical Exercise Q1: Task 2

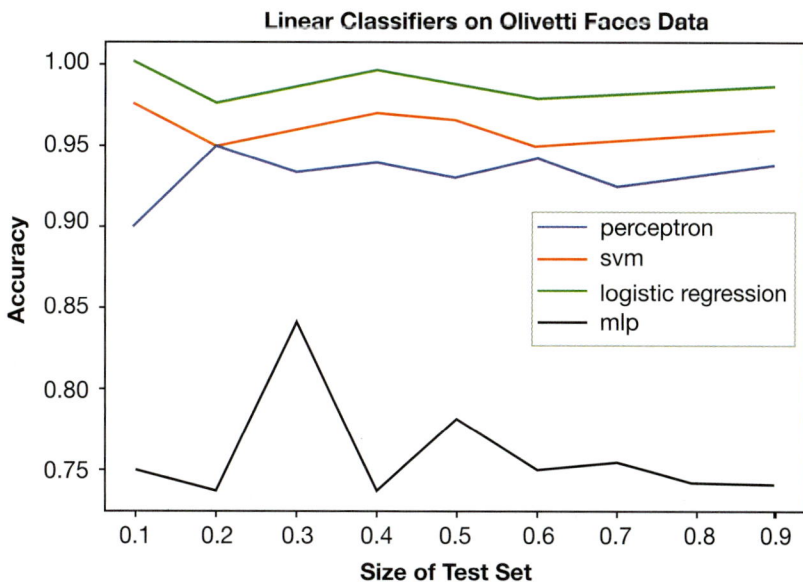

FIG. A.16 Accuracy of linear classifiers for the Olivetti Face data set

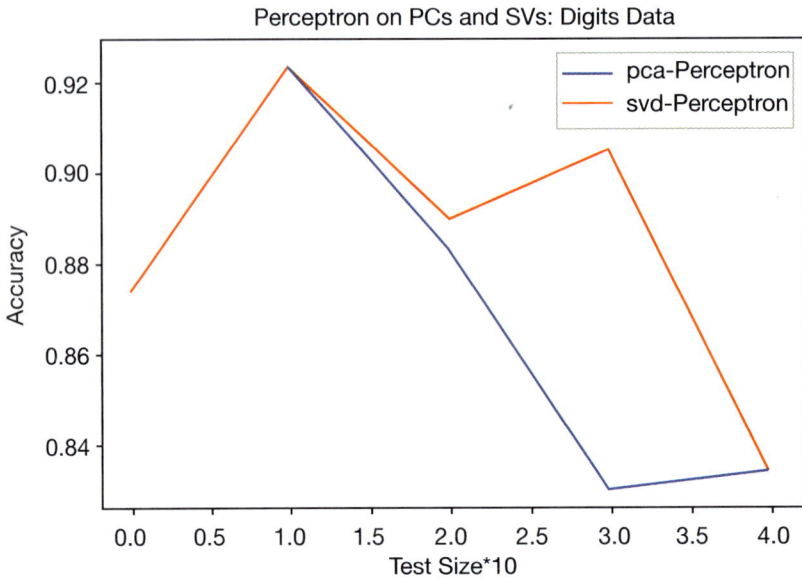

FIG. A.17 Accuracy of perceptron on principal components and singular vectors for the Digits data set

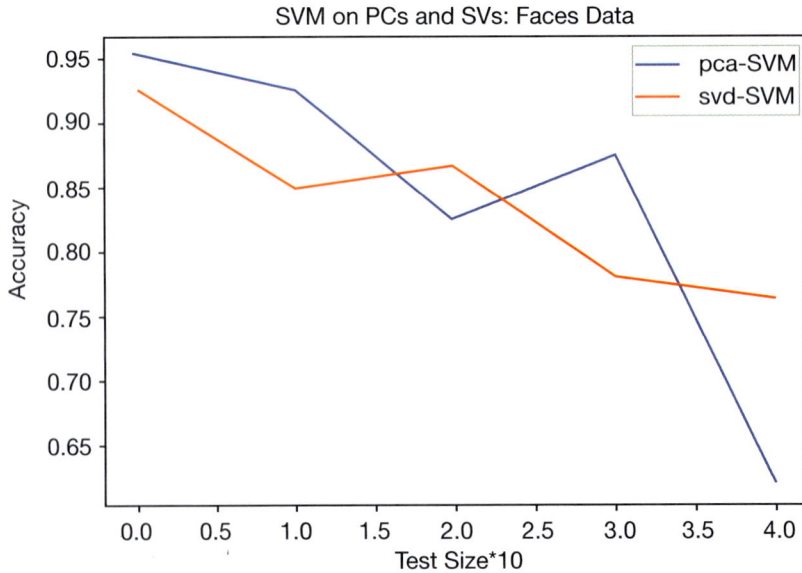

FIG. A.18 Accuracy of support vector machine on principal components and singular vectors for the Olivetti Face data set

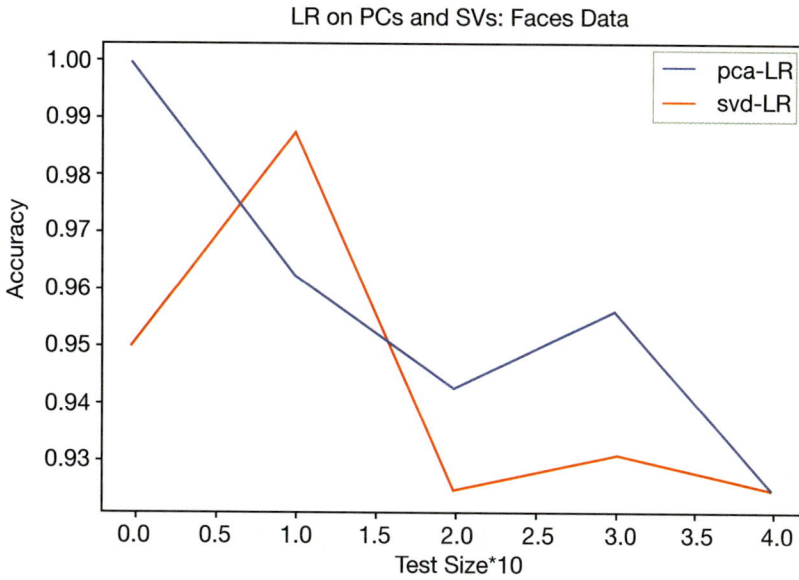

FIG. A.19 Accuracy of linear regression on principal components and singular vectors for the Olivetti Face data set

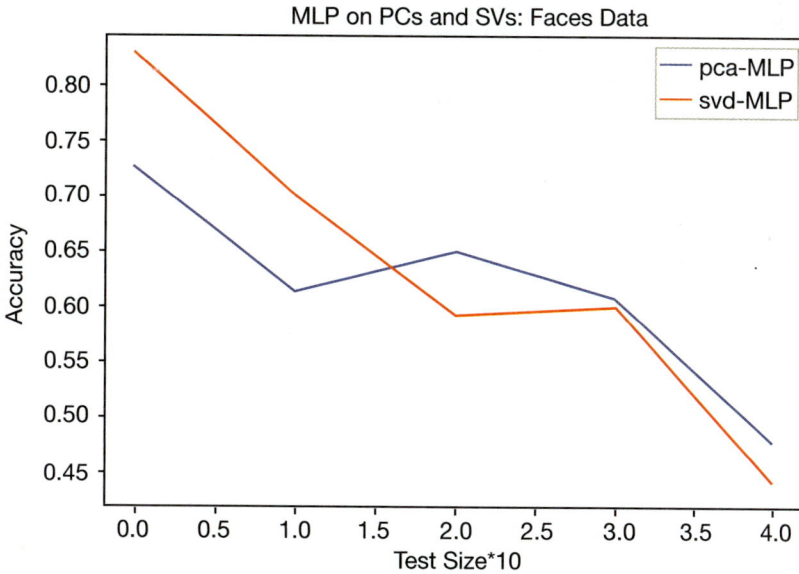

FIG. A.20 Accuracy of multi-layer perceptron on principal components and singular vectors for the Olivetti Face data set

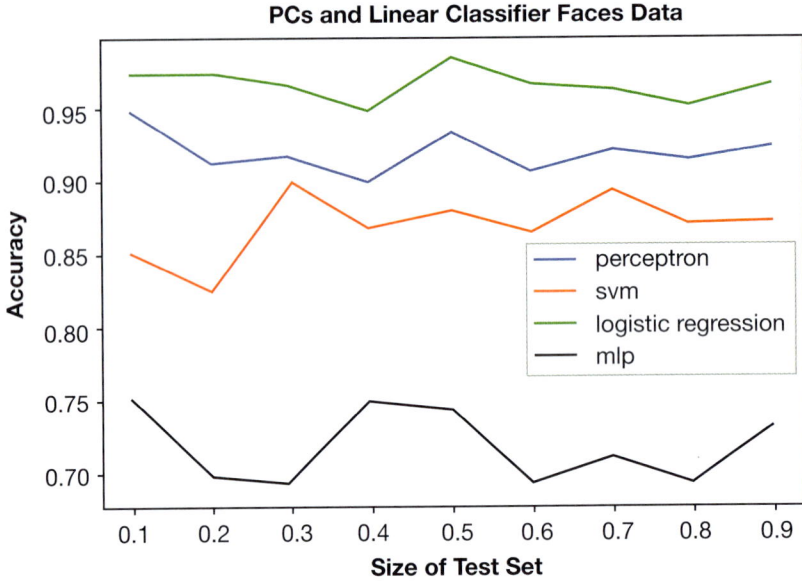

FIG. A.21 Accuracy of principal components and linear classifiers for the Olivetti Face data set

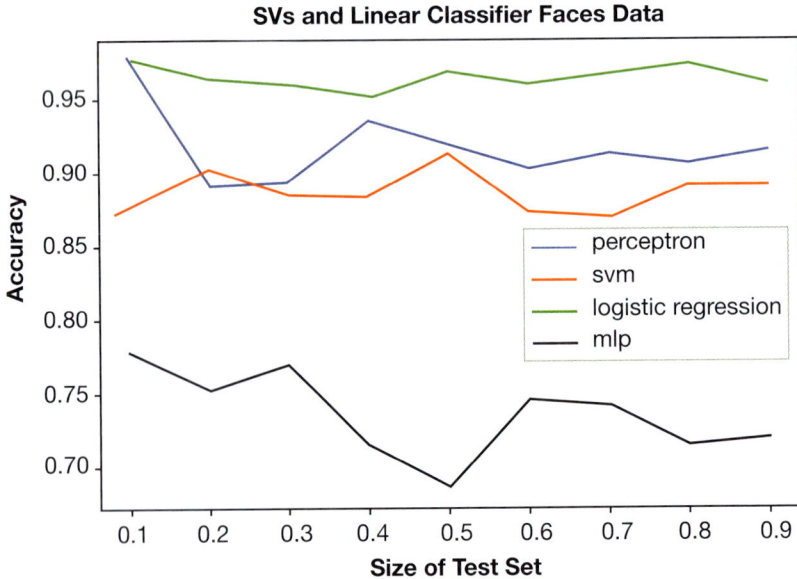

FIG. A.22 Accuracy of singular vectors and linear classifiers for the Olivetti Face data set

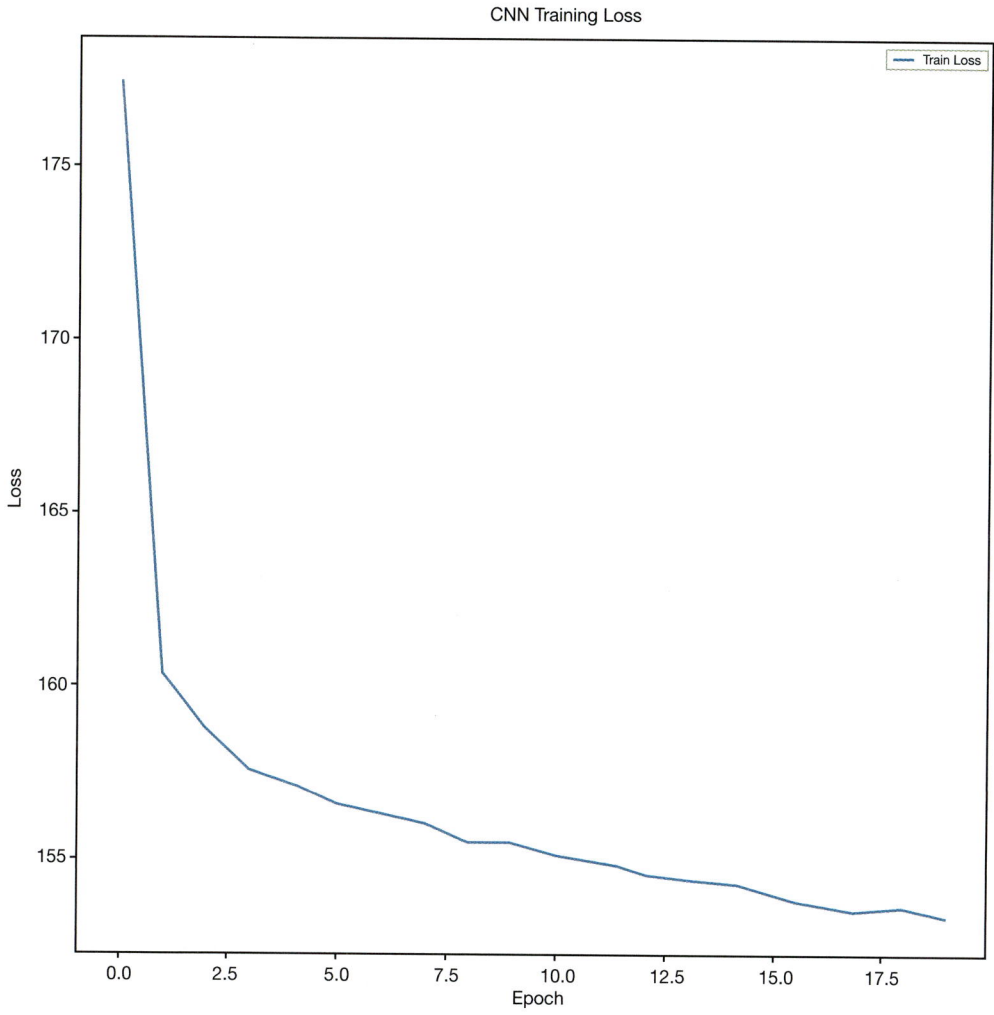

FIG. A.24 CNN training loss

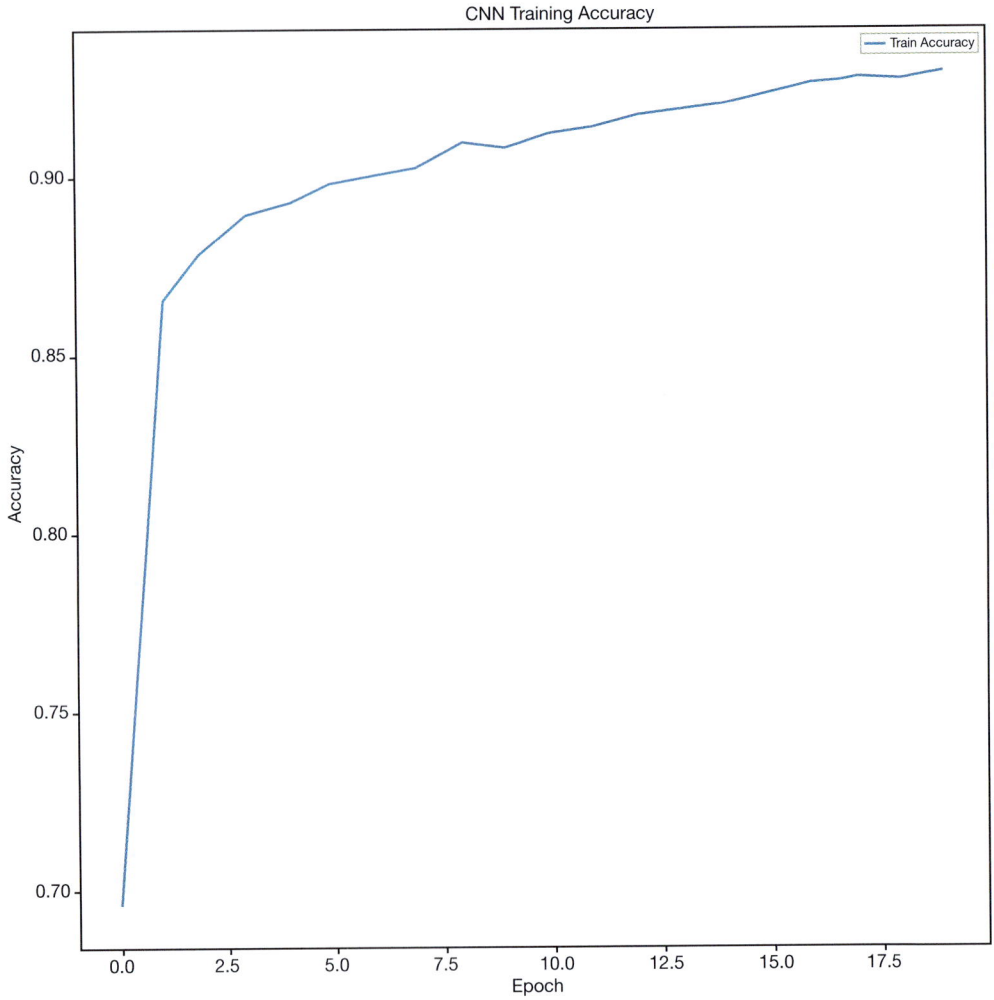

FIG. A.25 CNN training accuracy

Index

A

abductive learning 4
abstraction-based clustering 193
activation function 223, 234, 235, 238
AdaBoost
    about 50, 70
    application 77
    features 73
    in regression 74
AdaGrad 249
adaptive moment estimation (ADAM) 249
adaptive optimization
    about 248
    adaptive moment estimation (ADAM) 249
    root mean square propagation (RMSProp)
        249
adding noise
    to input 253
    to output 255
agglomerative clustering 171
airline passengers data set 18
Apriori algorithm
    about 112
    antimonotone property of support 112
    issues 115
    principle 112
    steps in 112
area under the curve (AUC) 42, 44
artificial intelligence (AI)
    about 1
    high-level view 2
    importance of background topics 2
    problems with conventional systems 17
artificial neural network (ANN) 202
association rule 110
association rule mining
    about 100
    steps in 100
augmented vector 206, 207
Australian weather data set 18, 271
autoencoder 234
Automatic Theorem Proving (ATP) 1

B

backpropagation
    about 17, 224, 225, 235
    backward pass 225, 227
    for training an MLP 225
    forward pass 225, 226
    gradient descent scheme 225, 235
    through hidden nodes 228
    through time 269
backward pass 225, 227
bagging 68
Bayes' classifier
    and its optimality 95
    Bayes' rule
        and classification 86
        and inference 85
    Bayesian estimation (BE) 81, 100, 103
        comparison with MLE 102
    cumulative distribution function (CDF) 90
    density estimation
        non-parametric 103
        parametric schemes 98
    introduction to 81
    multi-class classification 97
    naïve Bayes classifier (NBC) 105
        class-conditional independence 105, 106
    properties 81
    probability
        conditional 83
        event 82
        prior 82
        total 84
    probability mass function (PMF) 88
    random variable 87
        binomial 89
        continuous 91
        expectation 92
        normal distribution 93
        variance 93
Bayes' rule
    and classification 86
    and inference 85
Bayesian estimation (BE) 81, 100, 103

bias
    about 202
    variance trade-off 50, 65
binary classifier 211
binary cross entropy loss (BCE) 252
binary patterns 25
binomial
    coefficient 90
    random variable 89
Boolean
    AND 202
    OR 202, 208
boosting 70
bootstrapping 68
Boston housing data set 18, 36, 44

C
categorical data 6
categorical feature 15, 57, 81
centroid 5, 162, 164
chain rule 86
change in
    knowledge 122
    support value 122
characteristic
    equation 143
    value 143
    vector 143
city-block distance 22, 37, 152, 172
class association rule (CAR) 123
class-conditional independence 102, 105, 106
class imbalance 166
class label 5, 58
classification
    about 4, 58
    accuracy 40
    rule mining 123
classification algorithms
    about 26
    nearest neighbor (NN) 26
        k 27
        weighted k 28
        radius distance 29
        tree-based 30
        branch and bound 32
        leader clustering 34
        KNN regression 35
classifier performance
    about 40
    accuracy 40
    area under the ROC 41
    confusion matrix 40
clustering
    about 13, 44
    agglomerative 171
        steps in 171
    divisive 167
        monothetic 170

expectation maximization based 189
introduction to 157
    data compression 159
    data partitioning 157
    data reorganization 158
    data summarization 159
    matrix factorization 159
large data sets 192
    abstraction based 193
    divide-and-conquer 193
    incremental 192
of patterns 160
    algorithms 166
    data abstraction 162
partitional 173
    k-means ++ 177
    k-means 173
    soft partitioning 178, 180
spectral 190
combinational model 50
concentration effect 37, 38
conditional probability 83, 86
confusion matrix 40
convolutional neural network (CNN)
    about 1, 257
    convolution 257
    padding zero rows and columns 259
    pooling to reduce dimensionality 261
cosine similarity 7, 21
covariance matrix 97, 145
cross entropy loss 227, 252, 268
cross validation 60
cumulative distribution function (CDF)
    90

D
data
    abstraction 162
    acquisition 8
    compression 159
    from different domains 11
    preprocessing 8
    re-organization 158
    representation 13
data set
    about 17
    augmentation 254
    for classification 17
        Fashion MNIST 17, 246, 250
        MNIST handwritten digits 4, 17, 128,
            229, 237, 263, 264
        Olivetti face 18, 68, 77, 147, 150
        Wisconsin breast cancer 18, 38, 43, 60
    for regression 18
        airline passengers 18
        Australian weather 18, 271
        Boston housing 18, 36, 44

data types
    categorical 6
        nominal 6
        ordinal 6
    numerical 6
decision boundary 204
decision tree (DT)
    about 49
    applications 60
    bias–variance trade-off 65
    comparison with RF 68
    construction 54
    for classification 50
        Shannon's entropy 50
        Gini index 54
        misclassification impurity 54
    models 49
    properties 49
    properties of classifiers 56
    regression 63
decision tree construction, impurity measures
    about 54
    Gini index 54
    misclassification 54
decision tree classifiers, properties
    about 56
    binary or non-binary 57
    class label 58
    classification 58
    criterion for splitting 57
        information gain 57
    eliminating irrelevant features 59
    mixed data types 59
    pruning 59
    splitting rule 56
        axis-parallel 56
        oblique 56
    termination condition 58
    transparency 58
decision tree pruning 49, 59
decoder 15, 234, 235
deductive learning 3
deep learning
    about 1, 233
    convolutional neural network 257
    deep neural network 237
    difficulties associated with 17
    generative adversarial network (GAN) 272
    introduction to 233
    non-linear feature extraction using
        autoencoder 234
    properties 233
    recurrent neural network 265
deep neural network
    about 237
    activation functions 238
        leaky ReLU 243

        linear 238
        linear threshold 239
        rectified linear unit (ReLU) 242
        sigmoid 240
        softmax 243
        tanh 242
    adaptive optimization 248
    improved optimization methods 246
    loss functions 251
    regularization 252
    weight initialization 244
density estimation
    non-parametric 103
    parametric schemes 98
dendrogram 167
determinant 142
diagonal matrix 143
digits data set 229, 236
dimensionality reduction 13, 14
discriminator 272
disjoint events 82
dissimilarity 157
distance measure
    about 7, 21
    classification algorithms 25
    Euclidean 7, 21
    Hamming 21, 25
    Levenshtein 23
    Mahalanobis 22
    metric 22
    Minkowski 22
    mutual neighborhood 24
    non-metric similarity functions 23
    proximity between binary patterns 25
    weighted 22
divide-and-conquer method 193
divisive clustering 167
dot product 15
dropout 254
dynamic mining
    about 120
    change in knowledge 122
    change in support value 122
    incremental 120
dynamic range 44

E
early stopping 253
eigenpair 142
eigenvalue 141, 145
eigenvector 142, 145
elbow method 175
embedded feature selection 62
embedded methods 136
encoder 15, 234, 235
encoder–decoder model 270
entropy 3, 49

Euclidean distance 7, 21
Euclidean norm 38
event 82
expectation maximization-based clustering 189
explanation 16, 19, 45
exploding gradient 257, 269

F
F1 score 41
false
    negatives 41
    positives 41, 42
Fashion MNIST 17, 246, 250
feature engineering
    about 8, 19, 277
    data preprocessing 8
        data from different domains 11
        missing data 9
        outliers in data 12
feature extraction
    about 13, 14
    linear 14
        non-negative matrix factorization (NMF) 14
        principal components (PCs) 14
    non-linear 15
feature selection 13, 14, 132
feature space 202
feature vector 202
feed-forward neural network 223
forward pass 225, 226
forward propagation 223
fractional norm 22, 37, 39
filter methods
    about 132
    dependency 133
    mutual information (MI) 134
    variance 132
frequent itemset
    about 109
    dynamic mining
        about 120
        change in knowledge 122
        change in support value 122
        incremental 120
    generation strategy 112
frequent itemset approach
    classification rule mining 123
        class association rule (CAR) 123
    frequent itemset 109
        Apriori algorithm 112
        association rule 110
        generation 111
        market basket analysis 110
        support 110

frequent pattern (FP) tree 115
    comparison with PC tree 119
    frequent itemset generation 115
    introduction to 109
    pattern count (PC) tree 117
        comparison with FP tree 119
        construction 118
        for classification 125
        for clustering 126
        frequent itemset generation 117, 130
frequent pattern (FP) tree
    -based nearest neighbor 32
    about 115
    about 30, 115
    comparison with PC tree 119
    frequent itemset generation 115
    steps to construct 30
frequent sequential pattern 109
function value estimation 44
fuzzy C-means clustering 180

G
gated recurrent unit
    about 270
    reset 270
    update 270
Gaussian random projection 152
Gaussian RV 93
General Problem Solver (GPS) 1
generative adversarial network (GAN) 272
generator 272
Gini index 54
gradient boosting (GB)
    about 50, 74
    steps to estimate function 74
    properties 76
gradient descent
    about 17, 202, 225, 235
    adaptive (AdaGrad) 249
    momentum-based 247
    Nesterov accelerated 247
    stochastic 246
gradient clipping 270

H
Hamming distance (HD) 21, 25
hard clustering 161
He initialization 246
Hessian matrix 51
hierarchical clustering algorithm 167
high-dimensional data
    about 14, 49
    difficulties with using 14
    dimensionality reduction approach 14
        feature extraction 14
        feature selection 14
        hybrid clustering 197

I

identity matrix 141
identity transformation 141
impurity measure 54
incremental clustering algorithms
        192
incremental mining 120
independent events 84
inductive learning 4, 6
information gain
        about 57
        ratio 157
infrequent itemset 113
invertible matrix 142

J

Jaccard
        coefficient (JC), 25
        similarity 21

K

k-means clustering
        about 173
        rough 186
k-means++ clustering 177
k-median distance 23
k-nearest neighbor (KNN)
        classifier 27, 44, 147, 152
        regression 35
kernel 258
kernel trick 202, 218
Kullback–Leibler divergence 3

L

L_1 norm 22
L_2
        norm 22
        regularization 252
large data sets
        about 192
        abstraction based 193
        divide-and-conquer 193
        incremental 192
leader clustering 34, 44
learning by
        abduction 4
        deduction 3
        induction 4
                from examples 4
                from observations 4
        reinforcement 6
        rote 3
learning w and b 205, 221
LeCun initialization 245
Levenshtein distance 23

linear
        activation function 224, 238
        classifier 204
        feature extraction 136
        regression 219
        schemes 14
        threshold function 202, 239
linear discriminant function
        about 202
        learning w and b 205
        parameters involved in 203
linear discriminants
        for classification 202
                artificial neural network (ANN) 202
                linear discriminant function 202
                logistic regression 219
                multi-layer perceptron (MLP) 222
                perceptron 207
                support vector machine (SVM) 212
        introduction to 201
linearly non-separable
        case 215
        classes 210
linearly separable 201, 202, 206
Lisp 1
logistic
        function 222
        regression 202, 219
logistic regression
        about 202, 219
        learning w and b 221
        linear regression 219
        sigmoid function 220
long short-term memory (LSTM) 269
loss functions
        about 251
        binary cross entropy (BCE) 252
        mean absolute error (MAE) 252
        squared error 251

M

machine learning
        AdaBoost 50, 70
        based on frequent itemsets **109–130**
        clustering **157–200**
        deep learning **233–276**
        definition 1
        evolution of 1
        gradient boosting (GB) 50, 74
        linear discriminants **201–232**
        matching 7
        nearest neighbor (NN) 21, 22, 26
        paradigms for 3
        random forest (RF) 50, 67
        representation **131–156**
        search 16
        stages in 8

machine learning paradigms
    about 3
    learning by
        abduction 4
        deduction 3
        induction 4
        reinforcement 6
        rote 3
machine learning stages
    about 8
    data acquisition 8
    data representation 13
    feature engineering 8
    model evaluation 15
    model explanation 16
    model learning 15
    model prediction 16
    model selection 15
Mahalanobis distance 22, 97
market basket analysis 110
matching
    about 7
    proximity measure 7, 21
    applications 7
matrix factorization 16, 159
maximum likelihood estimation (MLE)
    about 81, 98, 103
    comparison with BE 102
maximum margin 202
mean absolute error (MAE) 43, 252
mean squared error (MSE) 10, 43, 251
medoid 163, 164
minimal distance classifier (MDC) 97
minimum confidence 110
minimum error-rate classifier 95
minimum support 110
Minkowski
    distance 22
    metric 21
    norm 38
MNIST handwritten digits data set 4, 17, 128, 229,
        237, 263, 264
model
    evaluation 15
    explanation 16
    learning 15
    performance measure
        about 40
        of classifiers 40
        of regression algorithms 43
    prediction 16
    selection 15
moments 250
monothetic clustering 170
multi-class classification 97
multi-class problems
    about 211

one-vs-one 211
one-vs-rest 211
multi-layer perceptron (MLP)
    about 222
    backpropagation for training 225
multiple membership approach 179
mutual information 3, 133
mutual neighborhood distance (MND) 24

N
naïve Bayes classifier (NBC)
    about 81, 105
    class-conditional independence 105, 106
nearest neighbor classifier (NNC) 21, 22, 26
negative class 204
negative half space 204
neural network 15
neuron
    about 13, 15, 202, 222, 223, 234, 238
    dead 243
    saturated neuron
noisy data 13
nominal data 6
non-linear
    decision boundary 201, 202, 216
    dimensionality reduction 234
    feature extraction 15, 234
    SVM 216
non-metric similarity functions 23
non-negative matrix factorization (NMF) 14
non-parametric schemes 103
normal distribution 93
normalization
    about 12, 22
    schemes 12

O
oblique split 56
odds ratio 221
Olivetti face data set 18, 68, 77, 147, 150
optimal classifier 95
optimization 3
ordinal data 6
orthogonal matrix 149
orthonormal 139
outliers
    about 12, 163, 164
    reasons for occurrence 13
    schemes for detecting 13
overfitting 14, 16, 49, 59, 234

P
parametric density estimation
    about 98, 103
    Bayesian estimation (BE) 103
    maximum likelihood estimation (MLE) 98
part of speech (POS) tagging 267

partitional clustering 173
pattern count (PC) tree 117
    comparison with FP tree 119
    construction 118
    for classification 125
    for clustering 126
    frequent itemset generation 117, 130
peaking phenomenon 14
perceptron
    about 1, 202
    classifier 207
        convergence 209
        learning algorithm 207
        linearly non-separable classes 210
        multi-class problems 211
        multi-layer 222
piece-wise linear boundaries 201
positive class 204
positive half space 204
positive rate
    true 41
    false 42
positive reflexivity 22
principal component analysis (PCA) 145
principal components 14
prior probabilities 81, 82
Prolog 1
probability
    about 3
    conditional 83
    event 82
    of error 81, 95
    posterior 81
    prior 82
    total 84
probability mass function (PMF) 88
proximity measures
    about 7, 21
    cosine 7, 21
    Euclidean 7, 21
    Hamming 21, 25
    Jaccard 21, 25
    simple matching coefficient 25
PyTorch 1

R
radius distance nearest neighbor algorithm 29
random forest (RF)
    about 50, 67
    bootstrapping 68
    comparison with DT 68
    features 68
    in regression 74
random projections 152
random variable 87
    binomial 89
    continuous 91

    expectation 92
    normal distribution 93
    variance 93
receiver operating characteristic (ROC) 42
rectified linear unit (ReLU) 242
recurrent neural network (RNN)
    about 1, 265
    auto completion of words 267
    encoder-decoder model 270
    properties 265
    training 268
        exploding gradient 269
        long short-term memory (LSTM) 269
        vanishing gradient 269
regression
    about 4, 44, 74
    using decision trees 63
    using DT-based models 74
regression algorithm performance
    about 43
    mean absolute error (MAE) 43
    mean squared error (MSE) 43
regularization
    about 252
    adding noise to inputs 253
    adding noise to output 255
    data set augmentation 254
    dropout 254
early stopping 253
    L_2 252
    ridge 253
reinforcement learning 6
representation in machine learning
    about 8, 19
    feature selection
        about 132
        embedded methods 136
        filter methods 132
        wrapper methods 135
    introduction to 131
    issues with classifiers 131
    linear feature extraction
        about 132, 136
        column vector 139
        eigenvalue 141
        eigenvector 142
        linear transformation 140
        matric rank 144
        principal component analysis (PCA) 145
        random projections 152
        row vector 139
        singular value decomposition 147
        symmetric matrix 143
        vector spaces 136
    non-linear feature extraction
    about 132
reset gate 270

reward function 6
root mean square propagation (RMSProp) 249
rote learning 3
rough k-means clustering algorithm 186

S
sample space 82
scalar multiplication 137
scaling 12
search 16
Shannon's entropy 50, 54
sigmoid activation function 200, 220, 240
similarity measure
    about 7, 157
    cosine 7, 21
    Jaccard 21
simple matching coefficient (SMC) 25
single-link algorithm 172
singular matrix 142
singular value decomposition 147
singular values 149
sklearn 62, 152
soft partitioning clustering
    about 178, 180
    multiple membership 179
        fuzzy 179, 180
        neural-network based 180
        rough 179, 186
    split and merge 179
    stochastic 180
        evolutionary algorithms 180
        simulated annealing 180
        Tabu search 180
support vector machine (SVM)
    about 202, 212
    kernel trick 218
    linearly non-separable 215
    non-linear 216
softmax activation function 243
spectral clustering 190
squared error loss 222, 251
squared error 63
sum-of-squared-error criterion 175
supervised learning 4, 277
symmetric matrix 143

T
tanh activation 242
TensorFlow 1
termination condition 58
training data 15, 81

training pattern 26
transfer learning 233
tree-based nearest neighbor algorithm 30, 44
triangular inequality 22
triangular matrix 143
true
    negative rate 85
    negatives 41
    positive rate 41, 85
    positives 41

U
unit variance 12
unsupervised learning 277
update gate 270

V
validation data 15, 16
vanishing gradient 257, 269
variance 93, 132
vector addition 137
vector spaces
    about 136
    addition 137
    basis of 138
    scalar multiplication 137
    subspace 138
Venn diagram 84

W
weak learner 70
weight vector 202, 204
weighted
    k-nearest neighbor (WKNN) 28
    distance measure 22
    entropy 53
    impurity 53
Wisconsin breast cancer data set 18, 38, 43, 60
wrapper methods
    about 135
    sequential 135
    sequential floating forward selection (SFFS)
        135

X
Xavier initialization 246
XGBoost 77

Z
zero mean 12, 145